# Power Systems

More information about this series at http://www.springer.com/series/4622

Ali Arefi · Farhad Shahnia
Gerard Ledwich
Editors

# Electric Distribution Network Management and Control

Springer

*Editors*
Ali Arefi
School of Engineering
and Information Technology
Murdoch University
Perth, WA
Australia

Farhad Shahnia
School of Engineering
and Information Technology
Murdoch University
Perth, WA
Australia

Gerard Ledwich
Faculty of Science and Engineering
Queensland University of Technology
Brisbane, QLD
Australia

ISSN 1612-1287 ISSN 1860-4676 (electronic)
Power Systems
ISBN 978-981-13-3905-9 ISBN 978-981-10-7001-3 (eBook)
https://doi.org/10.1007/978-981-10-7001-3



Printed on acid-free paper

This Springer imprint is published by the registered company Springer Nature Singapore Pte Ltd.
part of Springer Nature
The registered company address is: 152 Beach Road, #21-01/04 Gateway East, Singapore 189721, Singapore

# Preface

Electric distribution networks are critical parts of power delivery systems. In recent years, many new technologies and distributed energy resources have been integrated into these networks. To provide electricity at the required quality and at the lowest possible cost, an appropriate control and management of future electric grids is indispensable. The management system continuously monitors and controls various equipment in distribution network to reduce long-term operation and maintenance costs, to satisfy technical constraints of utilities and customers, and to improve security and reliability of power delivery. Since the integration of new technologies such as renewables and demand response is increasing the uncertainties in distribution networks, one of the critical challenges of control and management systems is to tackle such uncertainties. A wide range of approaches along with optimization algorithms including mathematic- and heuristic-based are utilized for finding the best configuration of control systems in distribution networks.

Over the last decades, several researches have been carried out around the world on electric distribution management and control, whose results are available as journal articles, conference papers, or technical reports. However, to the best of the editors' knowledge, no single book has covered the different aspects of distribution networks' management and control so far. The interested readers had to search among several hundreds of papers on this topic through various databases in order to build up their knowledge on the subject. This book is the first one entirely focused on the distribution networks' management and control and is an effort to provide a research-oriented and a coherent book on the subject for postgraduate students and researchers.

This book is benefited from the inputs and comments of a large number of researchers and experts from the academia and industry. It contains 12 chapters. The breakdown of the chapters is as follows:

- Chapter 1 explores existing and emerging flexibility options for facilitating the integration of large-scale variable renewable energy sources in electric distribution networks.

- Chapter 2 presents a model for balanced and unbalanced electric distribution network and then studies their effects on the power flow analysis of practical networks.
- Chapter 3 presents some advanced methods for demand simulation with the increased uptake of low-carbon technologies using limited number of measurements.
- Chapter 4 presents coordinated voltage control methods in active electric distribution networks by coordinating various control options including power factor control, on-load tap changers, and generation curtailment.
- Chapter 5 reviews promising concepts for distribution network-oriented demand response to address local challenges that distribution network operators face in daily operation.
- Chapter 6 presents the feasibility of using customer coupon demand response in meshed secondary networks to achieve the objective of optimal operation cost during network peak periods.
- Chapter 7 describes a transactive approach to the optimal scheduling for prosumers in coupled electric and natural gas distribution networks to improve the integration of distributed energy resources.
- Chapter 8 introduces the optimal switch deployment in distribution networks including the mathematical formulation in the form of a mixed integer program.
- Chapter 9 presents a multi-time scale model predictive control approach which is stochastically applied in the cooperative distributed energy scheduling problem for microgrids.
- Chapter 10 introduces methodologies to assess the impact of distribution generation on protection systems within electric distribution networks for an integrated network planning.
- Chapter 11 studies modeling of various types of generators in wind farms for fulfilling the rules and requirements, set by network operators, to guarantee the continuity and stability of the interconnected grid.
- Chapter 12 reviews a unified integration solution that enables distribution management systems to flexibly adapt to various advanced metering infrastructure systems with different communication protocols and meter data models. Also, it discusses graphic processing unit technologies and explores their applications in distribution networks.

As the editors of this book, we would like to thank all the contributors for their support and hard work. We also would like to thank the reviewers who provided valuable comments for improving the quality of this book. Also, we are grateful to the publisher Springer Nature for agreeing to publish this book. Last but not least, we would like to thank our families—Ali thanks his wife and son (Behnaz and Amin), Farhad thanks his parents (Nahideh and Ali) and his spouse (Negar), and Gerard thanks his family for their continuous encouragement and support.

Perth, Australia — Ali Arefi
Perth, Australia — Farhad Shahnia
Brisbane, Australia — Gerard Ledwich
January 2018

# Reviewers

Ahmed Rashad, Aswan University, Egypt, and University of Jaén, Spain
Amir Safdarian, Sharif University of Technology, Iran
Azah Mohamed, University Kebangsaan Malaysia, Malaysia
Carlos F. Meschini Almeida, University of São Paulo, Brazil
Desta Z. Fitiwi, University of Beira Interior, Portugal
Fang Yang, University of Wisconsin Platteville, USA
Francisco Jurado, University of Jaén, Spain
Gabriel Quiroga, University of São Paulo, Brazil
Georgios Giasemidis, University of Reading, UK
Henrique Kagan, University of São Paulo, Brazil
Jing Qiu, Commonwealth Scientific and Industrial Research Organization, Australia
João P. S. Catalão, University of Porto, Portugal
Kazem Zare, University of Tabriz, Iran
Ke Meng, University of Sydney, Australia
Marco R. M. Cruz, University of Beira Interior, Portugal
Mehdi Rahmani-Andebili, Clemson University, USA
Miadreza Shafie-khah, University of Beira Interior, Portugal
Milad Izadi, Sharif University of Technology, Iran
Mohammad Farajollahi, University of California, USA
Mustafa Alparslan Zehir, Istanbul Technical University, Turkey
Mustafa Bagriyanik, Istanbul Technical University, Turkey
Nelson Kagan, University of São Paulo, Brazil
Sérgio F. Santos, University of Beira Interior, Portugal
Stephen Haben, University of Oxford, UK
Salah Kamel, Aswan University, Egypt
Shady H. E. Abdel Aleem, 15th of May Higher Institute of Engineering, Egypt
Tengku J. Tengku Hashim, University Tenaga Nasional, Malaysia

Xuejun Zheng, Huazhong University of Science and Technology, China
Zhao Li, LY Grid Innovation, USA
Zhao Yang Dong, University of New South Wales, Australia
Zhechao Li, Huazhong University of Science and Technology, China

# Contents

# About the Editors

**Ali Arefi** received his Ph.D. in Electrical Engineering in 2011. He is currently a Senior Lecturer in Power Engineering at Murdoch University, Perth, Australia. Prior to that, he was a Lecturer and Research Fellow at the Queensland University of Technology. He also has 6 years' experiences with electric distribution industry and has been the supervisor for five industry-funded research projects. His research interests are in the areas of electric delivery planning, state estimation, power quality, and energy efficiency. He has published more than 80 research articles.

**Farhad Shahnia** received his Ph.D. in Electrical Engineering from Queensland University of Technology, Brisbane, Australia, in 2011. Currently, he is a Senior Lecturer in Power Engineering at Murdoch University, Perth, Australia. His professional experience includes 3 years at Research Office-Eastern Azarbayjan Electric Power Distribution Company, Tabriz, Iran. Prior to joining Curtin University, he was a Research Fellow at the Queensland University of Technology, Brisbane, Australia, and a Lecturer at Curtin University, Perth, Australia. He has authored 1 book, 9 chapters, and over 120 research articles as well as editing 4 books.

**Gerard Ledwich** received his Ph.D. in Electrical Engineering from University of Newcastle, Newcastle, Australia, in 1976. He has been the Chair Professor in Power Engineering at the Queensland University of Technology, Brisbane, Australia, since 1998. Previously, he was with the University of Queensland, Brisbane, Australia, from 1976 to 1994. His research interests are in the areas of power system operation and control. He is the supervisor of more than 35 Ph.D. graduates and has published 1 book, 3 chapters, and over 350 research articles.

# About the Editors

# Abbreviations

| | |
|---|---|
| ADI | Meter data integration |
| ADMD | After diversity maximum demand |
| AMI | Advanced metering infrastructure |
| ANN | Artificial neural network |
| AR | Automatic recloser |
| ARIMA | Autoregressive integrated moving average |
| AVRS | Automatic voltage reference setting |
| BESS | Battery energy storage system |
| CB | Circuit breaker |
| CCDR | Customer coupon demand response |
| CDF | Customer damage function |
| CG | Conjugate gradient |
| CGNR | Conjugate gradient normal residual |
| CHP | Combined heat and power |
| CPP | Critical peak pricing |
| CPR | Critical peak rebate |
| CS | Counter sectionalizer |
| CUDA | Compute unified device architecture |
| CVaR | Conditional value at risk |
| CWF | Combined wind farm |
| DA | Day-ahead |
| DER | Distributed energy resource |
| DFIG | Doubly fed induction generator |
| DG | Distributed generation |
| DisCo | Distribution company |
| DLC | Direct load control |
| DLMP | Distribution locational marginal price |
| DMS | Distribution management system |
| DNO | Distribution network operator |
| DNR | Dynamic network reconfiguration |

| | |
|---|---|
| DR | Demand response |
| DSE | Distribution state estimation |
| ES | Expected shortfall |
| ESB | Enterprise service bus |
| EU | European Union |
| EV | Electric vehicle |
| FSB | Front-side bus |
| G2V | Grid-to-vehicle |
| GA | Genetic algorithm |
| GC | Gas compressor |
| GHG | Greenhouse gas emission |
| GPU | Graphics processing unit |
| GS | Gas storage |
| GSC | Grid-side converter |
| GSF | Generator shifter factor |
| HEMS | Home energy management system |
| HPC | High-performance computing |
| ISO | Independent system operator |
| KKT | Karush–Kuhn–Tucher |
| LCE | Loosely coupled event |
| LMP | Locational marginal price |
| LP | Linear programming |
| LSE | Load serving entity |
| LV | Low voltage |
| MCS | Monte Carlo simulation |
| MCWF | Modern combined wind farm |
| MDMS | Meter data management system |
| MG | Microgrids |
| MILP | Mixed integer linear programming |
| MIP | Mixed integer programming |
| MM | Minimum melting |
| MPC | Model predictive control |
| MS | Manual switch |
| NEM | Net energy metering |
| NN | Neural network |
| OLTC | On-load tap changer |
| OMS | Outage management system |
| OpenCL | Open computing language |
| OPF | Optimal power flow |
| PC | Personal computer |
| PCC | Point of common connection |
| PFBS | Probabilistic fitness-based selection |
| PFC | Power factor control |
| PMSG | Permanent magnet synchronous generator |
| PMU | Phasor measurement unit |

| | |
|---|---|
| PSO | Particle swarm optimization |
| PV | Photovoltaic |
| RCS | Remote-controlled switch |
| RMAE | Relative mean absolute error |
| RPDE | Relative peak demand error |
| RSC | Rotor-side converter |
| RT | Real time |
| RTP | Real-time pricing |
| SCADA | Supervisory control and data acquisition |
| SCB | Switchable capacitor bank |
| SCIG | Squirrel cage induction generator |
| SG | Synchronous generator |
| S-MILP | Stochastic mixed integer linear programing |
| SP | Shortfall probability |
| SPD | Symmetrical positive definite |
| SS | Sectionalizing switch |
| STTM | Short-term trading market |
| SVC | Static VAR compensator |
| SVM | Support vector machine |
| TC | Total clearing |
| ToU | Time of use |
| UBLF | Unbalanced power flow |
| ULTC | Under-load tap changer |
| V2G | Vehicle-to-grid |
| VaR | Value at risk |
| VC | Voltage control |
| VolIn | Volatility index |
| VPP | Virtual power plant |
| vRES | Variable renewable energy source |
| VSC | Voltage source converter |
| WES | Wind energy system |
| WLS | Weighted least squares |
| WRIG | Wound rotor induction generator |
| WT | Wind turbine |

# Chapter 1
# Managing Risk in Electric Distribution Networks

**Marco R. M. Cruz, Desta Z. Fitiwi, Sergio F. Santos, Miadreza Shafie-khah and Joao P. S. Catalao**

**Abstract** This book chapter explores existing and emerging flexibility options that can facilitate the integration of large-scale variable renewable energy sources (vRESs) in next-gen electric distribution networks while minimizing their side-effects and associated risks. Nowadays, it is widely accepted that integrating vRESs is highly needed to solve a multitude of global concerns such as meeting an increasing demand for electricity, enhancing energy security, reducing heavy dependence on fossil fuels for energy production and the overall carbon footprint of power production. As a result, the scale of vRES development has been steadily increasing in many electric distribution networks. The favorable agreements of states to curb greenhouse gas emissions and mitigate climate change, along with other technical, socio-economic and structural factors, is expected to further accelerate the integration of renewables in electric distribution networks. Many states are now embarking on ambitious "clean" energy development targets. Distributed generations (DGs) are especially attracting a lot of attention nowadays,

---

M. R. M. Cruz
Department of Electrical and Computer Engineering,
University of Beira Interior, Covilhã, Portugal
e-mail: marco.meneses.cruz@ubi.pt

D. Z. Fitiwi (✉) · S. F. Santos · M. Shafie-khah
Department of Electromechanical Engineering,
University of Beira Interior, Covilhã, Portugal
e-mail: desta.fitiwi@esri.ie

S. F. Santos
e-mail: sdfsantos@gmail.com

M. Shafie-khah
e-mail: miadreza@ubi.pt

J. P. S. Catalao
Department of Electrical and Computer Engineering,
University of Porto, Porto, Portugal
e-mail: catalao@fe.up.pt

D. Z. Fitiwi
The Economic and Social Research Institute, Dublin, Ireland



A. Arefi et al. (eds.), *Electric Distribution Network Management and Control*,
Power Systems, https://doi.org/10.1007/978-981-10-7001-3_1

and planners and policy makers seem to favor more on a distributed power generation to meet the increasing demand for electricity in the future. And, the role of traditionally centralized power production regime is expected to slowly diminish in future grids. This means that existing electric distribution networks should be readied to effectively handle the increasing penetration of DGs, vRESs in particular, because such systems are not principally designed for this purpose. It is because of all this that regulators often set a maximum RES penetration limit (often in the order of 20%) which is one of the main factors that impede further development of the much-needed vRESs.

The main challenge is posed by the high-level variability as well as partial unpredictability of vRESs which, along with traditional sources of uncertainty, leads to several technical problems and increases operational risk in the system. This is further exacerbated by the increased uncertainty posed by the continuously changing and new forms of energy consumption such as power-to-X and electric vehicles. All these make operation and planning of distribution networks more intricate. Therefore, there is a growing need to transform existing systems so that they are equipped with adequate flexibility mechanisms (options) that are capable of alleviating the aforementioned challenges and effectively managing inherent technical risk. To this end, the main focus of this chapter is on the optimal management of distribution networks featuring such flexibility options and vRESs. This analysis is supported by numerical results from a standard network system. For this, a reasonably accurate mathematical optimization model is developed, which is based on a linearized AC network model. The results and analysis in this book chapter have policy implications that are important to optimally design ad operate future grids, featuring large-scale variable energy resources. In general, based on the analysis results, distribution networks can go 100% renewable if various flexibility options are adequately deployed and operated in a more efficient manner.

**Keywords** Demand response · Electric distribution networks · Energy storage systems · Flexibility options · Mixed integer linear programming Network reconfiguration · Stochastic programing · Variable renewable energy sources

## 1.1 Chapter Overview

It is now widely accepted that integrating variable renewable energy sources (vRESs) in electric distribution networks is inevitable to meet a growing demand for electricity, enhance energy security and diminish the heavy dependence on fossil fuels to produce electricity, which are associated with high carbon footprint. Many states are now forging ahead with ambitious vRES integration targets aiming to achieve a substantial reduction of greenhouse gas emissions (GHGs), as in the European Union (EU). Integration of vRES technologies are expected to lead to 80–95% GHG emissions by 2050 [1]. One eminent fact about these technologies is that they depend on the availability of primary energy resources such as wind speed and solar irradiation, which are unevenly distributed over a wide geographical area. This means distributed (rather than centralized) development of such resources could be more convenient, efficient and even cost-effective despite the economies-of-scale. The main reason for this is because distributed generations are installed in places closer to demand, which means they are often connected to distribution networks. If this is executed in a well-coordinated manner, vRESs can bring vast benefits to the systems as a whole in terms of improved efficiency, deferred transmission investments, reduced use of fossil fuels for energy production and therefore lower GHG emissions [2]. Hence, distribution networks are expected to accommodate more and more vRESs.

Current trends generally show that the share of vRESs in the overall energy consumption is rapidly increasing in many electric distribution networks globally amid a number of barriers. However, the intermittent nature of such resources means a large-scale integration creates technical problems in the systems. Electric distribution networks are especially experiencing unprecedented challenges due to the increasing penetration level of distributed power generation sources of variable in nature, particularly, wind and solar. In other words, distributed generations (DGs) are attracting a lot of attention from policy makers and planners to meet the increased demand for electricity in the future. There is nowadays a growing trend of adding more new DG capacities than centralized generation capacities. This brings serious concerns to grid operators, though. The partially unpredictable nature of power generation from the key renewable type DGs may endanger the stability and integrity of electric networks as a whole, and at a distribution level in particular. This may also deteriorate the quality of power delivered to consumers.

Because of these concerns, future distribution networks should be prepared to handle the ongoing transformation process of power generation from the traditionally centralized to a more distributed and small power productions. Nonetheless, conventional distribution networks are not designed to manage this, and as a result, regulators often impose a maximum penetration limit which does not help further development of distributed vRESs. But distribution networks are slowly evolving to smart grids, which are adequately equipped with the necessary tools and mechanisms to accommodate large-scale vRESs while minimizing their side-effects mentioned earlier. This chapter explores and discusses the flexibility options that can support the much-needed integration and efficient utilization of large-scale vRESs in the future distribution networks. The assessment also includes managing the negative impacts of vRESs, induced by their high variability and uncertainty, by means of various flexibility options. For this purpose, optimal management of distribution networks is performed via an appropriate mathematical optimization—a stochastic mixed integer linear programing (S-MILP) for deploy different flexibility options along with vRESs. This chapter aims to address the operation issues that can occur in distribution networks due to the high-level variability and uncertainty of vRESs. The analysis is made from the economic and technical point of view. In particular, this chapter makes an extensive analysis on the impacts of vRESs on the overall performance of the system such as voltage profile, losses, costs, system reliability stability and power quality. In addition, the contributions of different flexibility options in enabling high penetration of vRESs and their wide-range benefits are assessed.

The remainder of this chapter is as follows. The next section presents an overview of the need for increased flexibility in distribution networks. The subsequent section describes the developed mathematical formulation used to carry out the required analysis. This is followed by numerical results and discussions. Finally, the last section summarizes the main findings of this chapter.

## 1.2 The Need for Flexibility Options in Distribution Networks

Because of the reasons mentioned earlier, an increasing level of DGs is being connected to distribution networks. The fact that these are based on erratic power sources (wind and solar, for example) is creating technical problems in such systems. Grid operators are especially concerned as the conventional means of overseeing the network systems are now becoming insufficient to keep a healthy operation of such systems. The main reason for this boils down to the partially unpredictable nature of these energy resources. In such circumstances, proper management mechanisms need to be put in place so as to seamlessly accommodate large-scale vRES type DGs. This is critical to address a multitude of global concerns, partially described in the previous section.

In general, there is an increased need for flexibility in distribution networks counterbalance the continuous fluctuations in RES power production and even demand [3, 4]. Traditionally, demand-generation balancing is handled by conventional power plants. However, in the presence of high level vRESs, this approach may be prohibitively expensive or even not sufficient to provide the standard balancing service level. Therefore, the existence of vRES in the system decreases the effectiveness of existing flexibility mechanisms compared with the traditional system (without these resources), mainly due to the intermittent nature of renewables. In other words, the system needs a greater level of flexibility to be able to guarantee the system reliability as the variation increases (both in supply and demand). This is one of the key challenges integrating these energy sources. Therefore, new flexibility options are needed to manage the real-time imbalances in demand and power production. This way, the security of electric supply, stability and power quality can be guaranteed.

Flexibility can be defined as the ability electric distribution networks to efficiently manage its own resources in the event of continuous changes in power supply and demand sides. In this regard, voltage and frequency controls are the primary resources to face uncertainty and variability [5, 6]. In addition, another resource in electric distribution network useful for handling the imbalances as a result of unpredictable changes in the system (either from the supply, demand or both sides) is the network's reserve capacity. Nonetheless, flexibility in electric distribution networks can be affected by many factors such as the amount of reserve capacity, the ramp rates of generators, the type of generation, the availability of generation, interconnection with other electric distribution networks, capacity of interconnections, etc. [7]. These are traditional mechanisms to deal with imbalances mainly caused by traditional sources of uncertainty and variability. Conventional power plants can add reserve capacity to the system but the inherent variability and uncertainty of vRESs definitively change the operation of distribution networks. Under these circumstances, it may not be economical for conventional power plants to offer spinning reserves. This would be costly because of a possibly increased use of fossil fuels for providing the huge requirement of spinning reserves [8].

The fact that the energy sector is transforming to a new paradigm with improved energy efficiency and environmentally-friendly technologies to produce energy at reasonably priced tariffs [9] brings both opportunities and challenges. Flexibility options will be highly needed to address those challenges and reap the benefits. The system-wide reliability, efficiency, reduction of GHGs and affordability of energy can be achieved by deploying and coordinating different flexibility options such as energy storage systems (ESSs), switchable capacitor banks (SCBs), demand response (DR) and others. These technologies substantially enhance the flexibility of the system and its ability to continuously maintain a standard service in the face of large fluctuations in the supply and demand [10, 11].

Given the background given above, the question of having adequate renewables to meet the electricity demand requires one to have sufficient flexibility technologies to balance forecasting errors and fluctuations [12]. These flexibility options can be provided by the energy storage medium, electric distribution networks, demand and supply sides as shown in Fig. 1.1. For example, from the network side, the

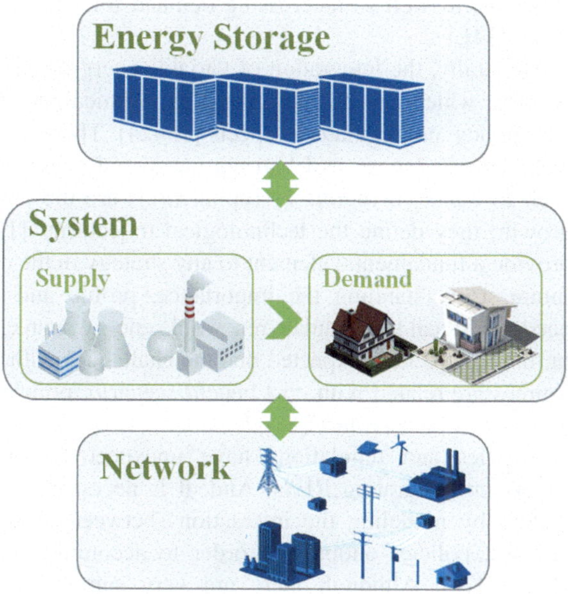


**Fig. 1.1** Identifying flexibility options in electric distribution networks

system can dynamically change its topology to adapt to changing operational situations. The more frequent the reconfiguration is, the better the contribution of such a flexibility mechanism will generally be. From the supply side, the traditional flexibility service in the form of spinning reserve provided by conventional generators is one example. Others include curtailment of variable power and reactive power control. On the demand side, some flexibility options are demand response, energy efficiency and electric vehicles.

### *1.2.1 Challenges of Variable Energy Sources Integration*

Traditionally, distribution networks are built to serve the peak demand, and fulfill reliability and quality requirements, in a radial structure [13]. The role of distribution operators has so far been mainly to construct, maintain and manage outages of their distribution network assets [13]. However, with advent of new technologies and new consumption forms as well as increasing penetration of DGs, this conventional business model needs to be structurally changed. Under this circumstance, distribution grids are expected to support bi-directional power flows, which is completely different from the way these are designed to. This is increasingly becoming a concern for grid operators as this new role complicates the operation of such grids. As a result, the architecture of distribution networks needs to change to effectively overcome the limitations and address the operators' concerns. The systems need to adopt modern technologies after careful planning and be equipped with necessary tools for their efficient operation. This is important to deal with compounded issues pertaining to the political, social, economic and environmental concerns, as well as meet rising demand for energy and sustainable development goals [14].

Generally, the integration of variable energy sources has several challenges and barriers, which can be categorized as technical, economic, social, political, financial, policy and regulatory aspects [15–24]. These are summarized in Fig. 1.2. The technical challenges and barriers are already discussed. There financial markets, such as banks, inventors or capital firms are the main contributors for economic growth; they define the technological trajectories [15]. Because of this, they can provide a fundamental element to any strategy in the direction of a more sustainable future. Understanding the importance, profile and information that an investor needs is critical to formulate renewable energy source (RES) policies and strategies. In this context, it is expected that the challenges with integration of variable energy sources are related with cost benefit scenario, policies and social acceptance analysis as can be seen in Fig. 1.2.

Policies and regulations have unexpected—sometimes counterproductive—effects on integrating RESs. And, it is necessary for policy makers to study the system by modeling the interactions between different parts of the system and different policies adopted in order to accommodate a large-scale integration of vRESs [25]. Although there are very supportive contributions from different

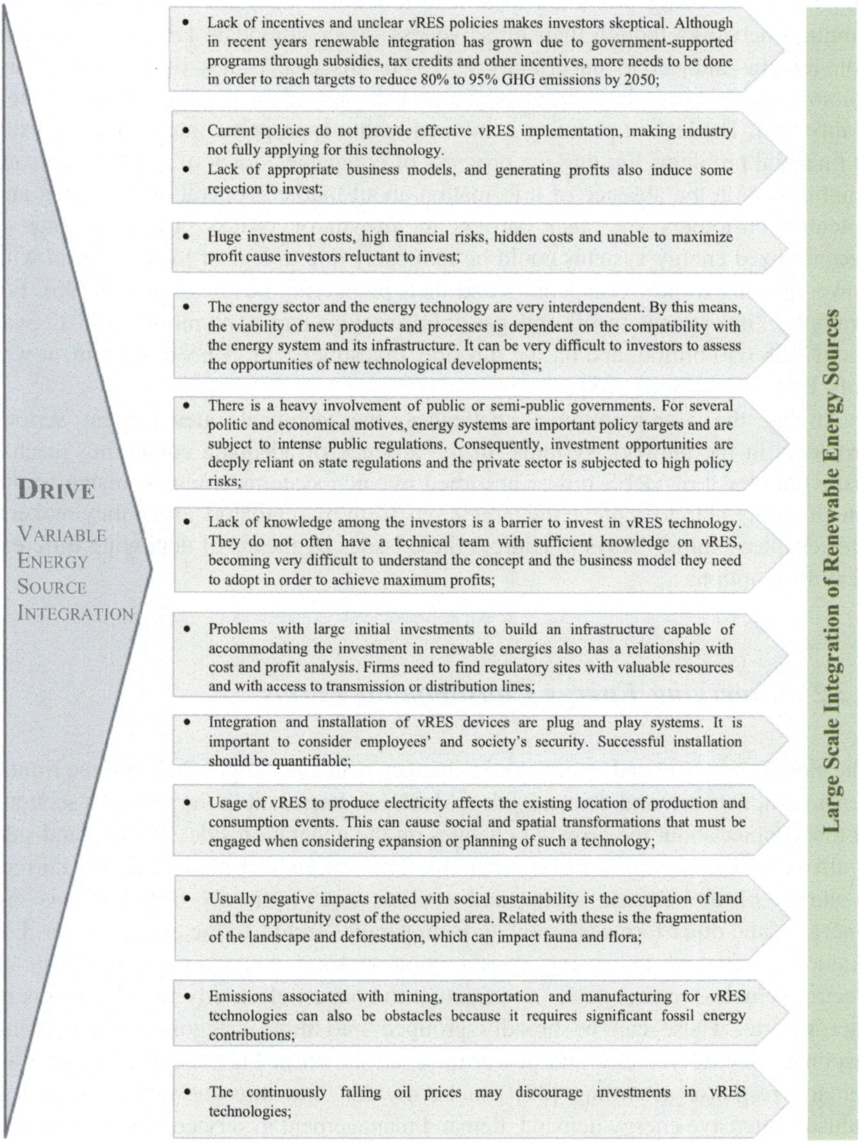


**Fig. 1.2** Challenges of integrating vRESs

nations, we face with a regulatory framework that comprises laws to overall support RESs but there is no long-term planning because the approaches and framework conditions are always changing [26]. As the network requires to build and operate complex systems involving many corporations, this changing conditions does not permit a system to function effectively [26]. Policies for renewable energy integration are being promoted to diffuse renewable energies within electric distribution networks though their effectiveness to accommodate large-scale integration remains subject to uncertainty [27]. For instance, states often try to assist countries that import laws from others and do not adapt the framework to their reality [26]. The lack of planning combined with inappropriate incentives can result in financial problems limiting the progress of companies. Lack of qualified persons combined with the absence of information about markets, operation, planning and potential customers are other barriers to growth of vRESs. The slow rate of decentralized energy systems could be purposely due to fear of losing control with power shifting to new competitors and their pioneering business models [26]. For example, "investment in oil and gas infrastructure and exploration in 2012 was about US$ 650 billion, and on the flip side, investments in vRES development was only US$ 244 billion" [28].

Among the aforementioned challenges, the technical ones present serious problems in the network systems. In the absence of adequate countering mechanisms, the level of vRES power absorbed by such systems could be insignificant, which hardly help to achieve the targets sets forth by regulators and policy makers. This chapter explores ways to address these issues by means of deploying different flexibility options.

### *1.2.2 Emerging Energy Consumption Forms*

The electric sector is undergoing rapid changes with a paradigm shift in three fronts: generation, network and demand sides. Much has been said in the previous sections of this chapter about the growing changes on the generation side. The demand side is also experiencing rapid transformations. This means that along with the current evolution of the electric sector and society, new forms of consumptions are emerging and other forms are moving from parallel sectors to the energy sector. For example, new and increasing consumption styles include e-mobility (such as electric vehicles), power-to-X (an initiative to convert electricity to other forms of energy), etc. These can be broadly grouped into three categories: the demand response, electric vehicles and power-to-X, as shown in Fig. 1.3. The category of demand response according [29] can be divided into three new sub categories, industry intensive energy demand, demand management in services and households and smart applications. The latter stems from the changes that are being made in the electricity sector by transforming the traditional networks into smart grid, taking advantage of the new communication capabilities that are being integrated into the system. The remaining subcategories arise from the electrification of other sectors such as, the transportation and the heating/cooling sectors [29]. A more detailed approach to each of these categories is made in the subsequent sections, where the main features of the new demand forms are presented together with the challenges associated with each one.

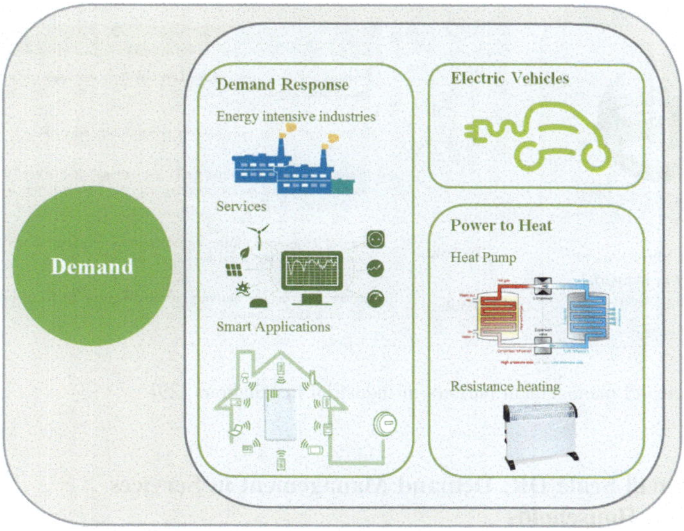


**Fig. 1.3** New emerging forms of energy consumption

#### 1.2.2.1 Demand Management in Industrial Installations

One of the subcategories within the demand is the demand management in industrial facilities. In this subcategory, demand is modeled by the specific industrial process characteristics, and can vary from one type of industry to another type. As a result, the demand is not uniform. However, as already explained before, the energy sector is changing and with the emergence of new concepts related to the smart grids, some ways can be found inside the demand to enhance system stability. Therefore, some types of industries have productive processes that offer a certain level of flexibility, that is, they can change the energy needs of the production process over a given period of time. Some examples of such processes are the ones that include electrolysis (very intensive installations), cement and paper industries, electric boilers, and electric arc furnaces [30]. However, a large number of industries do not have this flexibility.

In the industry, the factors that determine any action are the costs and gains. According to the reports in [29, 30], increasing flexibility at a low cost is generally possible in the cases where the primary process is not disturbed. These costs generally refer to the workers' shifts, the installation of communication and control equipment, and the additional potential storage of intermediate products on-site [31]. Therefore, this subcategory faces significant challenges that are presented in Fig. 1.4.

| Barrier | Description |
|---|---|
| Economic | Demand response is dependent on the electricity costs sensitivity and market price signals. |
| | In most markets, the extra capacity prevents large price spikes. |
| | In most industrial entities, the high organizational effort is not worthwhile due to the low level savings resulting from the demand shift to hours with lower electricity prices. |
| Technical | Some of the potential barriers are the product quality loss, shorter periods for production line changes, and the demand structure. |
| Political | Some markets punish temporary differences in demand, for example, higher rates on the network. |

**Fig. 1.4** Demand management barriers in industrial installations [29]

#### 1.2.2.2 Small-Scale DR: Demand Management in Services and Households

Another subcategory is related to the demand response in commercial and residential sectors, which can be accommodated in the same category because demand management can be applied to transversal processes such as heating and cooling. Including different demand electricity price levels, such as refrigeration timing for refrigerated warehouses, automatic adjustment of demand can be done by refrigerators [29].

Other technology types that are transversal to the two subcategories and with potential in the demand management are the air conditioned, air compression for mechanical use or even scheduling of washing processes in the dwellings. Several small load management programs are currently being installed up to 5 kW in several countries using two directions of communication (coming from the integration of smart grids) and the potential of these programs is very large [31]. However, existing IT infrastructures as well as primary device control constraints can present significant challenges. Therefore, demand management can reverse the game in the electricity markets, since this subcategory can also contribute to the creation of flexibility in the network systems, being no longer seen simply as a demand, in terms of flexible demand, by establishing the marginal prices in wholesale electricity markets [32]. The set of challenges pertaining to demand management in commercial and households are summarized in Fig. 1.5.

#### 1.2.2.3 Electric Vehicles (EVs)

Electric vehicles are one of the new energy consumption forms. For mobility purposes, they use energy stored in their EV batteries. The charging process is carried out by connecting the EV to the grid when the vehicle is parked at an EV parking lot [33]. Energy can also be transferred the EV to the electricity grid.

| Barrier Type | |
|---|---|
| Economic | The need to invest in information infrastructures and telecommunication (IT), and data processing. |
| | Few price tariffs are available in real time and market prices are not visible at the retail level. |
| Technical | The lack of communication infrastructure, costumer empowerment in most countries. |
| Political | Lack of acceptance or support in data security issues, change the paradigm of several different interest parts including the consumer. |

**Fig. 1.5** Demand management barriers in services and households

This effectively means that EVs can operate in two modes: power source and demand. The demand characteristics in the mobility sector makes the EV fleets be similar to the previous subcategories, as an option not only for demand but also for flexibility in the energy system that can be presented in two fundamental forms [34]:

- G2V (Grid-to-Vehicle mode, where the fleets of EVs are operated as a demand side management option, allowing a shift of load among different times).
- V2G (Vehicle-to-Grid mode, where in addition to charging the batteries of the electric vehicle, EVs could be discharged and feed power back to the grid).

Due to the fact that its primary use is for mobility, the provision of flexibility by EVs is subject to many constraints. In addition, it is highly uncertain supply source. However, several studies show that EVs may be competitive flexibility options [33–36] because they are expected to be largely available overnight (home charging). During the day, their availability depends on the charging infrastructures that exist elsewhere (for example, at work). The main advantage of EVs is that they are a parallel development, i.e., their investment comes from the transport sector [36]. EVs have a potential role to serve as a source of balancing and reserve requirements, as well as a solution to solve problems locally [34]. However, EVs face a significant set of challenges that are presented in Fig. 1.6.

#### 1.2.2.4 Power to Heat

Electricity can be used to replace other fuels such as gas or oil for residential heating purposes. One of the possible options is the direct heating in a housing, where the electric current through a resistor converts electrical energy into thermal energy [37]. Moreover, this subcategory has potential at the flexibility level, which can be created by selectively energizing heaters and storing the heat generated for later use. Thermal energy can be stored with relative efficiency in several ways [38], typically including insulated ceramic type containers and hot water tanks. The heat is then released as needed by the end users. Electric heat pump technology is one of the most efficient technologies that convert electricity into heat. The heat pumps

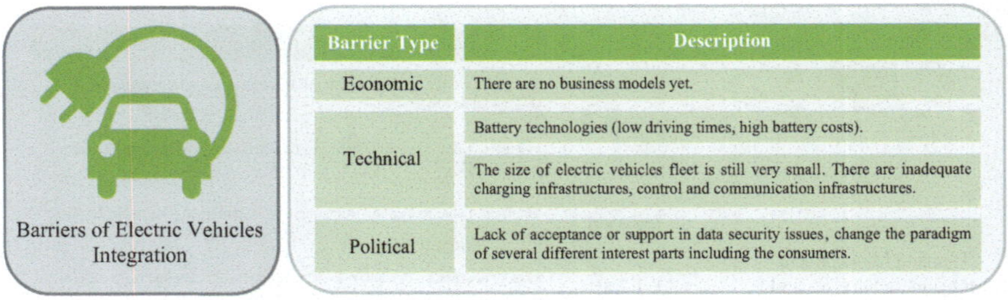


| Barrier Type | Description |
|---|---|
| Economic | There are no business models yet. |
| Technical | Battery technologies (low driving times, high battery costs).<br>The size of electric vehicles fleet is still very small. There are inadequate charging infrastructures, control and communication infrastructures. |
| Political | Lack of acceptance or support in data security issues, change the paradigm of several different interest parts including the consumers. |

**Fig. 1.6** Electric Vehicles main barriers

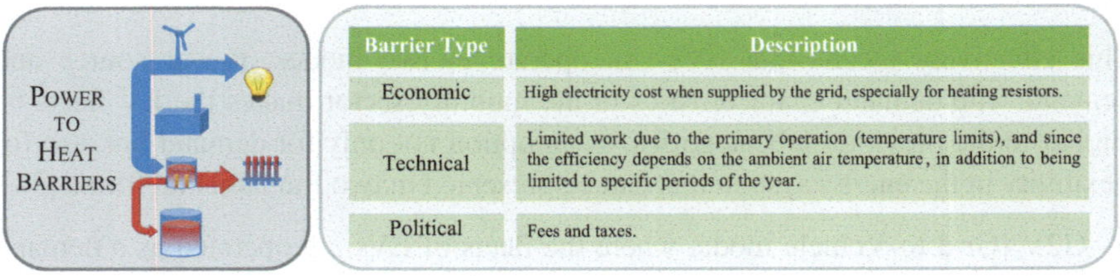


| Barrier Type | Description |
|---|---|
| Economic | High electricity cost when supplied by the grid, especially for heating resistors. |
| Technical | Limited work due to the primary operation (temperature limits), and since the efficiency depends on the ambient air temperature, in addition to being limited to specific periods of the year. |
| Political | Fees and taxes. |

**Fig. 1.7** Power to Heat barriers

efficiently move the stored heat energy from a heat source (e.g., ambient air) to the end use or storage.

Heat pump technology is a part of conditioned air and refrigerators. The principle is the same, but the direction of the heat flow is from outside, the ambient air from the conditioner in cooling applications, whereas in the heating is the inverse [37]. In fact, heat pumps are reversible and can perform as both heating and cooling functions simultaneously in some applications.

The electrification of the heating sector also shifts the demand from the heating sector to the electricity sector, and can add some important flexibility to the system. The combination of thermal storage with electric heating has the potential to increase the flexibility of the electric distribution networks as it builds an optional place to put temporary surges of vRES energy and reduce carbon emissions through the displacement of heat sources out of fossil fuels [12]. Power to Heat also has a set of barriers that are summarized in Fig. 1.7.

#### 1.2.2.5 Power to Gas

Power to Gas is a category commonly found in energy storage but can also be integrated into demand since Power to Gas refers to the storage of chemical energy, namely the use of electric power to create fuels that can be used in conventional power plants. The key fuel is synthetic methane (and hydrogen in some cases). The procedure consists of two steps [39]:

| Barrier Type | Description |
|---|---|
| Economic | High cost. |
| Technical | Low efficiency, an external $CO_2$ source or extraction of air (which leads to a reduction of efficiency) |

**Fig. 1.8** Power to Gas barriers

- Electricity is used in electrolysis to divide water into hydrogen and oxygen.
- Hydrogen is combined with carbon to obtain methane.

Methane is the main constituent of natural gas and therefore can be injected into the existing infrastructure of natural gas (network and storage). The high storage capacity in the network could then be used for medium and long term storage purposes.

A first demonstration project at the kW scale was built and operated in Germany, and a 6 MW German plant also started operating in 2013 [40]. The key to chemical storage compared to other technologies is their energy density (kWh/l) compared to most other technologies as well as the long period of change. The key barrier is low efficiency [41].

The great strength of this category lies in the seasonal storage, probably to be used in the transport sector in the first place. The technology perspective increases with the prospect of relying on 100% of renewable resources, storing the surplus of electric energy in the (central) gas infrastructure when generation from vRES is low [39]. Some of the challenges of the Power to Gas technology can be seen in Fig. 1.8 [29].

## *1.2.3 Risk Posed by Increasing Uncertainty and Variability*

Variable RESs are not always available when needed. They are subject to high level variability and uncertainty. Variability is related to the natural variation, for instance, of wind or sun to produce energy, meaning that the produced energy can fluctuate in certain quantity over regular time intervals. Uncertainty refers to the partially unpredictable nature of the uncertain parameters. As a result, daily and seasonal effects and limited predictability turns vRESs as highly intermittent generation sources [27]. Hence, as they are intermittent, they are not dispatchable and their output power cannot be controlled. Because of these reasons, in the absence of proper strategies, integration of vRESs can pose significant operational risk, making system voltage and frequency controls very difficult. This is because increasing penetration of vRESs increases fluctuations and creates big and uncertain generation-demand imbalances [42]. This leads to power quality and stability concerns. Grid disturbances, for instance, short-circuit faults can cause voltage sags and frequency variations, sending them both off the standard limits. Generally, increased levels of vRESs may cause more complex and uncertain operation situations [42]. Accordingly, there is a need for proper planning and decision making to face uncertainties for achieving optimal vRES integration [43].

Power quality issues when integrating vRES encompass the following important issues: (1) voltage and frequency oscillations triggered by non-controllable vRESs and by power grid disturbances, and (2) harmonics that are introduced by the electronic converters used in vRESs, that are necessary for adapting fluctuating production with grid requirements [42, 44]. Because of the intermittence of vRESs, one way to control power output is simply by curtailing the power production. Nonetheless, it is not an effective way since the curtailed energy could be stored and used on latter moments, not only for demand supply but also for voltage and frequency control of the power output.

In order to face voltage and frequency problems, utilities have introduced various grid codes for connecting vRESs to electric distribution networks. The regulatory framework of the grid codes are defined by the system operators to outline the duties and rights of all loads and power generation connected to the transmission and distribution networks [45]. Previously, the large-scale integration of vRESs, grid codes did not include regulations for wind and solar systems because the installed generation was very insignificant compared to the traditional generation systems. This situation has been changing in recent years as the level of vRESs integrated in distribution grids is on the rise. Such a massive integration of vRESs creates genuine stability concerns in the system due to the negative impacts of large solar and wind power plants. These concerns are related with voltage and frequency drops in the presence of a fault or high winds, making wind turbines to stall, that can lead to outages [45]. Accordingly, rigorous technical requirements are enforced to protect networks to contrast to these threats. As an example, wind power plants are required to withstand various grid disturbances and contribute to the stability of the system and provide ancillary services.

The technical challenges that vRES introduces to electric distribution networks increases the need for high level flexibility from other parts of the systems and flexibility through interaction with other energy sectors, like heating sector, natural gas and interaction between transportation and distribution networks [25].

### *1.2.4 The Path Towards More Flexible and Smarter Grids*

Given the new developments from the demand and supply sides, distribution network systems need to undergo the necessary transition to more flexible and smarter grids. Future grids will be equipped with different types of flexibility options such as ESSs, reactive power sources such as SCBs, DR and dynamic network reconfiguration (DNR). Moreover, a coordinated deployment and scheduling of flexibility options are needed to optimally manage an increased penetration of vRES in distribution networks. For example, energy storage systems can be added onsite for frequency control and add quick reserve capacity to the system. ESSs can also

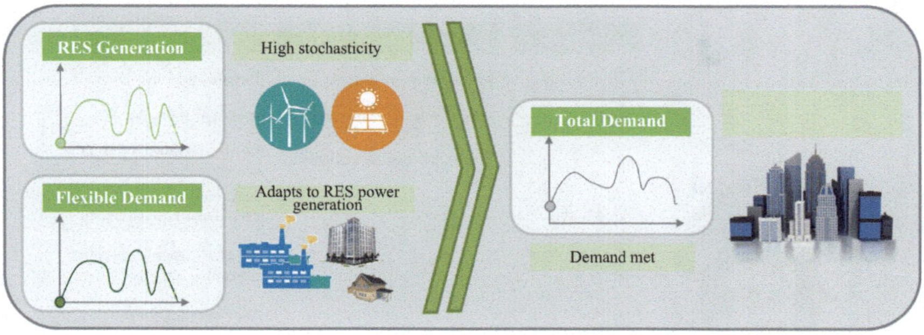


**Fig. 1.9** Flexibility options and smart grids

provide other services. Their fast response means that they can be part of the ancillary services (frequency control) and suited to black-out restart of the system. The operation principle of ESSs is to store excessive energy during the demand low period that will be utilized in periods of high demand.

Load flexibility options like demand response (DR) can also enhance the integration of vRESs, giving the control of operation of contracted services to a new competitor, named aggregator. From the network side, one example of potential flexibility option is dynamic reconfiguration of the distribution network. Dynamic reconfiguration can play substantial role in improving reliability, increasing RES penetration and minimizing power losses. Switchable capacitor banks can also provide adequate flexibility to the system, enhancing stability and RES integration level.

Flexibility options form important components of electric distribution networks and play important roles in the transformation of current electric distribution networks to smarter grids in the future. Most current systems are based on fossil fuels. Yet, the recent trend of system evolution shows that future grid systems will be based on the efficient accommodation of large scale variable renewable energy sources [32]. The existence of sufficient operational flexibility is a necessary prerequisite for the efficient large-scale integration RES energy in such network systems. Flexibility is not only necessary to mitigate supply variations due to increased uncertainties but also the variations in from demand side due to new and relatively unpredictable energy consumption forms. This is graphically illustrated in Fig. 1.9.

Therefore, future power grids need to become smarter, allowing multi-directional power flows, and allowing consumers to no longer have a passive role instead to play an active role in the electricity markets [11, 46, 47 p. 21]. Intelligent infrastructures are being developed both at the distribution and transmission levels. Intelligent network projects are being generalized around the world, where budgets have kept on increasing almost exponentially from 2006 [12].

However, the development of smart grids faces a significant set of challenges. In particular, standardization of communication and operational protocols, which will play a key role in future networks, is yet an ongoing process. Energy consumption

| Impact | Costs | Benefits |
|---|---|---|
| Direct | Investment in smart grids | Deferred grid investments |
| | Smart Grids and maintenance | Reduced grid losses |
| | Cost on location for equipment | Lower investment needs in storage |
| | | Lower need of investments in central generation capacity |
| | | More efficient use of central generating capacity |
| | | More energy savings |
| | | Reduced imbalances |

| Impact | Costs | Benefits |
|---|---|---|
| Indirect and External | Welfare losses due to adaptation to new energy demand patterns | Welfare gain due to new services |
| | | Reduced $CO_2$ emissions |
| | | Reduced outage time |
| | | Reduced curtailment of distributed generation |
| | | Increased grid hosting capacity for distributed generation |

**Fig. 1.10** Comparison of potential costs and benefits of developing smart grids and flexibility

optimization should be based on near-real time, which requires well-developed communication framework to facilitate the active interactions between producers and consumers. In order to select these communications individually, standardized protocols already exist. However, these are limited to a single domain [48]. With regard to the introduction of smart grids, one of the key tasks in the near future is the establishment of an interactive bidirectional communication system from the generation to the final consumer.

Having smart grids in perspective, the main ways to introduce flexibility into the electric distribution networks are through the introduction of fast markets, flexible generation (e.g., gas and water), demand side management, energy storage systems and interconnections. The smart grids in combination with all other forms of flexibility options mentioned previously will considerably increase the flexibility of the system, overcome congestion in the network systems, either by changing flexible loads from peak periods to periods with less congestion, or through the control of the network power flow due to the integration of large-scale renewables in the near future, among others. This leads to the creation of a more flexible and manageable network. However, the costs and benefits associated with the development of smart grids and network flexibility have direct and indirect effects, as can be seen in the scheme of Fig. 1.10.

With regard to integrated solutions for low carbon emissions, Smart Grids will be a key element in the implementation of modern technologies. The need for flexibility resulting from the integration of renewable energies, demand and contingencies can be met in different ways, including through flexible generation, response to demand, energy storage and interconnections of the electric distribution networks. All this makes it a key component for the emergence of Smart Grids.

## 1.3 Managing Distribution Networks Featuring Large-Scale Variable Energy Sources

### *1.3.1 General Problem Description*

This chapter develops an optimization model for carrying out detailed analysis of optimally operating distribution network systems featuring large-scale intermittent power sources with the help of various flexibility mechanisms. These mechanisms include dynamic network reconfiguration, energy storage systems, reactive power sources and demand response. A coordinated use of these technologies should lead to increased benefits in distribution network systems such as reduced costs, increased utilization of renewables and others.

The uncertainty inherent to the problem addressed in this chapter is handled by means of stochastic programming. In order to ensure solution exactness and enhance problem tractability, the entire problem is formulated as a stochastic mixed integer linear programming (MILP) optimization model. The accuracy of the analysis is guaranteed because this chapter proposes a model that employs a linearized AC power flow model, which strikes the right balance between accuracy and computational requirement.

### *1.3.2 Algebraic Formulation*

The objective of the formulated DNR problem is to minimize the sum of relevant cost terms, namely, costs of switching, expected costs of operation, unserved power and emissions in the system. This is given as

$$Minimize\, TC = TSC + TEC + TENSC + TEmiC \tag{1.1}$$

where *TC* refers to the expected total cost in the system.

The first term in (1.1), *TSC*, is related to the switching costs as a result of dynamic network reconfigurations. A switching cost is incurred when the status of a given feeder changes from 0 (open) to 1 (closed) or 1 (closed) to 0 (open). This leads to the absolute value of difference in successive switching variables. In order to linearly represent such a module, two non-negative auxiliary variables $x^{+}_{l,h}$ and

$x_{l,h}^{-}$ are introduced. Thus, *TSC* can be expressed as a function of the sum of these variables:

$$TSC = \sum_{k \in \Omega^k} \sum_{h \in \Omega^h} SW_k * \left( x_{k,h}^{+} + x_{k,h}^{-} \right) \tag{1.2}$$

where

$$u_{k,h} - u_{k,h-1} = y_{k,h}^{+} - y_{k,h}^{-}; y_{k,h}^{+} \geq 0; y_{k,h}^{-} \geq 0 \tag{1.3}$$

$$u_{k,0} = 1; \forall k \in \Omega^1 and\, u_{k,0} = 0; \forall k \in \Omega^0 \tag{1.4}$$

It should be noted that $\Omega^1$ and $\Omega^0$ refer to the sets of normally closed feeders and tie lines (which are normally opened), respectively. However, the status of any of these feeders can change during the course of the day depending on the optimality of the dynamic network reconfiguration.

The second term in (1.1), *TEC*, represents the expected costs of producing power using DGs, operating ESSs and importing power from upstream, which is given as in (1.5).

$$TEC = EC^{DG} + EC^{ES} + EC^{SS} \tag{1.5}$$

Each term in (1.5) is calculated as

$$EC^{DG} = \sum_{s \in \Omega^s} \rho_s \sum_{h \in \Omega^h} \sum_{g \in \Omega^g} OC_g P_{g,i,s,h}^{DG} \tag{1.6}$$

$$EC^{ES} = \sum_{s \in \Omega^s} \rho_s \sum_{h \in \Omega^h} \sum_{es \in \Omega^{es}} \lambda^{es} P_{es,i,s,h}^{dch} \tag{1.7}$$

$$EC^{SS} = \sum_{s \in \Omega^s} \rho_s \sum_{h \in \Omega^h} \sum_{\varsigma \in \Omega^\varsigma} \lambda_h^\varsigma P_{\varsigma,s,h}^{SS} \tag{1.8}$$

The third term, *TENSC*, captures the expected cost of load shedding. This is calculated as the sum of the costs of unserved active and reactive power as

$$TENSC = \sum_{s \in \Omega^s} \rho_s \sum_{h \in \Omega^h} \sum_{n \in \Omega^n} \left( \upsilon_{s,h}^{P} P_{i,s,h}^{NS} + \upsilon_{s,h}^{Q} Q_{i,s,h}^{NS} \right) \tag{1.9}$$

where $\upsilon_{s,h}^{P}$ and $\upsilon_{s,h}^{Q}$ are penalty parameters corresponding to active and reactive power demand curtailment. These parameters should be sufficiently large to avoid undesirably high unserved power.

The last term, *TEmiC*, accounts for the expected cost of emissions as a result of generating power using DGs and importing power through the substation as in (10).

$$TEmiC = EmiC^{DG} + EmiC^{SS} \tag{1.10}$$

Each of the terms in (1.10) are determined by

$$EmiC^{DG} = \sum_{s\in\Omega^s} \rho_s \sum_{h\in\Omega^h} \sum_{g\in\Omega^g} \sum_{n\in\Omega^n} \lambda^{CO_2} ER_g^{DG} P_{g,i,s,h}^{DG} \tag{1.11}$$

$$EmiC^{SS} = \sum_{s\in\Omega^s} \rho_s \sum_{h\in\Omega^h} \sum_{\varsigma\in\Omega^\varsigma} \sum_{n\in\Omega^n} \lambda^{CO_2} ER_\varsigma^{SS} P_{\varsigma,s,h}^{SS} \tag{1.12}$$

There are a number of technical and economic constraints that need to be respected all the time to ensure a healthy operation of distribution networks. Kirchhoff's current law states that the sum of all incoming flows to a node should be always equal to the sum of all outgoing flows at any given time. This constraint applies to both active (1.13) and reactive (1.14) power flows, and should be respected all the time:

$$\begin{aligned}&\sum_{g\in\Omega^g} P_{g,i,s,h}^{DG} + \sum_{es\in\Omega^{es}} \left(P_{es,i,s,h}^{dch} - P_{es,i,s,h}^{ch}\right) + P_{\varsigma,s,h}^{SS} + P_{i,s,h}^{NS} + \sum_{in,l\in\Omega^l} P_{l,s,h} \\ &- \sum_{out,l\in\Omega^l} P_{l,s,h} = PD_{s,h}^i + \sum_{in,l\in\Omega^l} \frac{1}{2} PL_{l,s,h} + \sum_{out,l\in\Omega^l} \frac{1}{2} PL_{l,s,h}; \forall\varsigma \in \Omega^\varsigma; \forall\varsigma \in i; l \in i\end{aligned} \tag{1.13}$$

$$\begin{aligned}&\sum_{g\in\Omega^g} Q_{g,i,s,h}^{DG} + Q_{c,i,s,h}^{c} + Q_{\varsigma,s,h}^{SS} + Q_{i,s,h}^{NS} + \sum_{in,l\in\Omega^l} Q_{l,s,h} \\ &- \sum_{out,l\in\Omega^l} Q_{l,s,h} = QD_{s,h}^i + \sum_{in,l\in\Omega^l} \frac{1}{2} QL_{l,s,h} + \sum_{out,l\in\Omega^l} \frac{1}{2} QL_{l,s,h}; \forall\varsigma \in \Omega^\varsigma; \forall\varsigma \in i; l \in i\end{aligned} \tag{1.14}$$

As can be seen in (1.13), incoming flows include the active power injected by DGs, inward active power flows in associated feeders, power discharged from ESSs and the amount of power imported (if the bus under consideration is a substation). On the other hand, outgoing flows encompass demand, losses (which are treated here as fictitious loads), outward flows in feeders and charged amount of ESSs.

Power flows in any feeder should also be governed by Kirchhoff's voltage law. This is enforced by including linearized power flow equations, derived by considering two practical assumptions. The first assumption is related to bus voltage magnitudes, which is expected to be close to the nominal value $V_{nom}$ in electric distribution networks. The second one is related to the voltage angle difference $\theta_k$, which is often very small due to practical reasons. The second assumption leads to the trigonometric approximations $\sin\theta_k \approx \theta_k$ and $\cos\theta_k \approx 1$. Given these simplifying assumptions, the well-known AC power flow equations (which are naturally complex nonlinear and non-convex functions of voltage magnitude and angles) can be linearly represented. The linearized active and reactive flows in a line are given by the disjunctive inequalities in (1.15) and (1.16), respectively.

$$\left|P_{k,s,h} - \left(V_{nom}\left(\Delta V_{i,s,h} - \Delta V_{j,s,h}\right)g_k - V_{nom}^2 b_k \theta_{k,s,h}\right)\right| \leq MP_k\left(1 - u_{k,h}\right) \tag{1.15}$$

$$\left|Q_{k,s,h} - \left(-V_{nom}\left(\Delta V_{i,s,h} - \Delta V_{j,s,h}\right)b_k - V_{nom}^2 g_k \theta_{k,s,h}\right)\right| \leq MQ_k\left(1 - u_{k,h}\right) \tag{1.16}$$

where $\Delta V^{min} \leq \Delta V_{i,s,h} \leq \Delta V^{max}$.

Moreover, power flows in each line should not exceed the maximum transfer capacity, which is enforced by

$$P_{k,s,h}^2 + Q_{k,s,h}^2 \leq u_{k,h}(S_k^{max})^2 \tag{1.17}$$

The following constraints are related to the active (1.18) and reactive (1.19) power losses in line $k$.

$$PL_{k,s,h} = R_k\left(P_{k,s,h}^2 + Q_{k,s,h}^2\right)/V_{nom}^2 \tag{1.18}$$

$$QL_{k,s,h} = X_k\left(P_{k,s,h}^2 + Q_{k,s,h}^2\right)/V_{nom}^2 \tag{1.19}$$

Note that the quadratic flows in (1.17)–(1.19) are linearized using a piecewise linearization approach, which is widely used in the literature.

Constraints (1.20)–(1.25) represent the energy storage model employed in this chapter. The amount of power charged and discharged are limited as in (1.20) and (1.21). Constraint (1.22) ensures that charging and discharging operations do not happen at the same time. The constraint related to the state of charge is given by (1.23). The storage level should always be within the permissible range (1.24). Equation (1.25) sets the initial storage level, and makes sure the storage level at the end of the time period is equal to the initial level. For sake of simplicity, both $\eta_{es}^{dch}$ and $\eta_{es}^{ch}$ are often set equal.

$$0 \leq P_{es,i,s,h}^{ch} \leq I_{es,i,s,h}^{ch} P_{es,i,h}^{ch,max} \tag{1.20}$$

$$0 \leq P_{es,i,s,h}^{dch} \leq I_{es,i,s,h}^{dch} P_{es,i}^{ch,max} \tag{1.21}$$

$$I_{es,i,s,h}^{ch} + I_{es,i,s,h}^{dch} \leq 1 \tag{1.22}$$

$$E_{es,i,s,h} = E_{es,i,s,h-1} + \eta_{es}^{ch} P_{es,i,s,h}^{ch} - P_{es,i,s,h}^{dch}/\eta_{es}^{dch} \tag{1.23}$$

$$E_{es,i}^{min} \leq E_{es,i,s,h} \leq E_{es,i}^{max} \tag{1.24}$$

$$E_{es,i,s,h0} = \mu_{es} E_{es,i}^{max};\, E_{es,i,s,h24} = \mu_{es} E_{es,i}^{max} \tag{1.25}$$

Equations (1.26) and (1.27) impose the active and reactive power limits of DGs, respectively.

$$P_{g,i,s,h}^{DG,min} \leq P_{g,i,s,h}^{DG} \leq P_{g,i,s,h}^{DG,max} \tag{1.26}$$

$$-tan\left(cos^{-1}\left(pf_g\right)\right)P_{g,i,s,h}^{DG} \leq Q_{g,i,s,h}^{DG} \leq tan\left(cos^{-1}\left(pf_g\right)\right)P_{g,i,s,h}^{DG} \tag{1.27}$$

The reactive power supplied by switchable capacitor banks (SCBs) is limited by inequality (1.28):

$$0 \leq Q_{c,i,s,h}^{c} \leq x_{c,i,h} Q_c^0 \tag{1.28}$$

where $Q_c^0$ is the minimum deployable unit of a capacitor bank.

For stability reasons, the reactive power at the substation is subject to lower and upper bounds as

$$-\tan\left(cos^{-1}(pf_{ss})\right)P_{\varsigma,s,h}^{SS} \leq Q_{\varsigma,s,h}^{SS} \leq \tan\left(cos^{-1}(pf_{ss})\right)P_{\varsigma,s,h}^{SS} \tag{1.29}$$

In order to account for demand response, the following constraints corresponding to the responsive active and reactive power demand are added:

$$PD_{s,h}^{i} = PD_{s,h}^{i,0}\left(1 + \sum_{h'} \xi_{h,h'}\left(\frac{\lambda_{s,h'} - \lambda_s^{flat}}{\lambda_s^{flat}}\right)\right) \tag{1.30}$$

$$QD_{s,h}^{i} = QD_{s,h}^{i,0}\left(1 + \sum_{h'} \xi_{h,h'}\left(\frac{\lambda_{s,h'} - \lambda_s^{flat}}{\lambda_s^{flat}}\right)\right) \tag{1.31}$$

$$\lambda_s^{flat} = \frac{\sum_h \lambda_{s,h}^{\varsigma}}{24} \tag{1.32}$$

where $PD_{s,h}^{i,0}$ and $QD_{s,h}^{i,0}$ are the active and reactive power loads before demand response. Note that, for the sake of simplicity, the flat price is assumed to be equal to the average electricity price of the day as in (1.32).

Electric distribution networks are normally operated in a radial configuration. Hence, in addition to the aforementioned ones, the radiality constraints in [16] are included in the model developed here. It should be also noted that, in (1.15) and (1.16), the angle difference $\theta_{l,s,h}$ is defined as $\theta_{k,s,h} = \theta_{i,s,h} - \theta_{j,s,h}$ where $i$ and $j$ correspond to the same line $k$.

## 1.4 Case Study, Results and Discussions

A standard IEEE 41-bus test system, whose single-line diagram is shown in Fig. 1.11, is employed here to perform the required technical and economic analysis. The total active and reactive power demand of this system are 4.635 MW and 3.25 MVAr, respectively. The nominal voltage of the system is 12.66 kV. Further details and information of this test system can be found in [49, 50].

The optimal locations and sizes of various distributed energy resources such as wind and solar type DGs, ESSs and SCBs in [50] are considered in this chapter. The only exception is at bus 14, where, instead of the optimal DG size (3 MW) reported in [50], a 2 MW DG is considered throughout this analysis. To make this chapter self-contained, the input data with regards to reactive power sources, DGs and ESSs are presented in Tables 1.1, 1.2 and 1.3 [50]. Figure 1.11 also clearly shows the locations of the considered DGs and ESSs. In addition, the following considerations are made when carrying out the simulations:

- The operational analysis spans over a 24-h period, with the possibility of hourly network reconfiguration.
- The maximum allowable deviation of the nodal voltage at each node is set to ±5% of the nominal value (12.66 kV).
- For all simulations, the substation serves as the reference node, whose voltage magnitude and angle are set equal to the nominal value and 0, respectively.
- The power factor at the substation is set equal to 0.8, and this is held constant throughout the analysis. The power factor of all DG types is considered to be 0.95.

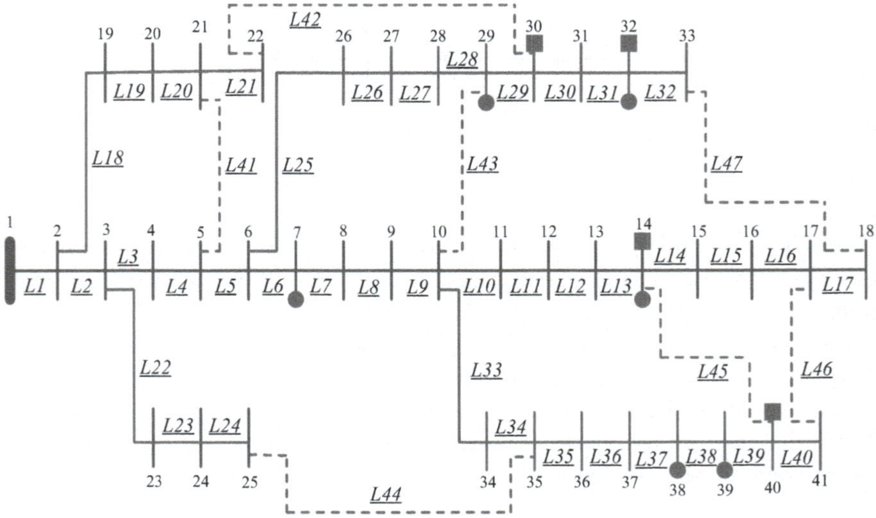


**Fig. 1.11** IEEE 41-bus distribution network with new tie-lines (square and circle dots represent the locations of ESSs and DGs, respectively)

**Table 1.1** Locations and sizes of capacitor banks

| Location (Bus) | Size (MVAr) |
|---|---|
| 7 | 0.9 |
| 14 | 1.3 |
| 24 | 0.1 |
| 25 | 0.3 |
| 29 | 0.3 |
| 30 | 1 |
| 31 | 0.2 |
| 32 | 0.5 |
| 37 | 0.1 |
| 38 | 2 |
| 39 | 0.1 |
| 40 | 0.6 |

**Table 1.2** Locations and sizes of DGs

| vRES type | Location (Bus) | Size (MW) |
|---|---|---|
| PV | 32 | 1 |
| PV | 38 | 1 |
| Wind | 7 | 1 |
| Wind | 14 | 2 |
| Wind | 29 | 1 |
| Wind | 32 | 1 |
| Wind | 38 | 1 |
| Wind | 39 | 1 |

**Table 1.3** Locations and sizes of ESSs

| Location (Bus) | Size (MW) |
|---|---|
| 14 | 2 |
| 30 | 1 |
| 32 | 1 |
| 40 | 1 |

- The emission rate at the substation is arbitrarily set to 0.4 $tCO_2e/MWh$ while those of solar and wind type DGs are assumed to be 0.0584 and 0.0276 $tCO_2e/MWh$, respectively.
- The price of emissions is considered to be 7 €/$tCO_2e$.
- The tariffs of solar and wind power generation are set equal to 40 and 20 €/MWh, respectively.
- Both charging and discharging efficiency of ESSs is 90%.
- The variable cost of operating ESSs is considered as 5 €/MWh.
- The cost of load shedding is 3000 €/MW, and any unserved reactive power is also penalized by the same amount.

- All feeders (including tie-lines) have a maximum transfer capacity of 6.986 MVA, which needs to be respected.
- All big-M parameters are set equal to 20, which is sufficiently large for the considered system.
- The number of partitions considered for linearizing quadratic terms in (1.17)–(1.19) is 5, which is set according to the findings in [51].
- The switching cost parameter is set to 10 €/switching.
- All self-elasticity parameters are set equal to −0.2 while the effect of cross-elasticities is not accounted for in this chapter. This means that cross-elasticity parameters are all considered to be zero.

In addition, for the sake of brevity, the energy intensities of solar and wind power sources is considered to be uniform throughout the system nodes. This means that the power generation profiles of solar and wind type DGs are the same in all the nodes where these resources are connected to. Moreover, it is assumed that the energy consumption patterns at all load nodes follow the same trend.

In order to account for the uncertainty pertaining to demand, wind and solar power outputs, six different scenarios are considered for each uncertain parameter, as shown in Figs. 1.12, 1.13 and 1.14. As can be seen in these figures, each scenario represents possible hourly realizations of the uncertain parameter over the 24-h period. The individual scenarios are obtained by clustering a larger number of scenarios (30 in this case). These scenarios are then combined to form a new set of 216 ($6^3$) scenarios that are considered in the analysis.

Electricity prices are assumed to follow a similar trend as demand, varying between 107 €/MWh during peak and 30 €/MWh during shallow hours. This is depicted in Fig. 1.15.

The potential of DR in the provision of flexibility for integrating vRESs is assessed by considering different self-elasticity values. Figure 1.16 demonstrates

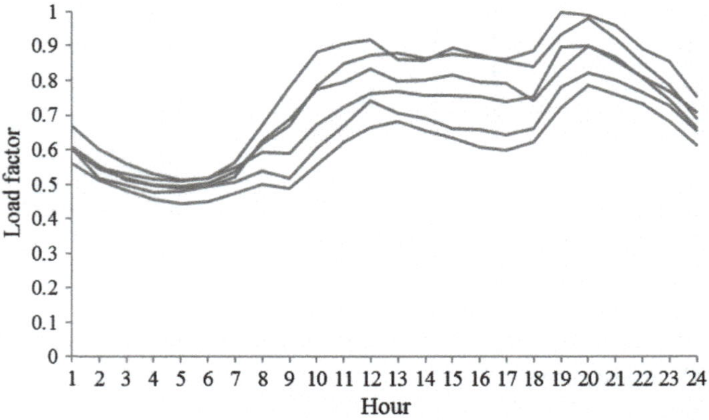


**Fig. 1.12** Considered demand scenarios

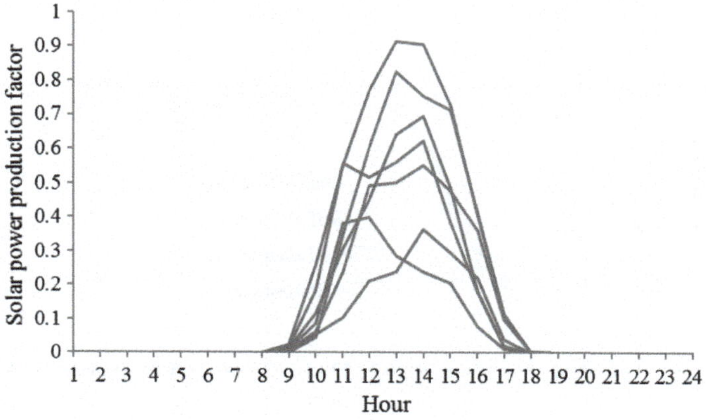


**Fig. 1.13** Considered solar PV power output scenarios

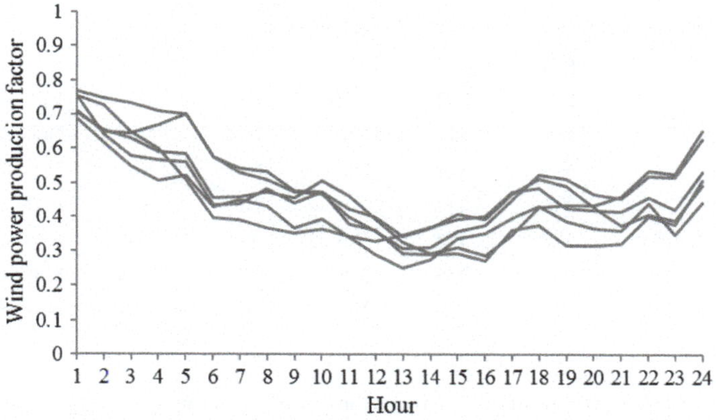


**Fig. 1.14** Considered wind power output scenarios

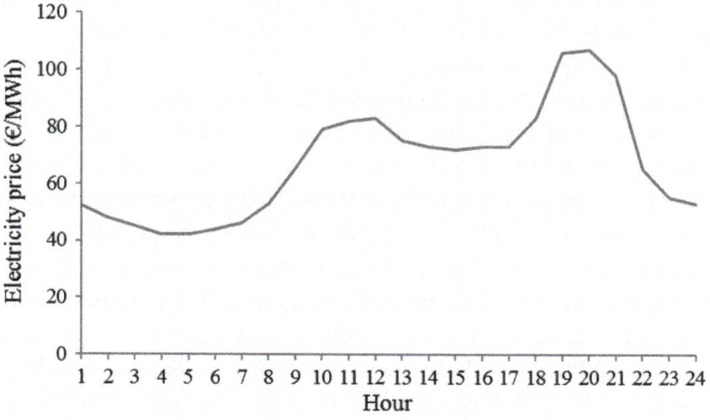


**Fig. 1.15** Hourly electricity prices

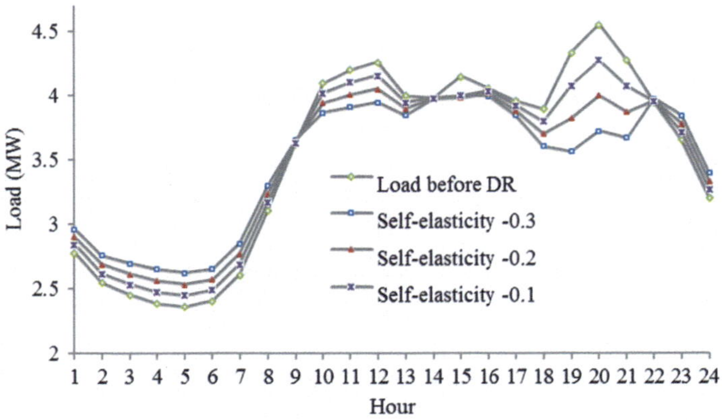


**Fig. 1.16** Flexibility via demand response

**Table 1.4** Details of the cases considered in the analysis

| Cases | Features | | | | | |
|---|---|---|---|---|---|---|
| | DNR | DGs | SCBs | ESSs | DR | Voltage limits |
| Base case | No | No | No | No | No | Not imposed |
| Only DNR | Yes | Yes | No | No | No | Imposed |
| Plus SCBs | Yes | Yes | Yes | No | No | Imposed |
| Plus SCBs & ESSs | Yes | Yes | Yes | Yes | No | Imposed |
| Full flex | Yes | Yes | Yes | Yes | Yes | Imposed |
| Plus SCBs & DR | Yes | Yes | Yes | No | Yes | Imposed |

the impact of DR in the hourly consumption profile. The results section presents the analysis results for self-elasticities of −0.2.

To ease the aforementioned analysis work, a total of six cases are considered here. Table 1.4 summarizes the distinctive features of each case. As can be observed in this table, all cases except the first case have two things in common—DNR and DG integration but differ in other aspects as clearly shown in Table 1.4.

The first case is related to the "do-nothing" scenario, where no distributed energy resource is connected and the entire load is met by importing power via the substation at bus 1. And, this is referred to as the "Base case". The second one considers DG integration with dynamic network reconfiguration, and is hereinafter referred to as "Only DNR". Note that DNR deals with the possibility of optimally changing the statuses of feeders (on an hourly basis) depending on the operational situation in the system. This case helps to understand the possible contribution of DNR in terms of enhancing system flexibility, and thereby increasing vRES utilization level. In addition to DNR, the third case considers switchable capacitor banks as a means of flexibility option. This is referred to as the "Plus SCBs" case in the remainder of this chapter. The fourth and the fifth cases are similar in that both

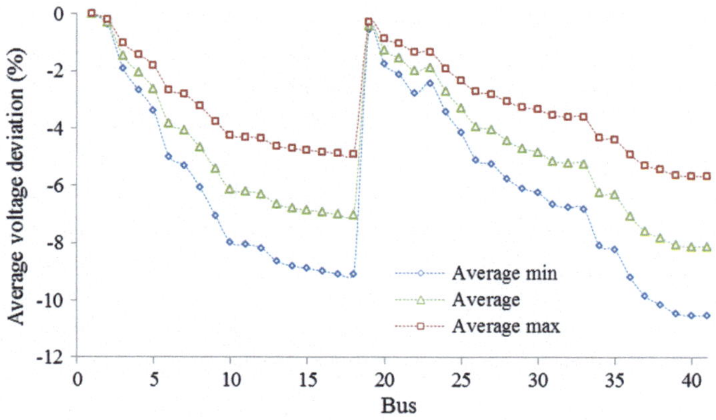


**Fig. 1.17** Average voltage deviation profiles with no flexibility options (base case)

consider the flexibility options provided by DNR, SCBs and ESSs. The only difference between these two cases is that the former does not have DR integrated as an additional flexibility mechanism. These cases are denoted as "Plus SCBs & ESSs" and "Full flex", respectively. The last case only considers the flexibility options: DNR, SCBs and DR, and is denoted by "Plus SCBs & DR". Note that lower bound of nodal voltage is relaxed in the base case to avoid infeasibility. This is due to the fact that the original system is poorly compensated. And, under this circumstance, it is not technically possible to meet the high reactive power requirement in this system while simultaneously imposing the voltage limits. For comparison purposes, the average voltage deviation at each bus is presented in Fig. 1.17. This also displays the minimum and maximum average values corresponding to different operational situations. It can be observed that most of the voltages fall outside the permissible range, particularly at the nodes located far away from the substation. The lowest voltage deviation occurs at bus 41, which can reach 18% in some operational situations.

Table 1.5 compares the objective function values and average losses corresponding to the different cases considered in the analysis. Compared to the base case, it can be seen that there are substantial improvements in the values of the designated function and variables. In the "Only DNR" case, for example, the total cost is reduced by about 9% and average losses by 24%. However, the vRES penetration level in this particular case (which stands at 12.2%) is not significant; solar PV and wind type DG utilization levels are only 0.4 and 11.8%, respectively. The wind and solar PV power sources are not being utilized because of technical constraints mainly related to the voltage limits. Since the system is not well-compensated, more power needs to be imported to support the high reactive power requirement in the system. Injecting more active power from the DGs, without proper compensation, would otherwise lead to voltage hikes which is not acceptable. Figure 1.18 shows the energy mix in the "Only DNR" case. Based on these results, it seems DNR alone may not contribute enough to enhance vRES

**Table 1.5** Total expected costs and average losses for the considered cases

| Cases | Total cost (€) | Average losses (MW/h) | | Voltage limits |
|---|---|---|---|---|
| | | Active | Reactive | |
| Base case | 6036.281 | 0.275 | 0.201 | Not imposed |
| Only DNR | 5512.385 | 0.208 | 0.158 | Imposed |
| Plus SCBs | 2677.782 | 0.073 | 0.058 | Imposed |
| Plus SCBs & ESSs | 2229.248 | 0.096 | 0.075 | Imposed |
| Full flex | **2151.926** | 0.093 | 0.073 | Imposed |
| Plus SCBs & DR | 2522.484 | **0.072** | **0.057** | Imposed |

Bold values indicate the lowest values computed in the optimization process

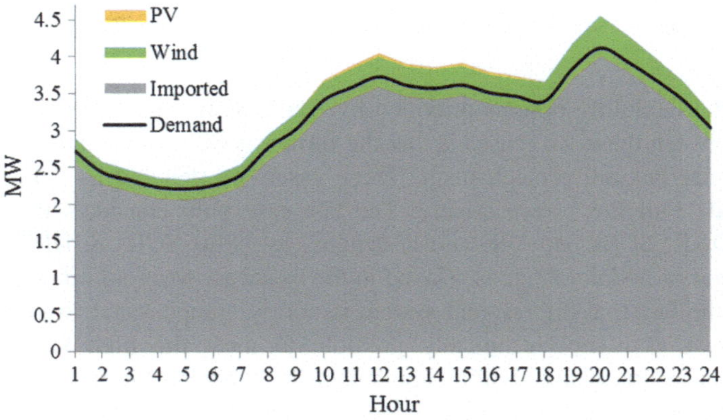


**Fig. 1.18** Aggregate energy mix in the system in the "Only DNR" case

penetration level in electric distribution networks. However, this may be case-dependent. Moreover, some of the assumptions made in this chapter may not reflect the real potential of DNR as a key flexibility option. For example, the assumptions on the uniform patterns of electricity consumptions and vRES power outputs may not encourage more frequent reconfigurations of the network so as to adapt to varying operational situations.

In the case of "Plus SCBs", the results in Table 1.5 show that the reduction in total cost and losses is simply dramatic, and so is the level of vRES penetration. Compared to the base case, costs are slashed by about 56% while the reduction of losses amounts to more than 73%. In this case, solar PV and wind cover about 12.6 and 66.8% of the aggregate demand in the system over the whole day. The energy-mix corresponding to this case is depicted in Fig. 1.19. As seen from this figure, there are hours where the system operates in island mode (see the first four hours). This mean the demand in these hours is fully met by locally produced renewable power. Generally, the results here reveal the substantial benefits of SCBs in enabling a large-scale penetration of variable energy resources. In other words,

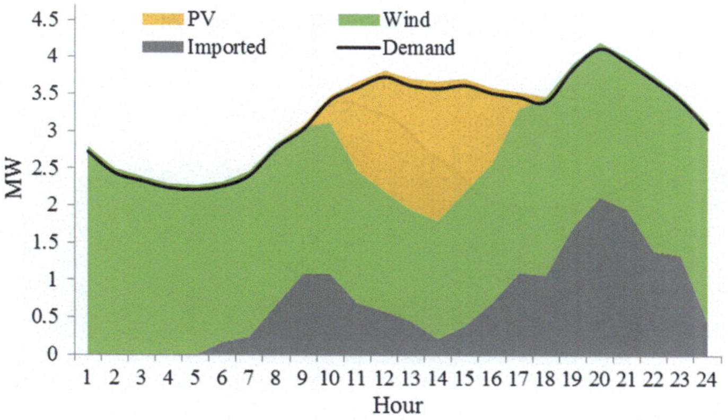


**Fig. 1.19** Aggregate energy mix in the system corresponding to the "Plus SCBs" case

a properly compensated distribution network can manage the technical risk posed by the intermittent nature of such resources.

As can be observed in Table 1.5, the overall cost is further reduced in the "Plus SCBs & ESSs" case by 63% in comparison to that of the base case. However, losses are slightly higher in this case than in the "Plus SCBs" one. This is mainly because of the fact that some feeders carry more power to charge/discharge the ESSs as opposed to the "Plus SCBs" case. It should be noted that the losses are yet substantially lower than that of the base case by 65%. The presence of ESSs in the "Plus SCBs & ESSs" case further increases the flexibility of the system, and allows a more efficient utilization of the "cleaner" DG power. This is can be seen in Fig. 1.20. One interesting observation in this figure is that the system operates autonomously during peak hours by releasing the cheaper energy stored in the ESSs

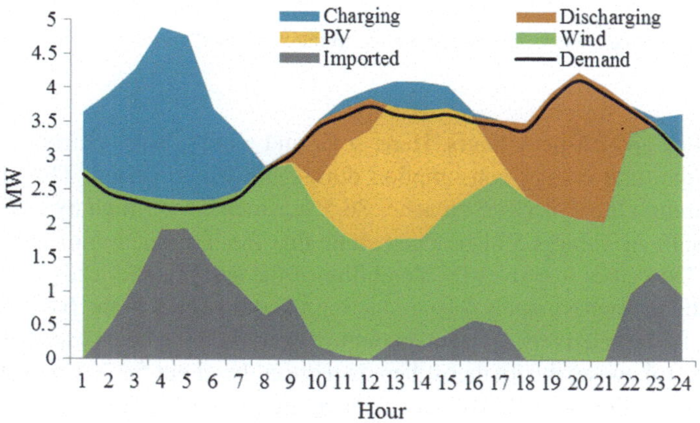


**Fig. 1.20** Aggregate energy mix corresponding to the "Plus SCBs & ESSs" case

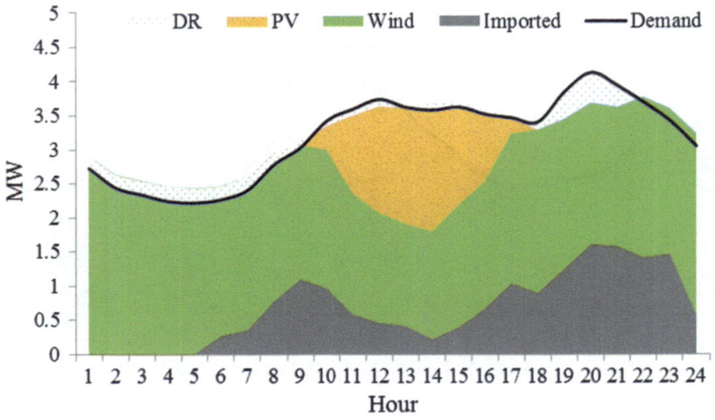


**Fig. 1.21** Aggregate energy mix corresponding to the “SCBs & DR” case

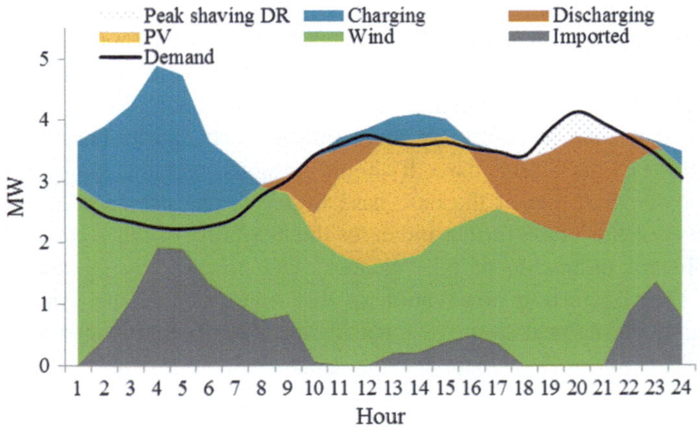


**Fig. 1.22** Aggregate energy mix corresponding to the “Full flex” case

during valley and off-peak hours. Here, solar and wind power contribute 14.3 and 72.2% to the total energy consumption during the whole period. This means the total penetration level of vRESs reaches 86.5%, which is very high by any standard.

The results in Table 1.5 also demonstrate that the introduction of DR, as in the “SCBs & DR” case, improves the flexibility of the system, and leads to the lowest losses (with an approximately 74% reduction in comparison to the base case). This is because of the relatively reduced amount of flows in the feeders especially during peak hours. Likewise, the total cost here is reduced by about 58%. This is higher by 2% than that of the “Plus SCBs” case. The aggregate energy mix corresponding to the “SCBs & DR” case is shown in Fig. 1.21. The shares of wind and solar PV power production over the whole period are 12.4 and 67.9%, respectively, which

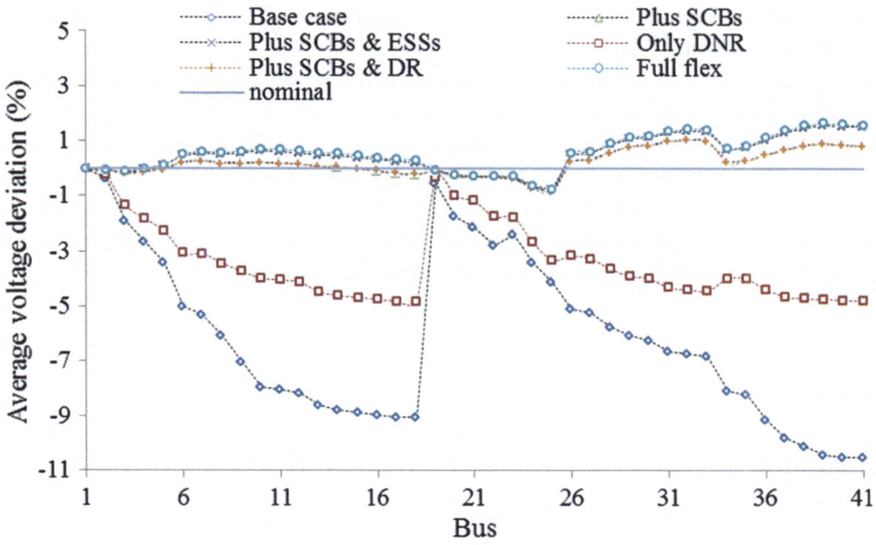


**Fig. 1.23** Comparison of average voltage profiles for the different cases

brings the total vRES penetration level to 80.3%. Because of the absence of a storage medium, this value is lower than the 86.5% share in the "Plus SCBs & ESSs" case.

As mentioned earlier, the "Full flex" case jointly deploys all four technologies that are capable of providing flexibility to the system: DNR, SCBs, ESSs and DR. As expected, this case leads to the lowest overall cost in the system (i.e., about 64% lower than that of the base case). As can be seen in Table 1.5, the benefit in terms of losses reduction is also evident even though this is slightly higher than that of the "Plus SCBs & DR" due to the same reasons as before. Because of the increased system flexibility in the "Full flex" case, the amount of imported energy is significantly lower than that of any other case. The total share of vRES power production reaches 86.6% (see Fig. 1.22). Wind and solar PV type DGs each contribute 14.4 and 72.2%, respectively.

So far, the analysis has been in terms of cost, energy mix and losses. Obviously, these are all relevant factors. However, it is also important to analyze the performance of the system from the technical point of view. To this end, the voltage profile is a good indicator. Ideally, voltage deviations in all nodes are desired to be close to the nominal value. But the nodal voltages often vary within certain permissible range (which in this case is $1 \pm 5\%$ of the nominal voltage). Figure 1.23 shows average deviations of voltages at every node in the system for all the cases considered in this chapter. This figure clearly shows that the introduction of flexibility mechanisms dramatically improve the voltage profile within the system. This is very critical to maintain the healthy operation of such a system. The "Only DNR" case alone keeps the voltages within the allowable range. For the remaining cases, the average voltage deviations for most of the nodes are practically insignificant, averaging at about 1%.

The benefits of all flexibility options considered in this chapter are evident with significant impact in achieving minimization of total costs of operation in the distribution networks. Analysis of jointly or separated operation of ESSs, capacitor banks, vRES and switching substantially improved voltage profiles. Operation of distribution networks with DR show the capability that this technology can have in the utilization of ESSs, making it a more valuable solution during operation, with less impact on total costs, increasing its utilization.

## Appendix

### Sets/Indices

| | |
|---|---|
| $c/\Omega^c$ | Index/set of capacitor banks |
| $es/\Omega^{es}$ | Index/set energy storages |
| $i/\Omega^i$ | Index/set of buses |
| $g/\Omega^g/\Omega^{DG}$ | Index/set of generators/DGs |
| $k/\Omega^k$ | Index/set of branches |
| $h', h/\Omega^h$ | Index/set of hourly snapshots |
| $s/\Omega^s$ | Index/set of scenarios |
| $\varsigma/\Omega^\varsigma$ | Index/set of substations |

### Parameters

| | |
|---|---|
| $SC_k$ | Cost of switching of branch k (€ per single switching) |
| $E^{min}_{es,i}, E^{max}_{es,i}$ | Energy storage limits (MWh) |
| $ER_g, ER^{SS}_\varsigma$ | Emission rates of DGs, and energy purchased, respectively ($tCO_2e$/MWh) |
| $g_k, b_k, S^{max}_k$ | Conductance, susceptance and flow limit of branch k ($\Omega, \Omega$, MVA) |
| $MP_k, MQ_k$ | Big-M parameters associated to active and reactive power flows through branch k |
| $OC_{g,i,s,h}$ | Operation cost of unit energy production by DGs (€/MWh) |
| $N_i, N_\varsigma$ | Number of buses and substations, respectively |
| $P^{ch,max}_{es,i}, P^{dch,max}_{es,i}$ | Charging and discharging power limits of storage system (MW) |
| $V_{nom}$ | Nominal voltage (kV) |
| $Z_k$ | Impedance of branch k ($\Omega$) |
| $\lambda^{CO_2e}_{s,h}$ | Price of emissions (€/$tCO_2e$) |
| $\lambda^\varsigma_{s,h}$ | Price of electricity purchased upstream (€/MWh) |
| $\overline{\lambda^\varsigma_s}$ | Average price of electricity purchased upstream (€/MWh) |
| $\lambda^{dch}_{es,i,s,h}$ | Cost of energy discharged from storage system (€/MWh) |

| | |
|---|---|
| $\eta_{es}^{ch}, \eta_{es}^{dch}$ | Charging and discharging efficiency (%) |
| $\rho_s, \pi_w$ | Probability of hourly scenario s and weight (in hours) of hourly snapshot group h |
| $\upsilon_{s,h}$ | Penalty for unserved power (€/MW) |
| $\xi_{h,h'}$ | Elasticity of electricity demand |

**Variables**

| | |
|---|---|
| $PD_{s,h}^i, QD_{s,h}^i$ | Active and reactive power demand at node $i$ (MW, MVAr) |
| $E_{es,i,s,h}$ | Reservoir level of ESS (MWh) |
| $I_{es,i,s,h}^{dch}, I_{es,i,s,h}^{ch}$ | Discharging/charging indicator variables |
| $P_{g,i,s,h}, Q_{g,i,s,h}$ | Active and reactive power produced by DGs (MW) |
| $P_{\varsigma,s,h}^{SS}, Q_{\varsigma,s,h}^{SS}$ | Active and reactive power imported from grid (MW) |
| $P_k, Q_k, \theta_k$ | Active and reactive power flows, and voltage angle difference of link k (MW, MVAr, radians) |
| $PL_k, QL_k$ | Active and reactive power losses (MW, MVAr) |
| $PL_{\varsigma,s,h}, QL_{\varsigma,s,h}$ | Active and reactive power losses at substation $\varsigma$ (MW, MVAr) |
| $P_{es,i,s,h}^{dch}, P_{es,i,s,h}^{ch}$ | Discharged/charged power (MW) |
| $P_{i,s,h}^{NS}$ | Unserved power at node $i$ (MW) |
| $Q_{i,s,h}^{c}$ | Reactive power produced by capacitor bank at node i (MVAr) |
| $Q_{i,s,h}^{NS}$ | Unserved power at node $i$ (MW) |
| $V_i, V_j$ | Voltage magnitudes at nodes $i$ and $j$ (kV) |
| $u_{k,h}$ | Utilization variables of existing lines |
| $x_{c,h}$ | Integer variable of switchable capacitor banks |
| $\theta_i, \theta_j$ | Voltage angles at node $i$ and $j$ (radians) |
| $\lambda_{s,h'}$ | Real-time price of electricity (€/MWh) |

**Functions (all units are in M€)**

| | |
|---|---|
| $EC_h^{SS}$ | Expected cost of energy purchased from upstream |
| $EC_h^{DG}$ | Expected cost of energy purchased from DG |
| $EC_h^{ES}$ | Expected cost of energy purchased from energy storage |
| $ENSC_h$ | Expected cost of unserved power |
| $EmiC_h^{DG}$ | Expected emission cost of DG power production |
| $EmiC_h^{SS}$ | Expected emission cost of purchased power |

## References


1. I.T. Papaioannou, A. Purvins, E. Tzimas, Demand shifting analysis at high penetration of distributed generation in low voltage grids. Int. J. Electr. Power Energy Syst. **44**(1), 540–546 (2013)
2. M. Junjie, W. Yulong, L. Yang, Size and location of distributed generation in distribution system based on immune algorithm. Syst. Eng. Procedia **4**, 124–132 (2012)

3. P.D. Lund, J. Lindgren, J. Mikkola, J. Salpakari, Review of energy system flexibility measures to enable high levels of variable renewable electricity. Renew. Sustain. Energy Rev. **45**, 785–807 (2015)
4. P. Crespo Del Granado, Z. Pang, S.W. Wallace, Synergy of smart grids and hybrid distributed generation on the value of energy storage. Appl. Energy **170**, 476–488 (2016)
5. R. Tulabing et al., Modeling study on flexible load's demand response potentials for providing ancillary services at the substation level. Electr. Power Syst. Res. **140**, 240–252 (2016)
6. E. Cutter, C.K. Woo, F. Kahrl, A. Taylor, Maximizing the value of responsive load. Electr. J. **25**(7), 6–16 (2012)
7. E. Lannoye, D. Flynn, M. O'Malley, Assessment of power system flexibility: a high-level approach, in *Power and Energy Society General Meeting, 2012 IEEE* (2012), pp. 1–8
8. J.L. Mathieu, M.G. Vayá, G. Andersson, Uncertainty in the flexibility of aggregations of demand response resources, in *Industrial Electronics Society, IECON 2013–39th Annual Conference of the IEEE* (2013), pp. 8052–8057
9. V. Calderaro, G. Conio, V. Galdi, G. Massa, A. Piccolo, Active management of renewable energy sources for maximizing power production. Int. J. Electr. Power Energy Syst. **57**, 64–72 (2014)
10. A. Ulbig, G. Andersson, Analyzing operational flexibility of electric power systems, in *Power Systems Computation Conference (PSCC), 2014* (Wroclaw, Poland, 2014)
11. A. Ulbig, G. Andersson, Analyzing operational flexibility of electric power systems. Int. J. Electr. Power Energy Syst. **72**, 155–164 (2015)
12. S.G. Task Force, Regulatory Recommendations for the Deployment of Flexibility, EG3 Report, Jan 2015
13. J.R. Aguero, E. Takayesu, D. Novosel, R. Masiello, Modernizing the grid: challenges and opportunities for a sustainable future. IEEE Power Energy Mag. **15**(3), 74–83 (2017)
14. A. Kumar et al., A review of multi criteria decision making (MCDM) towards sustainable renewable energy development. Renew. Sustain. Energy Rev. **69**, 596–609 (2017)
15. J.L. Christensen, D.S. Hain, Knowing where to go: the knowledge foundation for investments in renewable energy. Energy Res. Soc. Sci. **25**, 124–133 (2017)
16. Y. Fernando, S. Yahya, Challenges in implementing renewable energy supply chain in service economy era. Procedia Manuf. **4**, 454–460 (2015)
17. M. Bhattacharya, S.R. Paramati, I. Ozturk, S. Bhattacharya, The effect of renewable energy consumption on economic growth: evidence from top 38 countries. Appl. Energy **162**, 733–741 (2016)
18. P. Blechinger, C. Cader, P. Bertheau, H. Huyskens, R. Seguin, C. Breyer, Global analysis of the techno-economic potential of renewable energy hybrid systems on small islands. Energy Policy **98**, 674–687 (2016)
19. A. Botelho, L.M.C. Pinto, L. Lourenço-Gomes, M. Valente, S. Sousa, Social sustainability of renewable energy sources in electricity production: an application of the contingent valuation method. Sustain. Cities Soc. **26**, 429–437 (2016)
20. M. Engelken, B. Römer, M. Drescher, I.M. Welpe, A. Picot, Comparing drivers, barriers, and opportunities of business models for renewable energies: a review. Renew. Sustain. Energy Rev. **60**, 795–809 (2016)
21. C.-A. Gabriel, What is challenging renewable energy entrepreneurs in developing countries? *Renew. Sustain. Energy Rev.* **64**, 362–371, Out (2016)
22. M. Jamil, F. Ahmad, Y.J. Jeon, Renewable energy technologies adopted by the UAE: prospects and challenges—a comprehensive overview. Renew. Sustain. Energy Rev. **55**, 1181–1194 (2016)
23. A. Seetharaman, L.L. Sandanaraj, M.K. Moorthy, A.S. Saravanan, Enterprise framework for renewable energy. Renew. Sustain. Energy Rev. **54**, 1368–1381 (2016)
24. A. Zyadin, P. Halder, T. Kähkönen, A. Puhakka, Challenges to renewable energy: a bulletin of perceptions from international academic arena. Renew. Energy **69**, 82–88 (2014)

25. W. D'haeseleer, L. de Vries, C. Kang, E. Delarue, Flexibility challenges for energy markets: fragmented policies and regulations lead to significant concerns. IEEE Power Energy Mag. **15** (1), 61–71 (2017)
26. M. Engelken, B. Römer, M. Drescher, I.M. Welpe, A. Picot, Comparing drivers, barriers, and opportunities of business models for renewable energies: a review. Renew. Sustain. Energy Rev. **60**, 795–809 (2016)
27. M.S. Hossain, N.A. Madlool, N.A. Rahim, J. Selvaraj, A.K. Pandey, A.F. Khan, Role of smart grid in renewable energy: an overview. Renew. Sustain. Energy Rev. **60**, 1168–1184 (2016)
28. A. Zyadin, P. Halder, T. Kähkönen, A. Puhakka, Challenges to renewable energy: a bulletin of perceptions from international academic arena. Renew. Energy **69**, 82–88 (2014)
29. G. Papaefthymiou, K. Grave, K. Dragoon, Flexibility options in electricity systems, Ecofys 2014 by order of: European Copper Institute (2014)
30. Union internationale d'électrothermie and Electric load management in industry working group, in *Electric Load Management In Industry* (Paris-La Défense, UIE 1996)
31. M.H. Shoreh, P. Siano, M. Shafie-khah, V. Loia, J.P.S. Catalão, A survey of industrial applications of demand response. Electr. Power Syst. Res. **141**, 31–49 (2016)
32. E. Union, Research Challenges to Increase the Flexibility of Power Systems (Belgium, 2014)
33. K. Knezovic, S. Martinenas, P.B. Andersen, A. Zecchino, M. Marinelli, Enhancing the role of Electric Vehicles in the power grid: field validation of multiple ancillary services. IEEE Trans. Transp. Electrification **3**(1), 201–209 (2017)
34. D.T. Hoang, P. Wang, D. Niyato, E. Hossain, Charging and discharging of Plug-In Electric Vehicles (PEVs) in Vehicle-to-Grid (V2G) systems: a cyber insurance-based model. IEEE Access **5**, 732–754 (2017)
35. S.I. Vagropoulos, G.A. Balaskas, A.G. Bakirtzis, An investigation of plug-in electric vehicle charging impact on power systems scheduling and energy costs. IEEE Trans. Power Syst. **32** (3), 1902–1912 (2017)
36. N. Neyestani, M.Y. Damavandi, M. Shafie-khah, A.G. Bakirtzis, J.P.S. Catalao, Plug-In Electric Vehicles parking lot equilibria with energy and reserve markets. IEEE Trans. Power Syst. **32**(3), 2001–2016 (2017)
37. S. Hers, M. Afman, S. Cherif, F. Rooijers, *Potential for Power-to-Heat in the Netherlands* (CE Delft, Delft, Netherlands, 2015)
38. D. Böttger, M. Götz, N. Lehr, H. Kondziella, T. Bruckner, Potential of the Power-to-Heat technology in district heating grids in Germany. Energy Procedia **46**, 246–253 (2014)
39. M. Götz et al., Renewable Power-to-Gas: a technological and economic review. Renew. Energy **85**, 1371–1390 (2016)
40. F.D. Meylan, F.-P. Piguet, S. Erkman, Power-to-gas through $CO_2$ methanation: assessment of the carbon balance regarding EU directives. J. Energy Storage **11**, 16–24 (2017)
41. U. Mukherjee, S. Walker, A. Maroufmashat, M. Fowler, A. Elkamel, Development of a pricing mechanism for valuing ancillary, transportation and environmental services offered by a power to gas energy system. Energy **128**, 447–462 (2017)
42. X. Liang, Emerging power quality challenges due to integration of renewable energy sources. IEEE Trans. Ind. Appl. **53**(2), 855–866 (2017)
43. C.I. Ossai, Optimal renewable energy generation—approaches for managing ageing assets mechanisms. Renew. Sustain. Energy Rev. **72**, 269–280 (2017)
44. T. Strasser et al., A review of architectures and concepts for intelligence in future electric energy systems. IEEE Trans. Ind. Electron. **62**(4), 2424–2438 (2015)
45. S. Islam, Challenges and opportunities in grid connected commercial scale PV and wind farms, in *Electrical and Computer Engineering (ICECE), 2016 9th International Conference on*, 2016, pp. 1–7
46. M.I. Alizadeh, M. Parsa Moghaddam, N. Amjady, P. Siano, M.K. Sheikh-El-Eslami, Flexibility in future power systems with high renewable penetration: a review. Renew. Sustain. Energy Rev. **57**, 1186–1193 (2016)

47. J. Cochran et al., *Flexibility in 21st century power systems* (National Renewable Energy Laboratory (NREL), Golden, CO, 2014)
48. J.A. Schachter, P. Mancarella, A critical review of real options thinking for valuing investment flexibility in smart grids and low carbon energy systems. Renew. Sustain. Energy Rev. **56**, 261–271 (2016)
49. M. Wang, J. Zhong, A novel method for distributed generation and capacitor optimal placement considering voltage profiles, in *Power and Energy Society General Meeting, 2011 IEEE*, 2011, pp. 1–6
50. S.F. Santos, D.Z. Fitiwi, M. Shafie-khah, A.W. Bizuayehu, C.M.P. Cabrita, J.P.S. Catalao, New multi-stage and stochastic mathematical model for maximizing RES hosting capacity—part II: numerical results. *IEEE Trans. Sustain. Energy* (2016), pp. 1–1
51. D.Z. Fitiwi, L. Olmos, M. Rivier, F. de Cuadra, I.J. Pérez-Arriaga, Finding a representative network losses model for large-scale transmission expansion planning with renewable energy sources. Energy **101**, 343–358 (2016)

# Chapter 2
# Distribution Network Modeling and Management

**Fang Yang and Zhao Li**

**Abstract** This chapter first examines balanced and unbalanced electric distribution network modelling techniques and then tests their effects on power flow analysis with practical electric distribution networks. Test results show that distribution network operating conditions obtained from the balanced and unbalanced model-based power flows may have obvious difference in terms of voltage violations and overloaded equipment. These results demonstrate the importance of adopting unbalanced network modelling in improving the accuracy of power flow. Furthermore, this chapter presents an unbalanced model-based DMS application in loss reduction, an essential application for the energy efficiency improvement in electric distribution networks. The loss reduction often involves the control of reactive power (VAR) resources to optimize the VAR flow in the distribution network. Based on the unbalanced power flow model, an advanced loss reduction approach is developed to achieve the optimal control coordination among multiple capacitors and distributed energy resources such as solar and wind resources. The effectiveness of the presented approach is demonstrated on practical utility distribution networks with varying degree of unbalance and model complexity.

**Keywords** Electric distribution network modelling · Distributed energy resource · Distribution management system · Power flow · Loss reduction

F. Yang (✉)
Electrical Engineering Department, University of Wisconsin Platteville, Platteville, WI, USA
e-mail: yangfan@uwplatt.edu

Z. Li
LY Grid Innovation, Menasha, WI 54952, USA
e-mail: leezhao@gmail.com



A. Arefi et al. (eds.), *Electric Distribution Network Management and Control*, Power Systems, https://doi.org/10.1007/978-981-10-7001-3_2

## 2.1 Introduction

Electric distribution network modelling is crucial to various distribution management system (DMS) analysis and control applications. Among these DMS applications, power flow is the most fundamental one to implement many others, such as Voltage VAR control, network reconfiguration, demand response, and so on. Therefore, investigating the effects of different distribution network modelling techniques on power flow analysis plays an important role in improving DMS applications and achieving more economic and optimal control of smart distribution networks.

The existing electric distribution network modelling techniques fall into two basic categories: balanced equivalent single-phase and unbalanced multi-phase modelling. The balanced single-phase modelling approximates the unbalanced network with a balanced network and then derives either the positive sequence network or the decoupled three single-phase networks. The unbalanced multi-phase modelling explicitly represents the network components and their connection with single-, two- or three-phase models.

Based on the two types of electric distribution network modelling techniques, corresponding power flow methods have also been developed. The balanced model-based power flow mainly adopts the forward backward sweep technique [1], which takes advantages of the radial network structure and can achieve fast power flow solution. The unbalanced model-based power flow methods include extended forward backward sweep [2], Newton-Raphson [3, 4], decoupled [5], distributed methods [6], and so on.

Although the electric distribution network is well known for its inherent unbalanced nature caused by the unbalanced network structure and loading, most academic research and industrial practice still select the balanced single-phase model and assume it can provide acceptable accuracy in power flow analysis and other DMS applications [7–9]. In recent years, with the penetration of smart grid technologies in the distribution network, such as distributed energy resources (DERs), increasing research and real-world applications begin to adopt unbalanced multi-phase model to achieve more accurate distribution network analysis and control [10–12].

With the common recognition that unbalanced model-based DMS applications (such as power flow) can provide more accurate results than that of the balanced model, few literature, however, has been found to provide a quantitative comparison on the effects of adopting balanced and unbalanced models on the DMS applications. Ref. [13] investigates the voltage magnitude errors caused by the balanced model based on a small scale traditional distribution network without DERs, in which the oversimplified approximation is made to obtain the balanced model from the unbalanced model.

Furthermore, power flow analysis results provide basis for many other DMS applications, one of which is loss reduction. Considering about 4% of electric energy generated by power plants is lost in the distribution network, loss reduction is a critical function in the DMS to improve energy efficiency. Traditionally, distribution utilities place switchable capacitor banks at strategic locations to supply reactive power (VAR) and reduce loss. In recent years, more and more DERs are deployed, many of which allow the control of their VAR output within a certain range and thus become an important VAR resource besides capacitor banks in the distribution network.

Comparing to the on/off capacitor switching, the control of DER VAR output has advantages such as no operating time limits and the continuous control of reactive power to prevent and correct the over-/under-VAR compensation resulting from the discrete capacitor switching. These advantages make the DER VAR control a promising means for loss reduction in addition to the capacitor switching. The availability of DER VAR control presents both new opportunities and challenges as well.

Some research efforts have been devoted to integrating DERs in the loss reduction: Ref. [14] decouples the capacitor and DER VAR control into two sub-problems that are solved in separate steps. References [15, 16] integrate the impact of DER power injection on the network loss into the optimal power flow (OPF) model. References [17–21] adopt meta-heuristic techniques to solve Voltage VAR optimization problems for the loss reduction purpose. References [22, 23] propose a three-level hierarchical Voltage VAR control structure, in which DERs are used in both the primary and the tertiary control aiming for day-ahead planning. These approaches attempt to avoid the complex mathematic modelling and develop solutions using the oversimplified single-phase balanced network model. In addition, the developed approaches and solution methods are demonstrated on small test networks, their effectiveness and performance on practical large-scale utility networks are unclear.

This chapter examines balanced and unbalanced electric distribution network modelling techniques and then tests their effects on DMS applications including power flow analysis and loss reduction with practical electric distribution networks. The contribution of this chapter includes the following:

This chapter provides a comprehensive comparison for the effects of balanced and unbalanced models on the power flow analysis. In particular, the investigation is done for both the traditional network without DERs and the distribution network with DERs in the smart grid environment. Three practical distribution networks with representative characteristics in terms of distribution network dimension and complexity are selected as test platforms. Balanced/Unbalanced model-based power flow results are examined under different loading levels.

Furthermore, this chapter extends the application of the unbalanced model in one of DMS applications, i.e., loss reduction. Based on authors' experience in applying multi-phase model in the traditional loss reduction approach using capacitors only [24], this chapter presents an advanced technique to optimally coordinate the control among multiple capacitors and DERs and take into account the impact of control actions on the voltage profile. In particular, this chapter (1) integrates the practical multi-phase model in the loss reduction application to reflect the unbalanced nature of the practical distribution network, which enables more accurate loss reduction application and provides valuable experience for other DMS applications, (2) achieves optimal coordinated control among multiple capacitors and DERs that can take fully advantage of the continuous DER VAR control capability to compensate the over-/under-VAR compensation caused by the discrete nature of the capacitor switching, and (3) adopts real-world distribution networks to demonstrate the effectiveness and performance of the presented approach and solution method on the common personal computer platform.

## 2.2 Electric Distribution Network Modelling

This section first describes the distribution network multi-phase modelling technique in detail and then introduces the approximation of the unbalanced network with a balanced network model.

### *2.2.1 Electric Distribution Network Multi-phase Modelling*

Models for multi-phase distribution network devices, e.g., distribution lines, transformers with various configurations, loads/capacitor banks in wye/delta connections, and DERs are presented in this section.

#### 2.2.1.1 Distribution Lines

Distribution lines (such as feeders, cables, and laterals) are the backbone for a distribution network. Not all distribution line sections are built with balanced three phases, some laterals may include only single or two phases. The pi-equivalent model is used for both balanced and unbalanced line sections. Figure 2.1a shows a pi-equivalent model for a three-phase line section, in which each phase is modeled explicitly with self-impedance, shunt susceptance, and the mutual coupling with other phases. The detailed mathematical equations of the line model can be found in [1]. Similar model is developed for unbalanced line sections with only single or two phases.

#### 2.2.1.2 Transformers

Transformers in the distribution network include substation transformers such as on-load tap changers (OLTC), transformers along the line section such as voltage regulators, and service transformers that directly connect to customers. These transformers may be balanced or unbalanced and present various configurations in practical distribution networks. The main configuration types include: Wye/Wye,

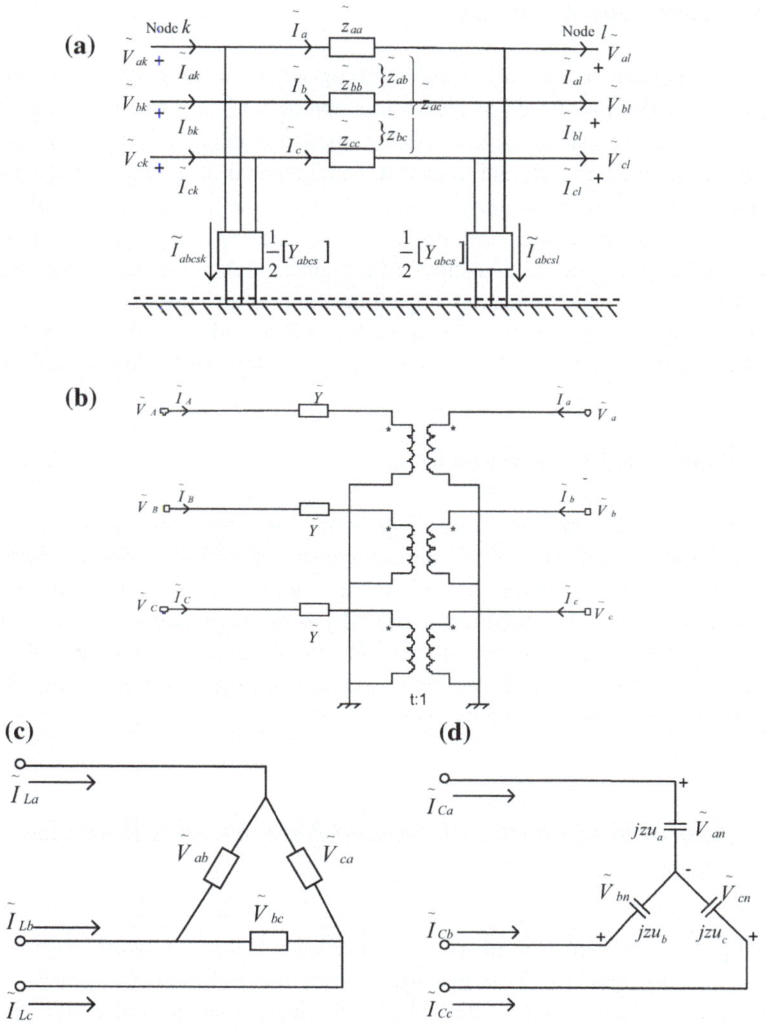


**Fig. 2.1** **a** A three-phase distribution line, **b** a transformer in wye/wye connection, **c** a three-phase delta-connected load, **d** a three-phase wye-connected capacitor bank

Wye/Delta, Delta/Wye, and Delta/Delta. Additional variations also derive from these types, such as grounded/ungrounded connection based on the grounding scheme of the wye connected side, leading/lagging Wye/Delta or Delta/Wye connection based on if the primary voltage is leading or lagging the secondary voltage, and open delta connection which provides three-phase voltages from two single-phase banks. All these transformer configurations are supported in this chapter. Figure 2.1b shows a Wye/Wye grounded transformer example.

#### 2.2.1.3 Loads/Capacitor Banks

Loads and capacitor banks are important components in the distribution network. They can be single-/two-/three-phase devices and in delta or wye connection. In this chapter, two types of load models, constant impedance and constant power loads [1], are adopted to represent different load voltage dependent characteristics. Capacitors may have either ganged or unganged control. For a capacitor with unganged control, each phase is modelled with an individual control variable to represent the unganged control capability. Figure 2.1c, d show a three-phase delta-connected load example and a three-phase wye-connected capacitor bank with three unganged control variables ($u_a$, $u_b$, and $u_c$) integrated with each phase impedance. These variables can take values of 0 or 1, corresponding to the open or close switching of each phase.

#### 2.2.1.4 Distributed Energy Resources

Under smart grid circumstances, distribution networks have increasing penetration of various types of DERs such as wind turbines, photovoltaic arrays, fuel cells, micro-turbines, diesel generators, and energy storage devices. Considering that power flow is a steady state application in DMS, only the steady state characteristics of DERs are of interest. In this chapter, DERs are represented with the multi-phase PQ models, and their real and reactive power can be adjusted to simulate different DER penetration levels in the network.

### *2.2.2 Approximation of Unbalanced Network with Balanced Model*

Regarding the approximation of an unbalanced distribution network with the balanced model, no standards exist in current academic and industrial practices. The approximation method used in Ref. [13] is to remove sections that only include single or two phases and then aggregate the downstream load to the first upstream bus that has three phases. One major disadvantage of this approximation method is that the power flow function based on such balanced model does not provide the voltage and any other information on the unbalanced nodes and lines that are removed.

In this chapter, a different approximation method that can overcome the drawback in [13] is adopted: any unbalanced section that only includes single or two phases is split into an equivalent three-phase balanced section, which still carries the same total amount of power/current flow as the original unbalanced section. In particular, in order to obtain the balanced model, following approximations are made to the unbalanced sources, lines, capacitors, transformers, and loads, respectively:

Unbalanced source/capacitor/load: The power generation or consumption associated with each phase of the unbalanced source/capacitor/load is aggregated first and then divided by three to obtain the power for each phase on the equivalent balanced model, as shown in Eqs. (2.1) and (2.2).

$$P_b = \frac{\sum_{i=1}^{n} P_{ub}^{i}}{3} \tag{2.1}$$

$$Q_b = \frac{\sum_{i=1}^{n} Q_{ub}^{i}}{3} \tag{2.2}$$

where

$P_b$ equivalent real power for each phase in the balanced model
$Q_b$ equivalent reactive power for each phase in the balanced model
$P_{ub}^{i}$ real power for phase i in the unbalanced device
$Q_{ub}^{i}$ reactive power for phase i in the unbalanced device
n total phase number of the unbalanced device.

Unbalanced line/transformer: The phase impedance of the equivalent balanced line section and transformer leg ($Z_b$) is calculated by Eq. (2.3), i.e., dividing each phase impedance in the unbalanced model ($Z_{ub}$) by the phase number ($\#Phase$) and then multiplying with three. Furthermore, the positive sequence impedance of the equivalent balanced model is then calculated based on $Z_b$.

$$Z_b = \frac{Z_{ub} \times 3}{\#Phase} \tag{2.3}$$

where

$Z_b$ equivalent phase impedance in the balanced model
$Z_{ub}$ unbalanced phase impedance in the unbalanced model.

## 2.3 Application of the Unbalanced Model in DMS Applications

The Newton-Raphson method is applied to solve both balanced and unbalanced model-based power flow. Balanced and unbalanced power flow results for three test networks are compared under two scenarios: traditional distribution networks without DERs and distribution networks with DERs in the smart grid environment.

In this chapter, three practical distribution networks (C1–C3) are selected to test the effect of different modelling techniques on power flow results. As these distribution networks originally do not include DERs, a number of PQ mode DERs are added to the networks at randomly selected locations. The power output of these

**Table 2.1** Power flow test network summary

| ID | #Feed | #Node | #Load | #Line | #DER | #Total |
|---|---|---|---|---|---|---|
| C1 | 2 | 363 | 165 | 374 | 3 | 980 |
| C2 | 8 | 2119 | 1331 | 2170 | 5 | 5904 |
| C3 | 18 | 3281 | 1696 | 3316 | 7 | 8789 |

DERs varies between 30 and 80% of the total network demand. For each network, the numbers of feeders (#Feeder), nodes (#Node), loads (#Load), lines (#Line), DERs (#DER), and total components are shown in Table 2.1. These are the numbers without counting actual phases. In other words, each component may have single, two, or three phases. Besides the components listed in the table, each network also includes multiple transformers, capacitors in various configurations, and many other devices.

All these test networks are typical real-world distribution networks with representative features in terms of distribution network dimension and complexity. Test results based on these networks provide realistic insight to the effects of different distribution network modelling on power flow results.

### 2.3.1 *Test Results on Power Flow Analysis*

In scenario I, both balanced and unbalanced power flows are executed under three loading levels: 30, 50, and 80%, respectively. In scenario II, the loading level of 50% is applied, while balanced and unbalanced power flows are executed under three different DER penetration levels: 30, 50, and 80% instead.

In each scenario, the impact of different loading levels or DG penetration levels on the voltage and current unbalance, calculated using Eqs. (2.4)–(2.7), is first investigated based on the unbalanced power flow results and provided in Tables 2.2 and 2.4. Furthermore, the impact of balanced and unbalanced model-based power flows on the network operating conditions, especially the operating constraint violations, is examined and the comparison are provided in Tables 2.3 and 2.5.

$$|V_k|_{ave} = \frac{\sum_{i=1}^{n} |V_k^i|}{n} \tag{2.4}$$

$$|V_k|_{ub} = \max\left\{ \left| \frac{|V_k^i| - |V_k|_{ave}}{|V_k|_{ave}} \times 100\% \right|, \quad i = 1\ldots n \right\} \tag{2.5}$$

$$|I_l|_{ave} = \frac{\sum_{i=1}^{n} |I_l^i|}{n} \tag{2.6}$$

**Table 2.2** Scenario I: voltage and current unbalance

| ID | Loading level (%) | Max \|V\| unbalance (%) | Range number (%) | Current unbalance |
|---|---|---|---|---|
| C1 | 30 | 0.285 | 12.4–13.98 | 2 |
| C1 | 50 | 0.478 | 12.4–13.89 | 2 |
| C1 | 80 | 0.772 | 12.4–13.82 | 2 |
| C2 | 30 | 0.185 | 0.54–96.96 | 6 |
| C2 | 50 | 0.314 | 0.96–98.17 | 6 |
| C2 | 80 | 0.518 | 1.27–98.86 | 6 |
| C3 | 30 | 0.907 | 1.6–26.1 | 16 |
| C3 | 50 | 1.557 | 1.63–26.05 | 16 |
| C3 | 80 | 2.614 | 1.68–25.9 | 16 |

**Table 2.3** Scenario I: operating constraints violations

| ID | Loading level (%) | Number of phase voltage violations | | #Overloaded equipment phases | |
|---|---|---|---|---|---|
| | | Balance | Unbalance | Balance | Unbalance |
| C1 | 30 | 0 | 0 | 0 | 3 |
| C1 | 50 | 0 | 0 | 6 | 6 |
| C1 | 80 | 0 | 0 | 6 | 6 |
| C2 | 30 | 0 | 3 | 0 | 0 |
| C2 | 50 | 911 | 3 | 0 | 0 |
| C2 | 80 | 2198 | 2041 | 6 | 18 |
| C3 | 30 | 0 | 0 | 3 | 1 |
| C3 | 50 | 0 | 112 | 6 | 4 |
| C3 | 80 | 701 | 1025 | 21 | 19 |

$$|I_l|_{ub} = \max\left\{\left|\frac{|I_l^i| - |I_l|_{ave}}{|I_l|_{ave}} \times 100\%\right|, \quad i = 1\ldots n\right\} \tag{2.7}$$

where

$|V_k^i|$ voltage magnitude at node-$k$ phase-$i$
$|V_k|_{ave}$ average voltage magnitude at node-$k$
$|V_k|_{ub}$ voltage unbalance at node-$k$
$|I_l^i|$ current magnitude at branch-$l$ phase-$i$
$|I_l|_{ave}$ average current magnitude at branch-$l$
$|I_l|_{ub}$ current unbalance at branch-$l$.

*Scenario I: Traditonal Distribution Networks without DERs*

Table 2.2 shows the maximum voltage magnitude unbalance, the range of current magnitude unbalance, and the number of branches with the current magnitude unbalance falling in this range for each network under three different loading levels: 30, 50, and 80%. For each test network, the maximum voltage unbalance increases as the loading level increases, while the current unbalance remains at the same level.

Table 2.3 shows the operating conditions including voltage violation and overloaded equipment information under different loading levels. When the loading level increases, more operating constraint violations occur to the network based on balanced and unbalanced power flow results. More importantly, the difference in the operating conditions reflected by the power flow results based on the two types of models is not trivial, such as the significant difference in the number of voltage violations for C2 under loading levels of 50%, the number of overloaded equipment for C2 under loading level of 80%, and the number of voltage violations for C3 under 50 and 80% loading levels.

*Scenario II: Distribution Networks with DERs*

In order to test the impact of different DER penetration levels on power flow results based on balanced and unbalanced models, various number of DERs are added to three test networks as shown in Table 2.4. All DERs in C1 and C2 and five DERs in C3 are three-phase balanced, while two DERs in C3 are single phase unbalanced. Power flows are executed with the total DER power output accounting for 30, 50, and 80% of the total demand for each network, given 50% loading level.

Table 2.4 shows the maximum voltage magnitude unbalance, the range of current magnitude unbalance, and the number of branches with the current magnitude unbalance falling in this range for each distribution network tested under three different DER penetration levels. For C1 and C2, maximum voltage unbalance values remain the same, while for C3, this value increases with the increasing of the

**Table 2.4** Scenario II: voltage and current unbalance

| ID | DER level (%) | Max \|V\| unbalance (%) | Range number (%) | Current unbalance |
|---|---|---|---|---|
| C1 | 30 | 0.474 | 16.387–28.343 | 2 |
| C1 | 50 | 0.471 | 18.62–123.62 | 2 |
| C1 | 80 | 0.467 | 23.4–24.75 | 2 |
| C2 | 30 | 0.316 | 3–98.17 | 6 |
| C2 | 50 | 0.316 | 0.84–98.16 | 6 |
| C2 | 80 | 0.317 | 0.4–98.16 | 6 |
| C3 | 30 | 1.55 | 3.01–115.3 | 16 |
| C3 | 50 | 2.36 | 2.1–72.02 | 16 |
| C3 | 80 | 3.56 | 2.5–25.9 | 16 |

**Table 2.5** Scenario II: operating constraints violations

| ID | Loading level (%) | Number of phase voltage violations | | #Overloaded equipment phases | |
|---|---|---|---|---|---|
| | | Balance | Unbalance | Balance | Unbalance |
| C1 | 30 | 0 | 0 | 6 | 6 |
| C1 | 50 | 0 | 0 | 6 | 6 |
| C1 | 80 | 0 | 0 | 6 | 6 |
| C2 | 30 | 0 | 3 | 0 | 0 |
| C2 | 50 | 0 | 3 | 0 | 30 |
| C2 | 80 | 0 | 3 | 0 | 120 |
| C3 | 30 | 0 | 99 | 6 | 4 |
| C3 | 50 | 0 | 86 | 6 | 7 |
| C3 | 80 | 0 | 35 | 11 | 12 |

DER penetration level. Current unbalance levels for C1 and C3 vary with the change in DER penetration levels.

The current unbalance level for networks with DERs also increases comparing to that of without DERs (at 50% loading level). Figure 2.2 shows current unbalance values (%) on two selected branches for three test networks under different levels of DER penetration. This figure indicates that the increasing DER penetration may result in the nontrivial and even significant changes in the current unbalance level.

Table 2.5 shows the number of voltage violations and overloaded equipment information with different DER penetration levels. When DER penetration level increases in the network, the number of operating constraint violations does not change for C1, increase for C2, and decrease for C3. When the DER penetration level reaches 50% or higher, balanced and unbalanced model based power flow results show significant difference for C2 and C3 in terms of overloaded equipment and voltage violations.

To summarize the results from above two scenarios, the voltage and current unbalance as well as the difference in operating conditions obtained from balanced and unbalanced model based power flows are influenced by following factors: network dimension and complexity, loading level, and DER penetration level. Tables 2.2 and 2.4 as well as Fig. 2.2 show that the nontrivial current unbalance is completely ignored when balanced network model is adopted. Furthermore, Tables 2.3 and 2.5 show the significant difference in the operating conditions obtained from different models, which will further affect many other DMS applications such as Voltage VAR control, reconfiguration, demand response and so on that are implemented on the basis of the power flow results.

(a)

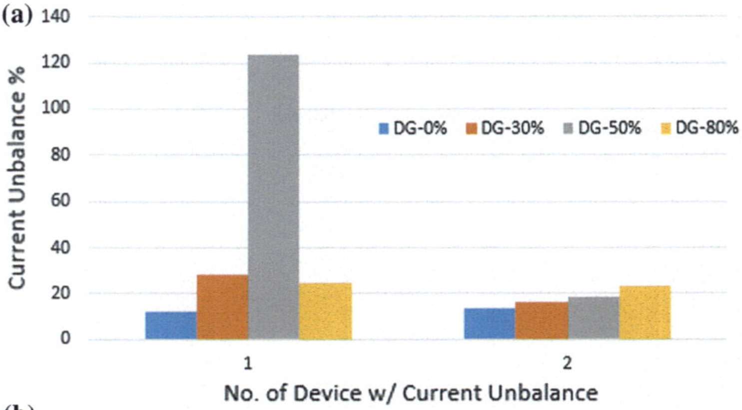


(b)

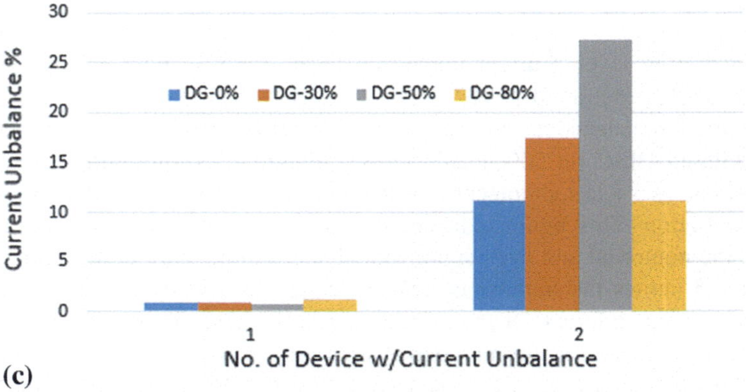


(c)

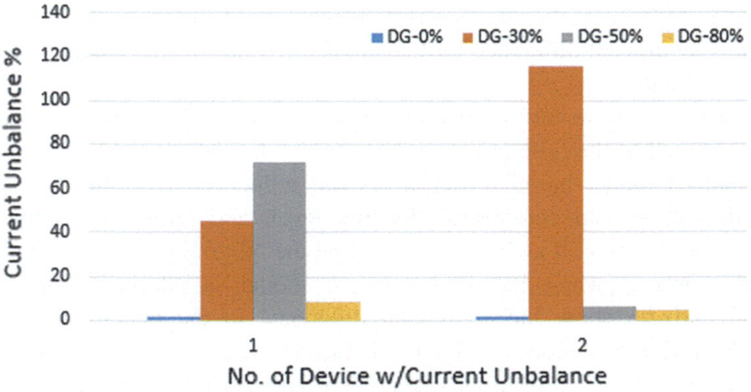


**Fig. 2.2** **a** Current unbalance under different DG penetration levels for C1, **b** current unbalance under different DG penetration levels for C2, **c** current unbalance under different DG penetration levels for C3

## 2.3.2 Unbalanced Model-Based Loss Reduction

This section describes a novel loss reduction approach and solution method developed on the basis of the multi-phase distribution network modeling and advanced optimization technology.

### 2.3.2.1 Methodology

Facilitated by the detailed multi-phase network model, an advanced loss reduction optimization approach is developed to achieve optimally coordinated control among multiple capacitors and DERs. The formulated optimization problem is shown in (2.8)–(2.15).

In this optimization problem, controllable devices include DERs besides switchable capacitor banks. Control variables are the continuous DER VAR output ($u_d$) and the discrete capacitor switching ($u_c$). Both ganged and unganged control of these controllable devices are included in this chapter.

In the objective function of the formulated optimization problem, the first term is to calculate total power loss using real and reactive current components and the branch resistance. While controlling capacitors and DERs to reduce loss, their impact on the voltage profile cannot be ignored. Distribution utilities often set higher priority for the voltage correction over loss reduction in the operating practice. That is, when the voltage violation exists, utilities desire to correct the voltage violations while reducing loss or even sacrificing the loss reduction for voltage correction; when no voltage violation exists, the loss reduction should not lead to any voltage violations. To address this practice, the objective function also includes a penalty function for each line-to-neutral/line-to-line voltage violation. By appropriately selecting the weight factor ($w_k$), the capacitor and DER VAR control targets to improve voltage violations while achieving the loss reduction.

The constraints in the formulated optimization problem include both current and voltage operating limits. The calculation of these parameters is based on their base values plus the estimated changes caused by the control actions. Equations (2.9)–(2.11) enforce the current magnitude. Equation (2.12) enforces the maximum and minimum DER VAR output limits. Equations (2.13)–(2.15) are the constraints for each line-to-neutral/line-to-line voltage magnitudes.

$$Minf = \sum_{i=1}^{n_b}\left\{\left(I_i^d\right)^2 + \left(I_i^q\right)^2\right\} * r_i + \sum_{k=1}^{n_v}\left\{V_k^+ + V_k^-\right\} * w_k \tag{2.8}$$

subject to

$$I_i^d = I_i^d(0) + \sum_{c=1}^{n_c} S_{u_c}^{I_i^d} * u_c + \sum_{d=1}^{n_d} S_{u_d}^{I_i^d} * u_d \quad i = 1, \ldots, n_b \tag{2.9}$$

$$I_i^q = I_i^q(0) + \sum_{c=1}^{n_c} S_{u_c}^{I_i^q} * u_c + \sum_{d=1}^{n_d} S_{u_d}^{I_i^q} * u_d \quad i = 1, \ldots, n_b \tag{2.10}$$

$$(I_i^d)^2 + (I_i^q)^2 \le (I_i^{max})^2 \quad i = 1, \ldots, n_b \tag{2.11}$$

$$Q_d^{MIN} \le Q_d(0) + u_d \le Q_d^{MAX} \quad d = 1, \ldots, n_{der} \tag{2.12}$$

$$V_k = V_k(0) - V_k^+ + V_k^- + \sum_{c=1}^{n_c} S_{u_c}^{V_k} * u_c + \sum_{d=1}^{n_d} S_{u_d}^{V_k} * u_d \quad k = 1, \ldots, n_v \tag{2.13}$$

$$V^{MIN} \le V_k \le V^{MAX} \quad k = 1, \ldots, n_v \tag{2.14}$$

$$V_k^+ \ge 0, V_k^- \ge 0 \quad k = 1, \ldots, n_v \tag{2.15}$$

where

| | |
|---|---|
| $I_i^d, I_i^q$ | active/reactive current component of branch-*i* |
| $n_b$ | number of network branches (lines and transformer windings) in the network |
| $r_i$ | resistance of branch-*i* |
| $V_k^+, V_k^-$ | slack variables associated with line-to-neutral/line-to-line voltage-*k* |
| $w_k$ | weight coefficient for voltage-*k* violation |
| $n_v$ | number of line-to-neutral/line-to-line voltages |
| $I_i^d(0), I_i^q(0)$ | base case value of the active/reactive current for branch-*i* |
| $S_{u_c}^{I_i^d}, S_{u_c}^{I_i^q}$ | sensitivity value of active/reactive current component of branch-*i* with respect to the switching of capacitor-*c* |
| $S_{u_d}^{I_i^d}, S_{u_d}^{I_i^q}$ | sensitivity value of active/reactive current component of branch i with respect to DER-*d* |
| $n_c$ | number of controllable capacitors |
| $u_c$ | discrete control variable of capacitor-*c* |
| $n_d$ | number of controllable DERs |
| $u_d$ | VAR output control variable of DER-*d* |
| $Q_d(0)$ | base case VAR output of DER d |
| $Q_d^{MAX}, Q_d^{MIN}$ | upper/lower limit of the VAR output of DER-*d* |
| $V_k$ | voltage-*k* magnitude |
| $V_k(0)$ | base case voltage k magnitude |
| $V^{MAX}, V^{MIN}$ | upper/lower limit for voltage-*k* magnitude |
| $S_{u_c}^{V_k}, S_{u_d}^{V_k}$ | sensitivity value of voltage k magnitude with respect to control variables (capacitor switching and DER VAR output) |
| $I_i^{max}$ | current magnitude limit of branch-*i*. |

All parameters in the objective function and constraints shown in (2.8)–(2.15), such as voltages, currents, resistances, are the parameters used in the multi-phase network model. Two examples are provided below to illustrate how the multi-phase distribution network model is adopted in the optimization problem formulation.

In the objective function, based on the detailed multi-phase network model, the power loss on each existing phase of the distribution line and transformer winding is explicitly taken into account. For instance, given a lateral section ($l$) that only consists of two phases (e.g., phases a and b), which is common in distribution networks, the two-phase resistances $r_{la}$ and $r_{lb}$, as well as the real and reactive current components on the two phases ($I_{la}^{d}$ and $I_{la}^{q}$ for phase a, $I_{lb}^{d}$ and $I_{lc}^{q}$ for phase b) are used to calculate the loss for the lateral section as shown in Eq. (2.16). No unrealistic three-phase balance assumption is made for this two-phase lateral.

$$P_l = \left\{ \left(I_{la}^{d}\right)^2 + \left(I_{la}^{q}\right)^2 \right\} * r_{la} + \left\{ \left(I_{lb}^{d}\right)^2 + \left(I_{lb}^{q}\right)^2 \right\} * r_{lb} \tag{2.16}$$

Similar to the objective function, constraints for currents and voltages are also derived based on the detailed multi-phase network model, i.e., each constraint reflects the limit for each phase or line variable. For example, if a single-phase load is connected between two phases of lateral $l$, the constraint for the line-to-line voltage at the load location is expressed in (2.17). Again, the impractical assumption for a three-phase balanced load is not necessary, and the constraint set only includes one line-to-line voltage for this load.

$$V^{MIN} \leq V_{ab} \leq V^{MAX} \tag{2.17}$$

In a nutshell, both objective function and constraints are formulated based on the multi-phase network model that can realistically reflect the nontrivial unbalanced characteristics of the distribution network. No approximation/assumption is made to represent any practical single-/two-phase component with a three-phase balanced component to obtain a balanced network model. The unbalanced distribution network model is fully utilized in the optimization formulation in this chapter.

The presented loss reduction approach is formulated as a mixed integer quadratic optimization problem. The optimization formulation is implemented in the Visual Studio C++ environment and interacts with the multi-phase distribution network model unbalanced power flow (UBLF). Third-party solver CPLEX is applied to solve the formulated optimization problem. The presented optimization technology can provide online solution speed.

#### 2.3.2.2 Test Distribution Networks

The advanced loss reduction approach and solution are tested with nine practical distribution utility networks. This section illustrates these networks and test results based on the presented coordinated control technology. Besides, this section also presents the test results from a benchmark approach that only includes capacitor control. The detailed benchmark formulation is provided in the Appendix, which is derived by removing DER VAR control from the advanced approach formulation in (2.8)–(2.15). The benchmark approach is implemented and solved in the same

**Table 2.6** Test network information summary

| ID | #Cap | #DER | #Node | #Load | #Line |
|---|---|---|---|---|---|
| C1 | 3 | 5 | 1569 | 538 | 1608 |
| C2 | 3 | 2 | 1740 | 543 | 1776 |
| C3 | 1 | 3 | 1312 | 402 | 1339 |
| C4 | 2 | 3 | 837 | 265 | 844 |
| C5 | 4 | 8 | 2085 | 644 | 2125 |
| C6 | 1 | 5 | 2465 | 770 | 2493 |
| C7 | 3 | 6 | 1016 | 307 | 1035 |
| C8 | 0 | 3 | 790 | 245 | 812 |
| C9 | 0 | 10 | 1760 | 566 | 1784 |

environment and tested with the same nine distribution networks as that of the advanced approach. Test results from two approaches are compared for the purpose of investigating the impact of DER VAR control on the loss reduction and voltage correction.

Table 2.6 lists the information of nine practical distribution utility networks (C1–C9) used to test both advanced and benchmark approaches. Each network has multiple feeders, and the average number of feeders is 10. Regarding distribution lines including both feeders and laterals, about 60% are three-phase and 40% are single-/two-phase. Voltage levels vary between 4.16 and 23.9 kV. In these networks, the number of switchable capacitors (#Cap) varies between zero to four, which are either ganged or unganged controllable. As these distribution networks originally do not include controllable DERs, a number of PQ mode DERs are added to the networks at randomly selected locations. The number of added controllable DERs (#DER) varies between two to ten. Load models include two types: constant power and constant impedance model.

The numbers of network nodes (#Node), loads (#Load), and lines (#Line) shown in Table 2.6 are the numbers without counting the actual phases. In other words, each node, load, or line may have single, two, or three phases. Each load and capacitor may connect in delta or wye connection. Besides the components listed in the table, each network also includes multiple transformers (on-load tap changer, voltage regulators, etc.) in various configurations.

All these test networks are typical practical distribution networks with representative features in terms of distribution network dimension and complexity. The test results based on these networks provide meaningful information in the effectiveness and performance of the presented loss reduction approach and solution for practical distribution networks.

#### 2.3.2.3 Test Results on Loss Reduction

The test results of advanced and benchmark loss reduction approaches on the nine networks are classified into three categories. The networks in each category exhibit certain common features in the loss reduction and voltage correction. Tables 2.7, 2.8 and 2.9 present test results for networks in each category, respectively. Each table includes the following information:

- Base Case: no loss reduction function is applied
- Cap Only: Apply the benchmark loss reduction approach based on capacitor control only
- Cap + DER: Apply the advanced loss reduction approach based on the coordinated control among capacitors and DERs
- Loss: the network power loss in kW
- Reduction %: the power loss reduction in percentage with respect to the base case power loss
- #VV: the number of voltage violations
- Max|VV|: absolute maximum voltage violation amount (in pu).

**Table 2.7** Category 1: test results on loss reduction and voltage violation correction

| ID | | Loss (kW) (Reduction %) | #VV | Max\|VV\| (pu) |
|---|---|---|---|---|
| C1 | Base case | 308.1 | 64 | 0.02 |
| C1 | Cap only | 308.1 (0%) | 64 | 0.02 |
| C1 | Cap + DER | 293.3 (4.8%) | 3 | 0.002 |
| C2 | Base case | 269.9 | 15 | 0.09 |
| C2 | Cap only | 215.4 (20%) | 10 | 0.002 |
| C2 | Cap + DER | 207.7 (23%) | 0 | 0 |
| C3 | Base case | 479.3 | 4 | 0.04 |
| C3 | Cap only | 479.3 (0%) | 4 | 0.04 |
| C3 | Cap + DER | 445.9 (7%) | 4 | 0.02 |

**Table 2.8** Category 2: test results on loss reduction and voltage violation correction

| ID | | Loss (kW) (Reduction %) | #VV | Max\|VV\| (pu) |
|---|---|---|---|---|
| C4 | Base case | 170.0 | 31 | 0.01 |
| C4 | Cap only | 170.0 (0%) | 31 | 0.01 |
| C4 | Cap + DER | 228 (−34%) | 0 | 0 |
| C5 | Base case | 410.3 | 199 | 0.04 |
| C5 | Cap only | 410.3 (0%) | 199 | 0.04 |
| C5 | Cap + DER | 442.9 (−8%) | 43 | 0.003 |
| C6 | Base case | 361.5 (0%) | 133 | 0.05 |
| C6 | Cap only | 361.5 (0%) | 133 | 0.05 |
| C6 | Cap + DER | 425.1 (−17%) | 6 | 0.007 |

**Table 2.9** Category 3: test results on loss reduction and voltage violation correction

| ID | | Loss (kW) (Reduction %) | #VV | Max\|VV\| (pu) |
|---|---|---|---|---|
| C7 | Base case | 49.9 | 0 | 0 |
| C7 | Cap only | 45.5 (8.9%) | 0 | 0 |
| C7 | Cap + DER | 42.1 (16%) | 0 | 0 |
| C8 | Base case | 51.6 | 0 | 0 |
| C8 | Cap only | NA | 0 | 0 |
| C8 | Cap + DER | 46.2 (11%) | 0 | 0 |
| C9 | Base case | 130 | 0 | 0 |
| C9 | Cap only | NA | 0 | 0 |
| C9 | Cap + DER | 120.4 (7.3%) | 0 | 0 |

Category 1 includes networks 1, 2, and 3 (C1–C3). All three test networks have voltage violations in the base case. Table 2.7 shows the test results from the presented and benchmark loss reduction approaches.

When applying the benchmark approach (Cap only): for networks 1 and 3, no capacitor switching action is available to improve both loss reduction and voltage violations or just correct voltage violations given its higher priority over the loss reduction; regarding to network 2, the capacitor switching action can achieve 20% loss reduction while improving the voltage profile by reducing the number of voltage violations and the maximum amount of the voltage violation. When applying the presented approach (Cap + DER): comparing to the benchmark approach, the voltage violations are further improved for networks 1 and 3 and completely eliminated for network 2. In addition more network loss reductions are achieved: loss reduces 4.8% for network 1, 23% for network 2, and 7% for network 3.

The test results in this category demonstrate that the coordinated control among capacitors and DERs is able to further reduce loss while eliminating/improving voltage violations, comparing to the benchmark approach with capacitor control only.

Category 2 includes networks 4, 5, and 6 (C4–C6). Similar to the networks in category 1, C4–C6 have voltage violations in the base case. Table 2.8 shows the test results from the presented and benchmark approaches.

When applying the benchmark approach (Cap only): for all the three networks, no capacitor switching action is available that can both correct voltage and reduce loss or just correct voltage violations. When applying the advanced approach (Cap + DER): given the higher priority of the voltage correction over the loss reduction, voltage violations are improved (for networks 5 and 6) or eliminated (for network 4) by sacrificing the network losses for all three networks. Loss increases 34% for network 4, 8% for network 5, and 17% for network 6.

Test results in this category show that the voltage correction and loss reduction cannot be achieved concurrently due to the contradict impact of capacitor and DER VAR control on the two targets. Given the high priority of voltage correction,

Table 2.10 Solution runtimes for benchmark approach

| ID | Solution time (s) | ID | Solution time (s) | ID | Solution time (s) |
|---|---|---|---|---|---|
| C1 | 1.38 | C4 | 0.9 | C7 | 1.4 |
| C2 | 2.43 | C5 | 2.1 | C8 | NA |
| C3 | 1.09 | C6 | 1.4 | C9 | NA |

Table 2.11 Solution runtimes for advanced approach

| ID | Solution time (s) | ID | Solution time (s) | ID | Solution time (s) |
|---|---|---|---|---|---|
| C1 | 2.87 | C4 | 2.0 | C7 | 1.6 |
| C2 | 2.51 | C5 | 6.0 | C8 | 0.9 |
| C3 | 1.85 | C6 | 3.8 | C9 | 2.4 |

the coordinated control between capacitors and DERs is able to eliminate or improve voltage violation, while cap only control cannot achieve similar results.

Category 3 includes networks 7, 8, and 9 (C7–C9). Different from the networks in the other two categories, C7–C9 have no voltage violation existing in their base cases. Table 2.9 shows the test results from the two approaches. Network 7 can achieve 8.9% loss reduction with the benchmark approach while 16% loss reduction with the advanced approach. Networks 8 and 9 do not include any capacitors and can achieve 11 and 7.3% loss reduction with the advanced approach, separately. Test results show that advanced approach can reduce loss without incurring new voltage violation.

In summary, test results from nine typical distribution utility networks show that the involvement of DER VAR control and especially the optimally coordinated control among capacitors and DERs can achieve further loss reduction and voltage correction comparing to the approach with capacitor control only. In particular, when the impacts of the control action on the loss reduction and voltage violation do not contradict, the presented approach can improve both aspects effectively. Otherwise, given the higher priority on the voltage correction, the presented approach can better improve the voltage profile. When no voltage violation exists in the base case, the presented approach can further reduce network loss without incurring new voltage violation.

The solution performance is tested on a desktop computer with following configuration: Intel Core 2 Quad Processor Q9400@2.66 GHz, 4 GB RAM, Microsoft Windows XP operating system.

Tables 2.10 and 2.11 provide the solution runtimes on the nine test networks based on two approaches, separately. Most of solution runtimes are less than or around 3 s. The solution runtime of network 5 is 6 s when applying advanced loss reduction approach. This network has relatively large dimension and number of controllable devices (four capacitors and eight DERs). Overall, the solution performance is suitable for online application to handle complex large-scale distribution networks in the real world.

## Appendix

The benchmark loss reduction problem formulation is derived based on the presented approach as illustrated in (2.8)–(2.15). The Benchmark approach only controls capacitor banks for loss reduction and voltage correction. The test results from this formulation are used as the reference to demonstrate the effectiveness of the advanced loss reduction approach. The detailed problem formulation is described in (2.18)–(2.24).

$$Minf = \sum_{i=1}^{n_b}\left\{\left(I_i^d\right)^2 + \left(I_i^q\right)^2\right\} * r_i + \sum_{k=1}^{n_v}\left\{V_k^+ + V_k^-\right\} * w_k \tag{2.18}$$

subject to

$$I_i^d = I_i^d(0) + \sum_{c=1}^{n_c} S_{u_c}^{I_i^d} * u_c \quad i = 1, \ldots, n_b \tag{2.19}$$

$$I_i^q = I_i^q(0) + \sum_{c=1}^{n_c} S_{u_c}^{I_i^q} * u_c \quad i = 1, \ldots, n_b \tag{2.20}$$

$$\left(I_i^d\right)^2 + \left(I_i^q\right)^2 \le \left(I_i^{max}\right)^2 \quad i = 1, \ldots, n_b \tag{2.21}$$

$$Q_d^{MIN} \le Q_d(0) + u_d \le Q_d^{MAX} \quad d = 1, \ldots, n_{der} \tag{2.12}$$

$$V_k = V_k(0) - V_k^+ + V_k^- + \sum_{c=1}^{n_c} S_{u_c}^{V_k} * u_c \quad k = 1, \ldots, n_v \tag{2.13}$$

$$V^{MIN} \le V_k \le V^{MAX} \quad k = 1, \ldots, n_v \tag{2.23}$$

$$V_k^+ \ge 0, V_k^- \ge 0 \quad k = 1, \ldots, n_v \tag{2.24}$$

## References

1. William H. Kersting, *Distribution System Modeling and Analysis*, 2nd edn. (CRC Press, Boca Raton, FL, 2007)
2. H. Liu, S. Cheng, C. Huang, Y. Hou, Unbalanced power flow calculation for low voltage distribution systems including DGs, in *Innovative Smart Grid Technologies-Asia, pp. 1–5* (2012)
3. A Saemni, A.B. Nassif, C. Opathella, B. Venkatesh, A Modified Newton Raphson Method for Unbalanced Distribution Systems, in *IEEE International Conference on Smart Grid Engineering*, pp. 1–7 (2012)
4. S. Bruno, S. Lamonaca, G. Rotondo, U. Stecchi, M. La Scala, Unbalanced three-phase optimal power flow for smart grids. IEEE Trans. Industr. Electron. **58**(10), 4504–4513 (2011)

5. R.D. Zimmerman, Hsiao-Dong Chiang, Fast decoupled power flow for unbalanced radial distribution systems. IEEE Trans. Power Syst. **10**(4), 2045–2052 (1995)
6. C.P. Nguyen, A.J. Flueck, A novel agent-based distributed power flow solver for smart grids. IEEE Trans. Smart Grid **6**(3), 1261–1270 (2015)
7. U. Leeton, T. Ratniyomchai, T. Kulworawanichpong, Optimal reactive power flow with distributed generating plants in electric power distribution systems, in *International Conference on Advances in Energy Engineering* (2010)
8. J.J. Zhao, X. Li, J.T. Hao, C.L. Zhang, J.P. Lu, Wind farm reactive power output optimization for loss reduction and voltage profile improvements, in *Power Electronics and Motion Control Conference* (2009)
9. M. Oshiro, T. Senju, A. Yona, N. Uraskaim T. Funabashi, C.H. Kim, Optimal operation strategy by battery energy storage systems in distribution system, in *IPEC* (2010)
10. Isha Sharma, Analysis of unbalanced distribution systems with solar PV penetration, in *IEEE PES GM* (2013)
11. J.F. Franco, A mixed-integer linear programming model for the electric vehicle charging coordination problem in unbalanced electrical distribution systems, in *IEEE Transactions on Smart Grid*, vol. 99 (2015)
12. F. Yang, Z. Li, Improve distribution system energy efficiency with coordinated reactive power control. IEEE Trans. Power Syst. **99**, 1–8 (2015)
13. K. Miu, M. Kleinberg, Impact studies of unbalanced multi-phase distribution system component models, in *IEEE Power and Energy Society General Meeting*, pp. 1–4 (2010)
14. Y. Zhu, K. Tomsovic, Optimal distribution power flow for systems with distributed energy resources. Electr. Power Energy Syst. **29** (2007)
15. A.A.S. Algarni, K. Bhattacharya, Disco operation considering DG units and their goodness factor. IEEE Trans. Power Syst. **24**(4) Nov. (2009)
16. A.A.S. Algarni, K. Bhattacharya, Utility-owned DG units' impacts on distribution system operation, in *Power Systems Conference and Exposition* (2009)
17. T. Niknam, A.M. Ranjbar, A.R. Shirani, Impact of distributed generation on volt/var control in distribution networks, in IEEE Power Tech Conference Proceedings (2003)
18. T. Niknam, A.M. Ranjbar, A.R. Shirani, A. Ostadi, A new approach based on ant colony algorithm to distribution management system with regard to dispersed generation, in 18th International Conference on Electricity Distribution (2005)
19. U. Leeton, T. Ratniyomchai, T. Kulworawanichpong, Optimal reactive power flow with distributed generating plants in electric power distribution systems, in International Conference on Advances in Energy Engineering (2010)
20. J.J. Zhao, X. Li, J.T. Hao, C.L. Zhang, J.P. Lu, Wind farm reactive power output optimization for loss reduction and voltage profile improvements, in Power Electronics and Motion Control Conference (2009)
21. M. Oshiro, T. Senju, A. Yona, N. Uraskaim T. Funabashi, C.H. Kim, Optimal operation strategy by battery energy storage systems in distribution system, in *IPEC* (2010)
22. F.A. Viawan, D. Karlsson, Coordinated voltage and reactive power control in the presence of distributed generation, in *Conversion and Delivery of Electrical Energy in the 21st Century* (2008)
23. F.A. Viawan, D. Karlsson, Combined local and remote voltage and reactive power control in the presence of induction machine distributed generation. IEEE Trans. Power Syst. **22**(4), Nov. (2007)
24. X. Feng, P. William, F. Yang, Implementation of control center based voltage and var optimization in distribution management system, in *IEEE Transmission and Distribution Conference and Exposition* (2010)

# Chapter 3
# Distribution Network Demand and Its Uncertainty

**Stephen Haben and Georgios Giasemidis**

**Abstract** This chapter presents some advanced tools for low voltage (LV) network demand simulation. Such methods will be required to help distribution network operators (DNOs) cope with the increased uptake of low carbon technologies and localised sources of generation. This will enable DNOs to manage the current network, simulate the effect of various scenarios and run load flow analysis. In order to implement such analysis requires high resolution smart meter data for the various customers connected to the network. However, only small amounts of individual smart meter data will be available and such data could be expensive. In likelihood, smart meter data is only going to be freely available at the aggregate level. Hence, in general, to implement LV network tools, customer loads will need to be simulated based on the assumption of limited amounts of monitored data. In addition, due to the high volatility of LV electric distribution networks, demand uncertainty must also be captured within a simulation tool. In this chapter, a number of methods are described for simulating demand on low voltage feeders which rely only on relatively small samples of smart meter data and monitoring. Firstly, a method called 'buddying' is described for assigning realistic profiles to unmonitored customers by buddying them to a customer who is monitored. Secondly, a number of methods are presented for capturing the uncertainty on the network. Finally the uncertainty models are incorporated into the buddying method and implemented in a load flow analysis tool on a number of real feeders. Both the buddying and the uncertainty estimation are presented for two different cases based on whether LV substation monitoring is present or not. This illustrates the different impacts of monitoring availability on the modelling tools. This chapter demonstrates the presented methods on a large range of real LV feeders.

S. Haben (✉)
Mathematical Institute, University of Oxford, Oxford, UK
e-mail: haben@maths.ox.ac.uk

G. Giasemidis
CountingLab Ltd, University of Reading, Reading, UK
e-mail: g.giasemidis@reading.ac.uk



A. Arefi et al. (eds.), *Electric Distribution Network Management and Control*,
Power Systems, https://doi.org/10.1007/978-981-10-7001-3_3

**Keywords** LV networks · Buddying · Network modeling · Uncertainty estimation Power flow analysis

## 3.1 Introduction

It is becoming increasingly necessary for distribution network operators (DNOs) to be able to model the demand behaviour at the low voltage (LV) level to improve network planning and management. With the uptake of low carbon technologies, such as electric vehicles and photovoltaics, demand is expected to increase and become more unpredictable [1]. Accurate modelling of network demands is essential to be able to predict and assess the stability and headroom on the LV network through power flow analysis [2]. Further, accurate simulation can also assess the validity of network demand solutions such as battery storage or demand side response [3, 4]. Such smart grids could reduce the costs from reducing additional traditional reinforcement, for example, in the UK the potential savings are around £2.5B–£12B [5].

With the roll out of smart meters, network operators can potentially populate network modelling tools with the actual load of the connected customer and then it is trivial to run power flow analysis tools [6, 7]. However, since smart meter data is proprietary, it is not guaranteed that it will be available to DNOs and if it is, it is likely to be expensive [1]. For example, in the UK, it has been stated that only aggregated data may be available, this will save a DNO from having to monitor at the substation/feeder level but will give no greater visibility of individual customers [8].

Since an LV network feeder typically represents the aggregation of around 10 up to 100 customers, the demand is relatively volatile compared to higher volatile networks, as illustrated in Fig. 3.1. Hence, any modelling of a network must capture the uncertainty to ensure that the diversity is incorporated and that a fuller understanding of the demands is available. Up until recently network operators have planned networks based around basic measures such as the after diversity maximum demand (ADMD), or estimated the aggregation of winter period profiles using assumption of Gaussianity, e.g., such as the specifications of ACE Report No. 49 (ACE49) for the design of LV radial distribution networks [9]. Unfortunately, although such metrics have been relatively successful, the increased uptake of low carbon technologies was not anticipated when these tools were developed and hence they may not be suitable for future network design in a low carbon economy.

In this chapter, simple and practical methods for modelling realistic profiles for residential customers on low voltage electric distribution networks are demonstrated using limited customer smart meter data. Two cases are considered: one where substation (i.e., aggregate) level monitoring is available and another where this monitoring is not available. Hence, even with less data resources an (albeit less accurate) modelling solution can be implemented. The second main topic of this

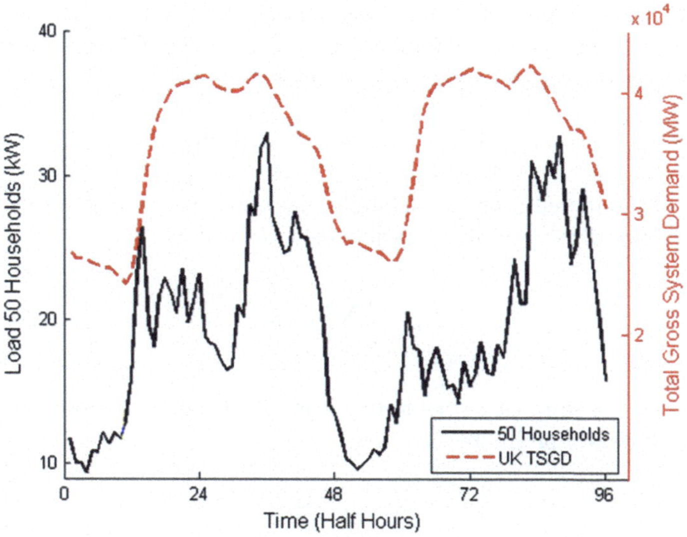


**Fig. 3.1** Half-hourly demand over 2 days for total gross system demand (red) and the aggregation of 50 residential customers (black)

chapter is about methods for quantifying LV demand uncertainty. Again, to take into account the scenario where little data is available, two methods are presented depending on different quantities of monitoring. Finally, both the customer network models and the uncertainty quantification are consolidated to show a complete modelling of an LV network taking into account the diversity of demand. This should enable a distribution network operator to run informative power flow analysis for any size LV network.

## 3.2 Buddying: Simulated Network Demand with Limited Monitoring

The basic aim of buddying is to simulate, as accurately as possible, the demand profiles for each customer point on the feeder of an LV network, as well as the aggregated feeder demand profile using a limited sample of monitored profiles.

Suppose $M$ customers, labelled $c_j$ for $j = 1, \ldots M$, are connected to an LV feeder (or phase) which has no available smart meter data (or other data from advanced metering infrastructures) but does have a known daily mean usage $U_j$ for

$j = 1, \ldots M$, (say, obtained from their quarterly meter readings which are typically available to network operators). Further, assume a sample of $N$ customers who are monitored with smart meters over $d$ days and let

$$\boldsymbol{P} = \{\boldsymbol{p}_k = (p_k(1), p_k(2), \ldots, p_k(48d))^T \in R^{48d} | k = 1, \ldots, N\}, \tag{3.1}$$

be the set of monitored profiles. Naturally, the mean daily usage for each profile $\boldsymbol{p}_k \in \boldsymbol{P}$ is known directly from the profile data, defined as

$$\widehat{U}_k = \frac{1}{d} \sum_{h=1}^{48d} p_k(h). \tag{3.2}$$

Finally, since in some cases monitoring is installed at the substation level, denote the feeder (or phase) profile by

$$\boldsymbol{s} = (s(1), s(2), \ldots, s(48d))^T \in R^{48d}, \tag{3.3}$$

for the same time period.

The aim is to assign, as accurately as possible, with the information available, a profile $\boldsymbol{p}_k \in \boldsymbol{P}$ to simulate the demand profile of each unmonitored customer $c_j$. This process is called *buddying* and the profile $\boldsymbol{p}_k$ is referred to as the *buddy* of the customer $c_j$. In the methodologies presented in the following sections the assignments are restricted so that both the unmonitored customer and the buddy come from the same group. The groups are explained in Appendix "Customer Groups" but are primarily determined by the Elexon profile class and the council tax band of the property (here viewed as a surrogate for assets) [10]. The groups serve two main purposes: (1) to ensure that similar customers are buddied and (2) to reduce the computational cost of the genetic algorithm employed in the optimization. It should be noted that many other groupings could be equally valid.

The buddying method described above can be considered as a method for creating *pseudo-measurements* which are utilized in state estimation methods when observation data is not available [11]. However, whereas pseudo-measurements are typically generated from standardized daily profiles, buddying uses the profile from a monitored household instead [12]. One advantage of the method is that, by definition, the customers are assigned realistic profiles; this is in contrast to the methods which use representative profiles, which are smooth due to the averaging of several similar customers (see Sect. 3.2.6 for descriptions of other potential methods).

### 3.2.1 Buddying Methodology

Initially it is assumed that the substation is monitored at the same resolution as the monitored customers (typically network operators own the substation and are free to monitor the load). In addition, although smart meter data may not be directly available, some countries are allowing network operators access to aggregated data. Hence, there are a number of options where substation/feeder level data may be available for use in optimizing a network model.

In this scenario, the assumption is that all customers are residential. In Sect. 3.2.5 an extension of the method is presented which allows a mix of both residential and commercial customers. Since only a low resolution view of individual customers' usage is available (i.e., their quarterly meter readings) the substation level demand is utilized to better match the intra-day (albeit aggregate) level demand on the network. The aim of the buddying is to find a set of profiles $\widehat{\boldsymbol{P}} = \{\boldsymbol{p}_{k_1}, \boldsymbol{p}_{k_2}, \ldots, \boldsymbol{p}_{k_M}\}$, where $\boldsymbol{p}_{k_j} \in \boldsymbol{P}$ is the buddy for unmonitored customer $c_j$ for $j = 1, \ldots, M$. This is achieved by minimising the cost function

$$F\left(\widehat{\boldsymbol{P}}\right) = (1 - w)\sum_{h=1}^{48d} \frac{\|a(h) - s(h)\|_1}{S} + w\sum_{j=1}^{M} \frac{\left\|U_j - \widehat{U}_{k_j}\right\|_1}{D}, \tag{3.4}$$

over all possible sets of buddied profiles $\widehat{\boldsymbol{P}}$. Here $a(h) = \sum_{j=1}^{M} p_{k_j}(h)$ is the aggregate demand of the buddy, $S = \sum_{h=1}^{48d} s(h)$ and $D = \sum_{j=1}^{M} U_j$. The weighting $w \in [0, 1]$ allows the buddy to either optimize fully to the substation ($w = 0$) or completely to the mean daily usages ($w = 1$), or a weighted sum of the two. The optimal choice in $w$ can also be interpreted as the trust in the accuracy of the quarterly meter readings (since they are generally estimates) as well as their importance in identifying an accurate buddying.

The optimal buddy is found by a genetic algorithm which optimizes the solution according to the fitness function (3.4). The genetic algorithm creates updates of several collections of monitored customers according to how well they score according to the fitness function. In subsequent generations, solutions are produced by crossbreeding between the fittest solutions and retaining common buddies. Mutations (exchanging of some random buddies) ensures that solution retains sufficient diversity. The genetic algorithm is iterated 100 times or until the solution converges. Complete details on the genetic algorithm method can be found in Appendix "Genetic Algorithm". For $w \neq 1$ the buddying is referred to as the GA or genetic algorithm buddy. In the special case of $w = 1$, which is described in more detail in the next section, the buddy is referred to as the simple algorithm or SA buddy.

### 3.2.2 The Simple Algorithm Buddy

The SA buddy is a special case of the algorithm which simply assigns the profile which has the closest mean daily usage and can be found quickly and does not require the implementation of the genetic algorithm. Since no half hourly information is used in the SA buddy the algorithm is expected to be less accurate at assigning buddying profiles than the GA buddy. However, this option is more attractive to a network operator since it uses the minimal amount of monitoring and hence reduces potential data costs and storage. The importance of the accuracy of this method is reduced if the uncertainty is appropriately estimated (see Sect. 3.3).

### 3.2.3 Assessing the Buddying

To assess the quality of the buddying the accuracy can be considered at both the aggregate, feeder level or at the individual, household level. This chapter considers the relative mean absolute error (RMAE), defined by

$$RMAE = \frac{1}{\|A\|_1}\sum_{i=1}^{T}\|A(i) - E(i)\|_1, \tag{3.5}$$

where $A$ is the actual profile (either individual residential household or substation) and $E$ is the estimated profile (either the buddy or the aggregate of the buddies). Since peaks in demand are also important for DNOs the relative peak demand error (RPDE) is also considered,

$$RPDE = \frac{\max\limits_{i=1,\ldots,T} A(i) - \max\limits_{i=1,\ldots,T} E(i)}{\max\limits_{i=1,\ldots,T} A(i)}, \tag{3.6}$$

Note that by using relative errors then substations of different sizes and magnitudes are easily compared.

### 3.2.4 Case Study for 122 LV Substation Feeders

Although this case study will be presented for low voltage electric distribution networks in the UK, the buddying method can be applied to any network as long as the mean daily demands of the individual customers (from quarterly meter reads), and the aggregate demand of all customers (say from substation monitoring) is available. A grouping of customers for implementation of the genetic algorithm is also required (see Appendix "Customer Groups").

#### 3.2.4.1 Data and Study Specifications

A case study of 122 low voltage substation feeders was considered from the Thames Valley Vision project, a low carbon network fund project based in Bracknell, UK [14]. The half hourly demand was monitored from the 20th March 2014 to 22nd September 2015 amounting to around a year and a half of data. An example of the feeders emanating from the Radcliffe Way substation is shown in Fig. 3.2. All feeders are connected to residential customers only with an average of 35 households per feeder and a range of 4–109 customers. A sample of $N = 242$ residential customers is monitored with half hourly smart meter data for the same time period as the feeder data. In the case study presented all feeders are buddied for

**Fig. 3.2** Radcliffe way low voltage substation, Basingstoke, UK. The substation (blue) feeds residential households using three phase feeders (solid black lines) [13]

different numbers of weeks and weights starting from the 29th September 2014 but similar results hold for testing other periods of the year [15]. The accuracy of the buddying is tested over the yearly period 1st September 2014 to 31st August 2015.

#### 3.2.4.2 Buddying Error Analysis

Figure 3.3 shows an example of the SA and GA ($w = 0$) buddy for a particular day (15th January 2015) for individual phases on a feeder. Clearly the genetic algorithm is more accurate in this example (and later in this section it is shown that this holds in general) but it also highlights some potential drawbacks to the method. On phase 1 the peak period is missed, this is because the cost function is expressed in terms of a 1-norm and hence the error is typically reduced by capturing the average behavior. To more accurately model peaks would require a p-norm for $p > 2$ so that the optimization focuses on reducing peak errors. However, this is a relatively

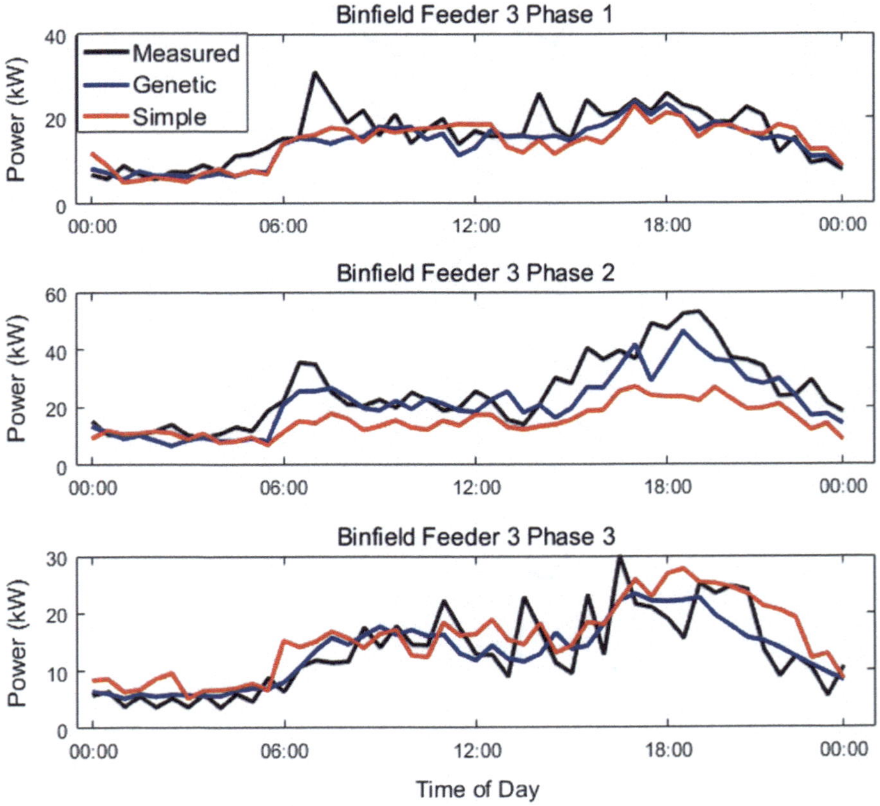


**Fig. 3.3** Example of the SA (red) and GA (blue) buddying versus the actual for individual phases of a particular Feeder

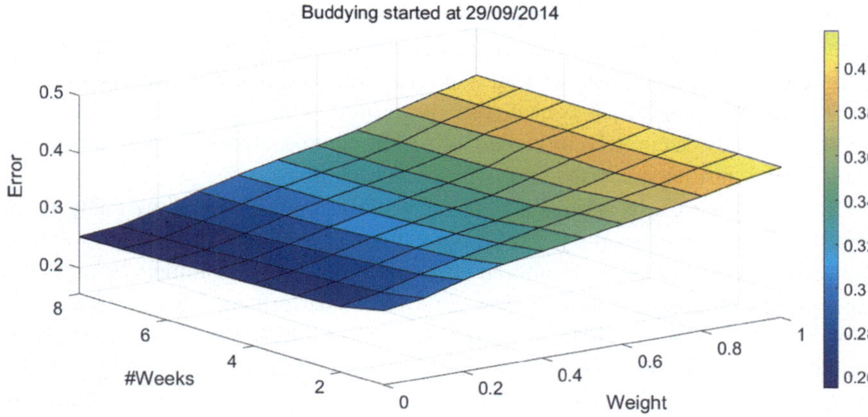


**Fig. 3.4** Average relative mean absolute errors for different numbers of weeks and weights in the GA buddying [15]

simple adjustment that can be made to the algorithm depending on the needs of the user. For more information on peak errors see [16].

A plot of the RAME errors of the match between the aggregated GA buddying and the substation feeder demand for the different numbers of weeks and weights is shown in Fig. 3.4. The more weeks in the buddying, and the higher the substation weighting (the smaller the $w$), the better the match to the substation, even over the entire year test period. This is perhaps unsurprising, since the buddying is optimized at the feeder level, however it can be further shown that some small weighting ($w = 0.1$) is required to ensure that there is accuracy at the individual level as well [15].

The accuracy, as may be expected, is a function of the number of customers, as shown in Fig. 3.5, for both the SA and GA ($w = 0$) buddying methods. The errors fit a power law distribution quite accurately as a function of the number of

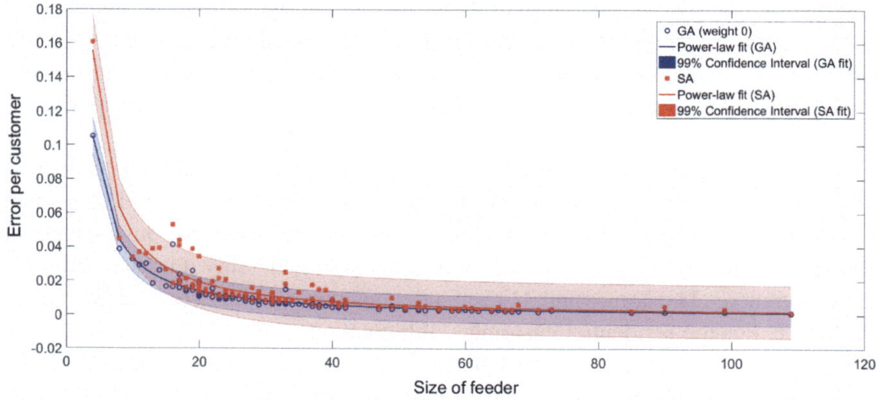


**Fig. 3.5** RMAE errors (normalized by number of customers) against number of customers for each feeder for both the GA and SA buddying methods [15]

customers on the feeder. Such plots can be useful for helping a distribution network operator make planning decisions, in particular understanding which LV electric distribution networks can be accurately modelled and which may need continuous monitoring. For example, it is clear that the feeders with more than 40 customers can be relatively accurately modelled using the simple algorithm and substation monitoring is not necessary to improve this modelling.

Often a Monte Carlo approach is taken when running power flow analysis on electric distribution networks [17, 18]. A random sample of the smart meter data is taken and assigned to the unmonitored customers (subject to customers being in the same group), and then a power flow analysis is performed. This is repeated a number of times (e.g., in [17] this is 1000 times) to estimate the potential spread of behaviors that can be expected on the network. When comparing against the buddying it is found that for 109 of the 122 feeders one run of the GA buddy is more accurate than all 1000 runs of random buddies (subject to customers being in the same group).

The choice of weighting, $w$, depends on the purpose of the buddying. If the accuracy of the individual profiles is less important than at the aggregate level the weighing should be skewed towards $w = 0$. However, if individual customers must be accurately modelled then some weighting should be skewed towards the individual setting $w = 1$ [15]. The inclusion of the uncertainty in demand in the modelling (see Sects. 3.3 and 3.4) means that the accuracy at either level is less important in the power flow analysis.

### 3.2.5 Incorporating Commercial Customers and Generation

The methods presented so far describe how to model unmonitored customers by assigning them a profile from a sample set of monitored profiles. This is suitable for residential customers since, given a large enough sample, their demand can be largely described by a finite set of behaviors (e.g., see [19]). For commercial customers this methodology no longer applies. For some larger commercial, that are already monitored, this will not be an issue since their actual profile can be subtracted from the aggregate/substation demand and buddying can be applied as normal.

For unmonitored commercial the buddying is more complicated since there is unlikely to be substantial monitored profiles of other commercial buildings with similar behavior to find a reasonable match. To deal with these customers, a slight adjustment is applied to the algorithm. Without loss of generality it is assumed there is one commercial customer and $M$ residential customers, who need buddies, on an LV feeder. For the residential customers, the notation of Sect. 3.2.1 are used. The commercial customer has a standard profile for the particular type of business (e.g., church, school, shop), which are part of current profile tools (such as WinDebut). This standard profile is normalized so that the daily mean usage is 1 kWh, denoted as $\boldsymbol{p}_{N+1}$. It is also assumed that the mean daily demand $U_{N+1}$ for the commercial customer is available (from the quarterly meter readings). As well as assigning each residential customer an appropriate profile from the sample of smart meter data, a scaling, $0.8 \leq \alpha \leq 1.2$, is also optimized for the normalized profile. Thus, a set of profiles $\widehat{\boldsymbol{P}} = \{\boldsymbol{p}_{k_1}, \boldsymbol{p}_{k_2}, \ldots, \boldsymbol{p}_{k_M}\}$ and a scalar $\alpha$ are now focused on to minimise the updated cost function of

$$F\left(\widehat{\boldsymbol{P}}\right) = (1-w)\sum_{h=1}^{48d} \frac{\|a(h) - s(h)\|_1}{S} + w\left(\sum_{j=1}^{M} \frac{\left\|U_j - \widehat{U}_{k_j}\right\|_1}{D} + \frac{\|U_{N+1}(1-\alpha)\|_1}{D}\right). \tag{3.7}$$

here $a(h) = \sum_{j=1}^{M} p_{k_j}(h) + \alpha U_{N+1}\boldsymbol{p}_{N+1}$ is the aggregate demand of the buddy, $S = \sum_{h=1}^{48d} s(h)$ and $D = \sum_{j=1}^{M+1} U_j$. This can be solved in a similar way to using a genetic algorithm. This time, the scalar $\alpha$ is sampled randomly from the range [0.8, 1.2] during buddying. At the crossover step, $\alpha$ is randomly chosen between the two parent genomes. At mutation, if the commercial customer is selected for mutation, a new $\alpha$ is sampled randomly in the range [0.8, 1.2]. For more details of the genetic algorithm see Appendix "Genetic Algorithm". Notice that $\alpha U_{N+1}$ is also an estimate for the daily mean usage for the commercial customer.

### 3.2.6 A Note on Other Approaches

This chapter only presents a particular approach for modelling network customers which has advantages in terms of its simplicity, accuracy and practicality. There are a number of other methods which could also be incorporated and they fit into roughly too cases: standardized profiles and bottom-up modelling [20].

For example, in the UK, electricity suppliers tend to use a two-level Elexon grouping of customers based on whether they have overnight storage heaters or not and this assigns them a Standard and Economy 7 profile respectively, which is scaled based on quarterly meter readings and weather data [10]. However, there are a large number of other standardised profiles which are typically created from clustering a sample of customer data and linking to household and demographic data ([19, 21], also see references within [22]). This approach depends on strong links to such properties but evidence suggests that such links to publically available (or easily accessible to a distribution network operator) data are not particularly strong [19, 23–25]. In addition, such profiles are typically quite smooth and unrealistic since they are based on average data and hence may not give accurate results via a power flow analysis.

For the bottom up approach realistic profiles can be generated by a statistical model which combines the profiles and usage of individual household appliances [26–28]. Hence realistic profiles are assigned but the drawback is that with rapidly changing appliances and technologies these profiles could become outdated rather rapidly and a large sample is required to create the models and ensure that correlations between appliance usages are also captured. In the buddying by using full profiles the correlations are naturally modelled.

## 3.3 Modelling Uncertainty on LV Electric Distribution Networks

In the previous section a method was described for assigning realistic profiles to all consumers on a low voltage electric distribution network without needing to monitor all customers. This can be used as a basis for understanding the impact of different uptakes of low carbon technologies and then power flow analysis can be performed to understand the current future network headroom [13].

However, in its current form the buddying does not take into account demand volatility, which can be particularly high at the low voltage network level. Traditionally, when planning for new building developments or planning new network connections standards, such as the ACE 49, have been used to estimate the potential 90% load on a network [29]. However, these standards may not be as robust as low carbon technologies are introduced on the network. In addition, the ACE49 framework is based on assuming the demands are Gaussian (or normally) distributed which is unlikely to be the case at the LV level and typically will only make sense at the higher voltages, i.e., larger aggregations of customers, where the law of large numbers becomes valid.

In this section methodologies for modelling network demand uncertainty are presented for two main scenarios, based on the amount of monitoring available. In the first scenario, it is assumed substation level monitoring (or equivalently, aggregations of the demand at the same level) is available. Methods used in this case will be more accurate, but more computationally expensive and require data and/or monitoring. The second scenario does not use substation monitoring but allows a network operator some insight into the demand variability with a much lower requirement for data and computational cost.

### *3.3.1 Monitoring Based Confidence Bound Generation*

The first scenario develops a model for specific quantiles using the historical information from monitors at the feeder level or from aggregations of the demand from all customers on the feeder. Only a particular model based on quantile regression is presented but there is a wide range of other possible methods that could also have been employed, such as kernel density estimation, general additive models, or any of a vast array of methods which are used in generating probabilistic load forecasts [30].

The advantages of the method presented here are that only specific quantiles (unlike say, kernel density estimation where the entire distribution must be specified) need to be generated, it is relatively quick, has been successfully implemented at higher voltage demands and the equations are also relatively simple and hence easy to interpret [31]. Each half hour is modelled as a separate time series and only models seasonality, trend and weekend identifiers as the main features of the system. Of course, further extensions can be developed including distinguishing between individual day types (Monday, Tuesday, etc.) and including autoregressive effects. However, this simple model was found to be quite successful for the purposes of quantile estimation and the considered test case.

Focusing on a particular half-hour of the day, say 5 pm, the individual load $L_d$, for $d = 1, \ldots, D$ is considered at the same hour for the $D$ days of the historical data set. Now assume a model for a particular quantile of the load data, say $\tau$ which is of the form

$$\widehat{L}_d^{\tau}(\boldsymbol{\alpha}^{\tau}) = a_0^{\tau} + a_1^{\tau} d + a_2^{\tau} W_d + \sum_{p=1}^{P} b_p^{\tau} \sin\left(\frac{2\pi p d}{365}\right) + c_p^{\tau} \cos\left(\frac{2\pi p d}{365}\right), \tag{3.8}$$

where $W_d$ is the weekend dummy for day $d$, having value one if the current day is on the weekend or zero otherwise. Here $P$ is the number of annual period terms to include in the model. $P = 2$ was found to be sufficient. In many cases, especially those countries with high air-conditioning use and hence more bi-annual behaviours, $P = 3$ or higher may be more suitable. Then, it is essential to find the corresponding coefficients, $\boldsymbol{\alpha}^{\tau} = (a_0^{\tau}, a_1^{\tau}, a_2^{\tau}, b_1^{\tau}, b_2^{\tau}, c_1^{\tau}, c_2^{\tau})^T$ for each half hour model and the specified quantile.

The parameters are found through quantile regression which is an analogue of the usual regression but instead of finding an estimate of the expected value, the specified quantile is estimated [32]. Much like standard regression the quantile is found by minimizing a cost function

$$\boldsymbol{\alpha}^{\tau} = \arg\min_{\boldsymbol{\alpha}} \sum_{d=1}^{D} \rho_{\tau}\left(L_d - \widehat{L}_d^{\tau}(\boldsymbol{\alpha})\right), \tag{3.9}$$

where

$$\rho_{\tau}(z) = \begin{cases} (\tau - 1)z & \text{if } z < 0 \\ \tau z & \text{if } z \geq 0 \end{cases}. \tag{3.10}$$

The minimization can be solved relatively quickly as a simple linear programming problem. Notice that the confidence bounds can be either estimated within sample (e.g., within the first $D$ days of historical data) or can be projected forward as a forecast. If there is insufficient historical information the out of sample forecasts may be inaccurate if projecting too far forward and hence it is suggested to use the in-sample estimates if trying to estimate the uncertainty on a particular day of the year.

### 3.3.2 Simple Confidence Bound Generation (Limited Monitoring)

If monitoring data at the aggregate level is limited (either through missing data or the monitoring has only just recently been installed) or not available then the method described in the previous section cannot be applied to estimate confidence bounds on the load. In this case a weaker estimate on the demand uncertainty at the feeder/aggregate level can be made by utilizing a relatively small sample of customer smart meter data from the same historical period. The method is based on a simple bootstrap method which repeatedly resamples from the sample of metered data to create several estimates of the aggregated demand [33]. From these bootstraps, quantiles can then be estimated empirically to give the confidence bounds.

Suppose an estimate of the $\tau$th quantile of the load on a feeder is required, on a particular time period of the year which is connected to $M$ customers, labelled $c_j$ for $j = 1, \ldots M$. Further, suppose these customers can be separated into $Q$ distinct groups, say based on profile types, household types, socio-demographics or whatever properties are good indicators/distinguishers of demand types. Next, assume a set of monitored profiles for each customer group for the time period specified is available, let these be defined as

$$\boldsymbol{P}^q = \{\boldsymbol{p}_k^q = (p_k(1), p_k(2), \ldots, p_k(48d))^T \in R^{48d} | k = 1, \ldots, N_q\}, \tag{3.11}$$

for $q = 1, \ldots, Q$ and $N = \sum_{q=1}^{Q} N_q$. Given there is $M_q$ customers from each group connected to the feeder, one bootstrap estimate can be created by sampling (with replacement) $M_q$ profiles $\boldsymbol{p}_{k_1}^q, \boldsymbol{p}_{k_2}^q, \ldots, \boldsymbol{p}_{k_{M_q}}^q$ from the set $\boldsymbol{P}^q$ for $q = 1, \ldots, Q$ given by

$$A^{(1)} = \sum_{q=1}^{Q} \sum_{l=1}^{M_q} \boldsymbol{p}_{k_l}^q. \tag{3.12}$$

This can be repeated a number of times, say 1000, to create several bootstrap estimates of the aggregate/feeder level demand $\boldsymbol{A}^{(1)}, \boldsymbol{A}^{(2)}, \ldots, \boldsymbol{A}^{(1000)}$ from which the empirical quantile can be used as an estimate for the confidence bounds. An advantage of the technique is that other quantities such as the peak demand or minimum demands can also be estimated for free from the same bootstraps.

The bootstrapping is a relatively cheap and effective method for estimating quantities (and the uncertainties associated to those quantities) with limited data available. There are of course several other ways of estimating quantiles from bootstraps, including bias correcting versions, but the version described here is amongst the simplest [33]. The quality of the bootstrap is very much determined by the representativeness of the sample of profiles available. However, in these cases there are methods to help increase the diversity of the sample, for example, decomposing each profile into several main components (e.g., mean daily, weekly and annual components as well as a residual component). Then bootstrapping can proceed as before but samples are taken with replacement on each component and combined to create new profiles.

### 3.3.3 Case Study

In the case study 223 residential-only feeders are considered from the Thames Valley Vision project varying from 4 to 109 customers. For the bootstraps, only two groups of customers are used based on profile class with 211 customers in profile class 1 and 30 in profile class 2. These are basic 2 classes for residential customers in the UK based on standard demand (profile class 1) or those with overnight storage heaters (profile class 2). See [10] and Appendix "Customer Groups" for more details on profile classes.

Examples of the half hourly 80% confidence regions (defined by the 90 and 10% quantiles) and the actual loads are shown for three feeders for the three days 14th–16th July 2014 in Figs. 3.6, 3.7 and 3.8 with increasing numbers of connected residential connections. The feeders service 22, 73 and 89 residential customers respectively with 22, 66 and 87 of those belonging to profile class 1 respectively. The substation confidence is clearly the better of the two and unsurprisingly the confidence bounds become smoother as the size of the feeders increase. The

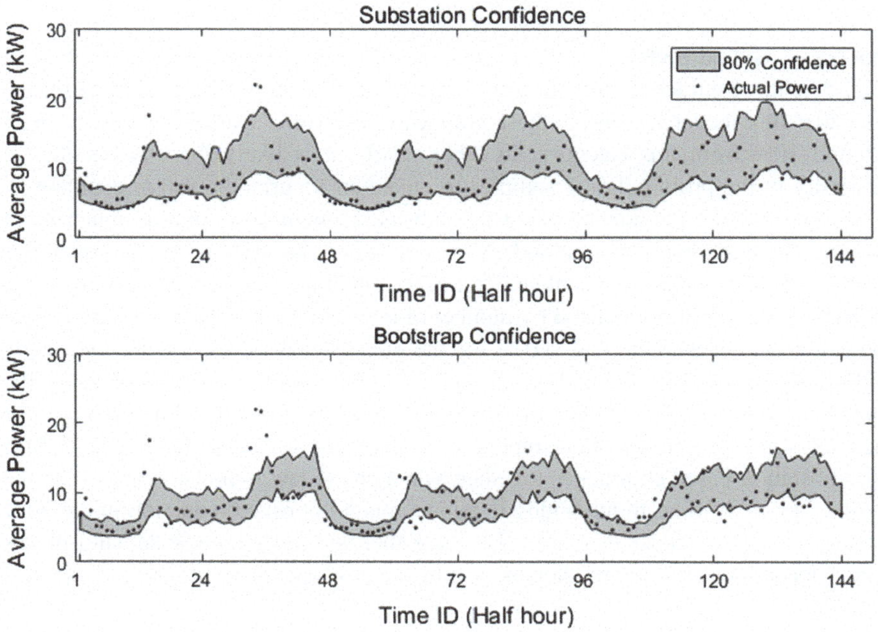


**Fig. 3.6** Three day demand (black dots) for a feeder with 22 residential customers and 80% confidence (grey) for both the substation method (top) and the bootstrap method (bottom)

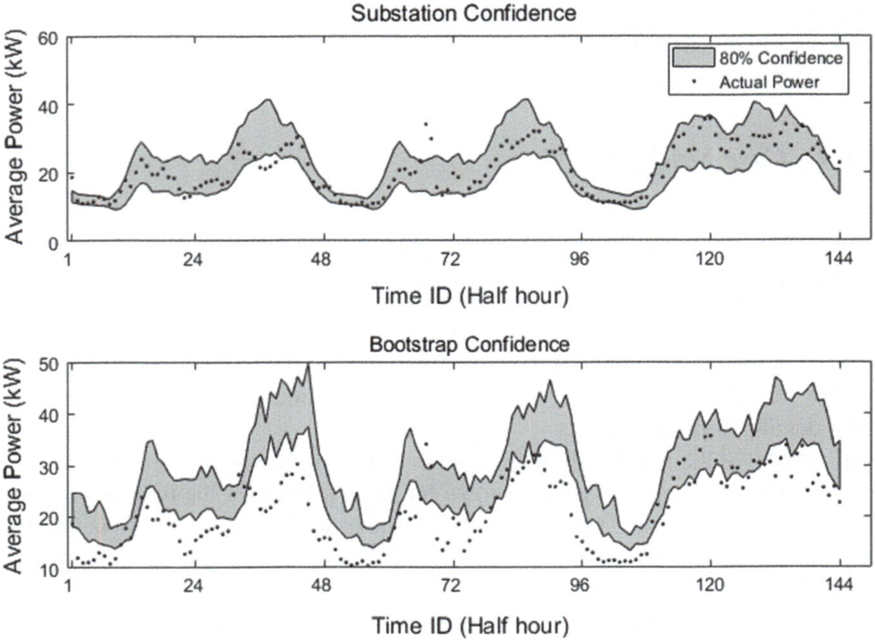


**Fig. 3.7** Three day demand (black dots) for a feeder with 73 residential customers and 80% confidence (grey) for both the substation method (top) and the bootstrap method (bottom)

bootstrap method performs well in a number of cases but occasionally, such as in Fig. 3.7, the confidence estimates are quite poor. This could be explained by the fact that this feeder has the highest proportion of profile class 2 customers, with around 10% from this category. In other words, it is likely there is insufficient diversity in the profile class 2 samples to create an accurate bootstrap estimate.

In general, the confidence using the substation monitoring data is much more accurate than the bootstrap method. A measure of the accurate of the two quantiles from each feeder can be calculated from the continuous ranked probability score (CRPS) [34]. When normalized by number of households the CRPS = 0.235 for the bootstrap-based estimate and 0.071 for the substation-based estimate, i.e., using substation data makes the estimates over 3.3 times more accurate than using the bootstrap estimate. Much like the buddying there is a relationship between the accuracy and the number of customers on a feeder. This is presented in Fig. 3.9 for the substation-based confidence together with a power law fit and confidence on the power law fit. Similar to the buddy there is a much improved reduction in the error as the size of the feeder increases. The smoother and more regular demand makes larger feeders much more predictable and hence produces more accurate confidence estimates.

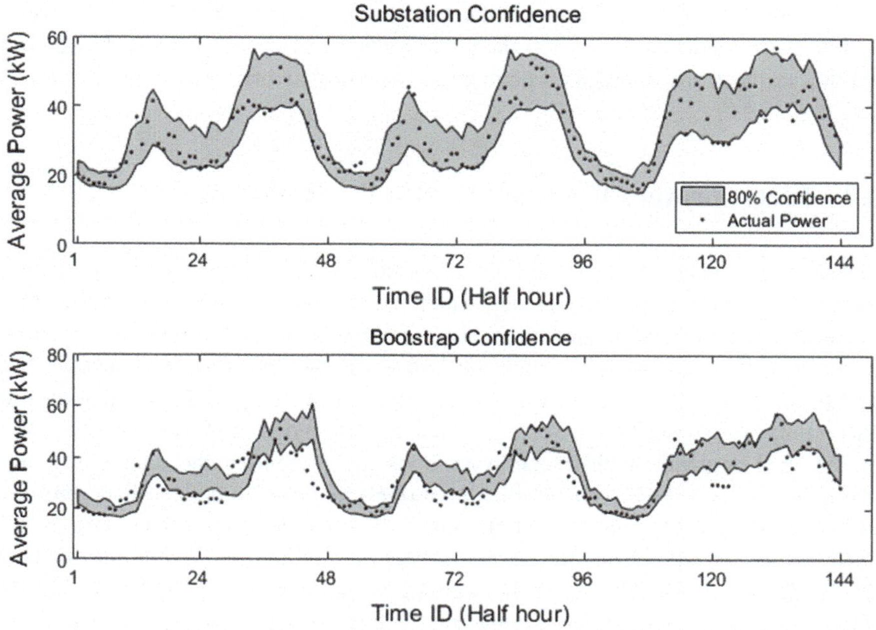


**Fig. 3.8** Three day demand (black dots) for a feeder with 89 residential customers and 80% confidence (grey) for both the substation method (top) and the bootstrap method (bottom)

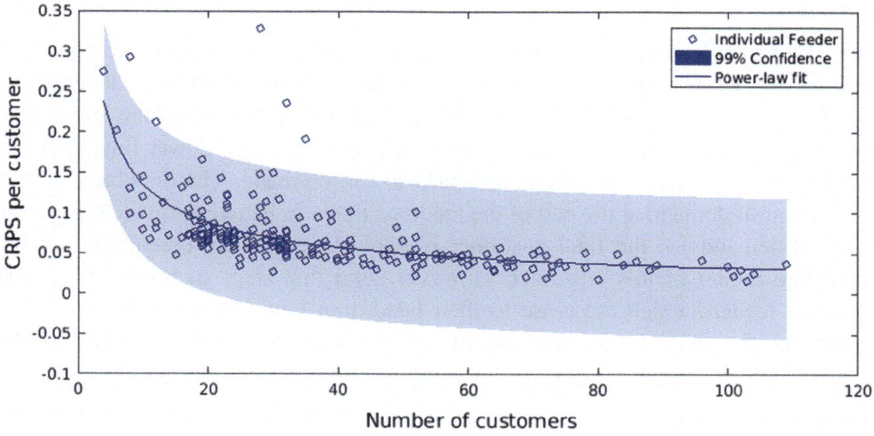


**Fig. 3.9** Normalized CRPS score per customer for the substation-based method as a function of number of customers on the feeder. Also included are the least squares power-law fit and the 99% confidence bounds fit

As with the buddying, although the case study is based on UK electric distribution networks the methods are valid for any electric distribution network and only require, at most, the aggregated demand of the full network.

## 3.4 Incorporating Uncertainty into the Buddying

In Sect. 3.2 methods were described for assigning realistic profiles to individual customers on a low voltage feeder which could then be used for power flow analysis. The drawback to the approach is the buddying only describes one possible scenario of demand behavior since the buddying does not take into account the historic diversity on the network. To remedy this, techniques for estimating the uncertainty were described in Sect. 3.3 via the creation of confidence bounds on the feeder level demand. In this section methods for combining the uncertainty and buddying are described to ensure that the models incorporate both realistic profiles and demand diversity. In the sections that follows, unless specified, most of the methods are independent of the buddying or confidence generation method employed. Hence for the rest of the section, assuming a time period of $d$ days is being considered, the confidence bound (at the feeder level) of interest, say the 90% quantile, is denoted as $\boldsymbol{C} \in \boldsymbol{R}^{48d}$.

### 3.4.1 A Simple Update

One of the simplest ways of incorporating the diversity is to look at the difference between the aggregated demand from the buddy and the confidence to create a so-called "diversity chunk" which can then be assigned to the end of the feeder. An advantage of this technique is that it can be simply added to a power flow analysis of a buddied network and a direct comparison can be made. The drawback is that the amount of demand at the end of the feeder is likely to be overestimated. Such an adjusted demand for the final customer is likely to be highly unrealistic and an inaccurate model of this customer. However, as a first step, such modelling can highlight feeders which are close to their headroom limit. Note that for the 10% confidence the confidence chunk would typically be negative. This update would only be valid for radial LV electric distribution networks.

### 3.4.2 Uniformly Distributed

To reduce the exaggerated demand of the customer at the end of a feeder the confidence bound profile could be uniformly divided amongst the $M$ customers, giving each customer a profile defined by $\boldsymbol{p} = \frac{1}{M}\boldsymbol{C}$. This guarantees that no household is modelled with an unlikely large demand. Unfortunately, now no customer has a realistic demand. However, the profiles should be sufficient to give a good indicator of the thermal constraints on the network after a power flow analysis.

### *3.4.3 Buddying to Confidence*

Another alternative is to implement the buddying algorithm to the confidence bound rather than the substation feeder demand. The advantage of this is that now all customers will be assigned a realistic profile which is constrained to their mean daily demand. Since the confidence bound is also smoother than the feeder level demand the optimization within the buddying algorithm (via the genetic algorithm) should also be a lot simpler to implement. For accurate application, the original sample of monitored customers must contain a representative collection of larger demand users. This is because larger customers are likely to be required due to buddying to larger values. Hence a larger initial sample of smart meter data would be required.

### *3.4.4 Uncertainty of Buddying from the Distribution of Smart Meter Data*

Uncertainty on individual buddies can be also estimated from the empirical distribution of the smart meter data of a given group. For each half-hour, empirical distributions of load demand can be created from the readings of the smart meters in a given group. This way, buddies in the same group are assigned empirical distributions for each half-hour, from which one can estimate confidence bounds. The empirical distributions may be replaced by parametric ones, e.g., Gaussian, whose parameters are estimated from the readings of the smart meters in the group. The parametric approach has the advantage that these distributions can be convoluted to give the aggregated distribution, hence the uncertainty of the substation/feeder.

## 3.5 Power Flow Analysis Examples

The main application of this chapter has been to model network loads so that they could be used as input for a network modelling environment to perform power flow analysis. Here, simple implementations of a power flow analysis are demonstrated for the presented methods using the CYME Secondary Network Analysis module which implements the full Newton-Raphson method for unbalanced load analysis [35].

Buddies for the electric distribution network were created using the GA version as outlined in Sect. 3.2.1 and 90 and 10% confidence bounds were generated using the method described in Sect. 3.3.1 using substation monitoring at the feeder level. An example of the confidence generated and the buddied (baseload) versus actual (Measured) is shown in Fig. 3.10.

First, load flow analysis was performed to show the effect of the upper confidence to the maximum loading factor for 43 feeders on 11 substations. The confidence in these examples was applied using the simple update presented in Sect. 3.4.1. As Fig. 3.11 shows no feeders are overloaded (i.e., exceed 100% loading) in this case which is to be expected since the current feeders are all in working order.

Next the effect of the maximum and minimum loadings (based on maximum and minimum confidence) on the minimum and maximum voltages respectively are shown in Figs. 3.12 and 3.13 respectively. The first plot shows that the lower voltage limit of 216 V could be violated for feeder 42 due to a high load at the end of the feeder. This is an artifact of the simple update procedure for incorporating the uncertainty with the buddying and demonstrates the potential complications with this method, as explained in Sect. 3.4.1.

In Fig. 3.13 the maximum voltage rose to close to the maximum limit of 253 V (red line) for three of the 43 feeders. This was caused by a large negative load

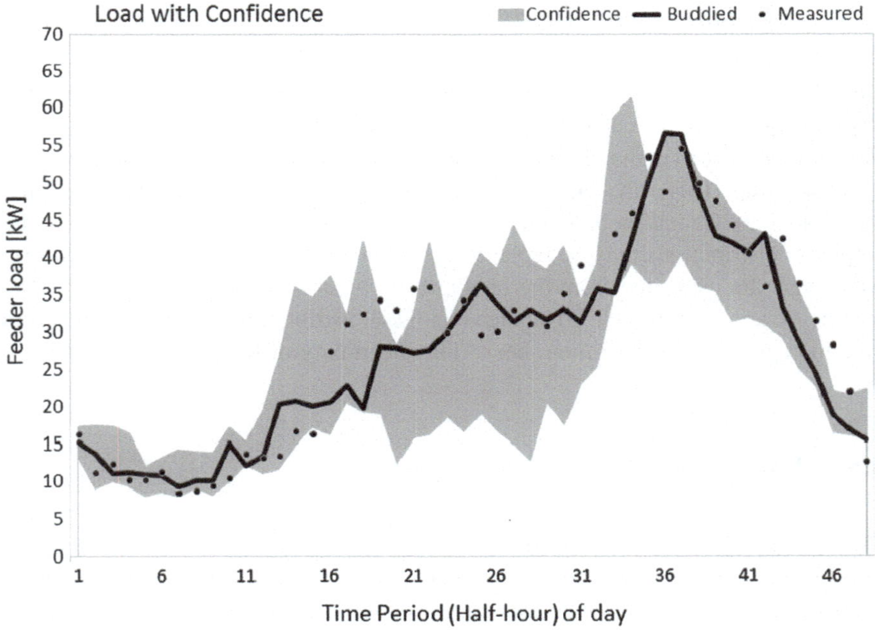


**Fig. 3.10** Confidence bound (grey area) for a daily load profile (dots) for a Binfield feeder. Also included are the buddies (black)

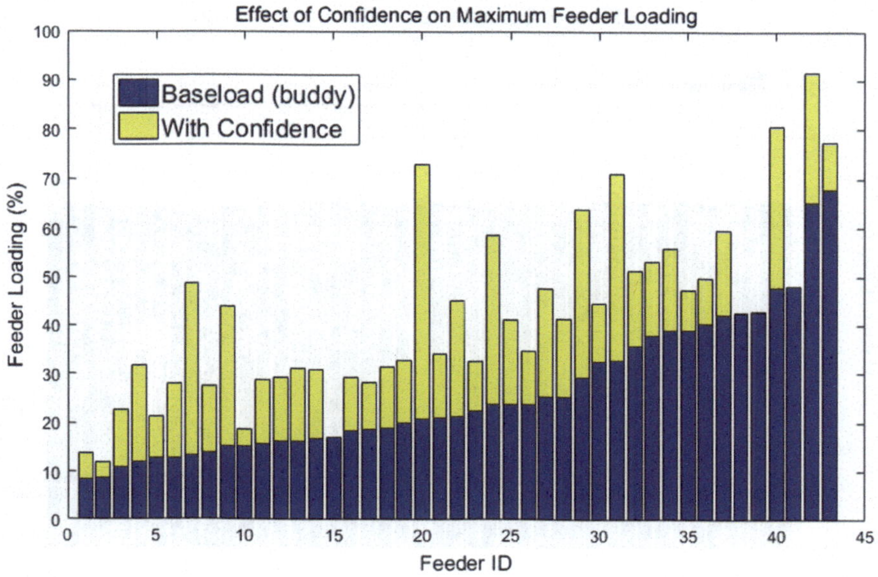


**Fig. 3.11** Feeder load analysis on the buddied load (blue) and the effect on the load of adding the positive confidence (yellow) on a range of feeders. The feeders have been ordered in increasing maximum loading of the buddied loads

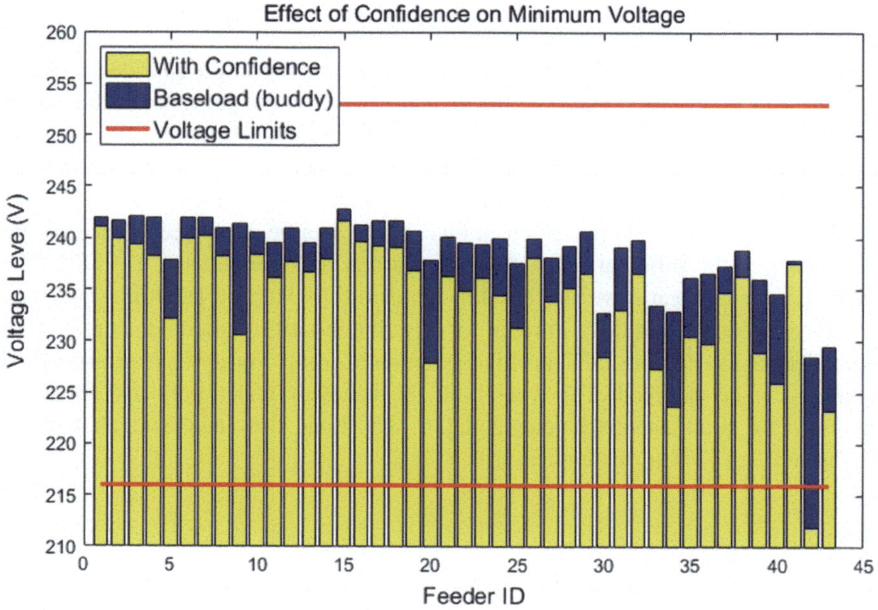


**Fig. 3.12** Minimum voltage analysis on a number of feeders which are buddied (blue), and comparison when the upper 90% loading (yellow) is applied at the end of the feeder. The maximum and minimum voltage limits are shown by the red horizontal lines. Feeders are ordered by maximum loading of buddied loads

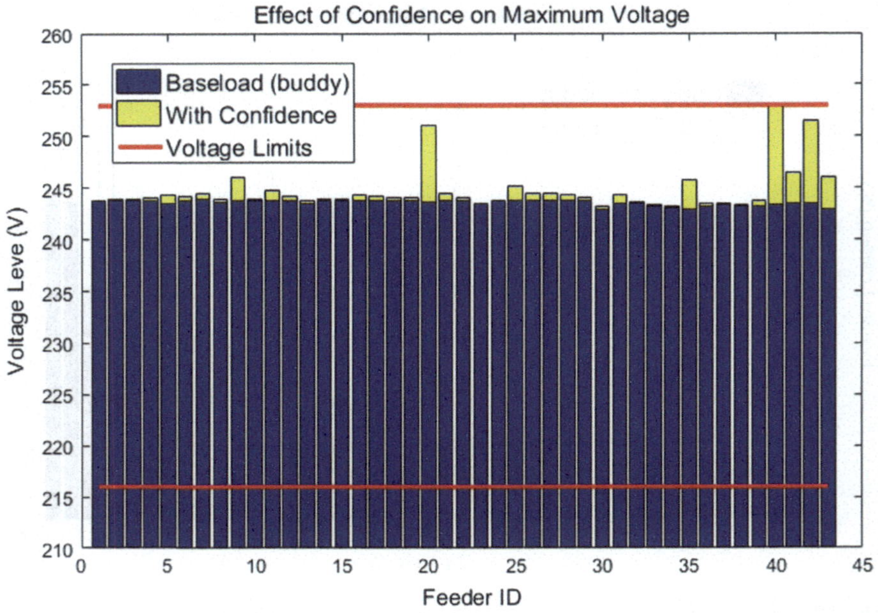


**Fig. 3.13** Maximum voltage analysis on a number of feeders which are buddied (blue), and when the 10% confidence loading (yellow) is applied at the end of the feeder. The maximum and minimum voltage limits are shown by the red horizontal lines. Feeders are ordered by maximum loading of buddied loads

(generation) at the end of the feeder. Again in this case the lower confidence chunk (Sect. 3.4.1) exaggerated how low the demands were on the customer at the end of the feeder which was registered as excessive generation. However, although the assigned profiles were unrealistic they have shown that these feeders could breech their voltage limits if large amounts of photovoltaics are installed.

The buddying method presented in this chapter can be expanded to incorporate the impact of the different uptakes of low carbon technologies (LCTs) [13]. The buddied networks can serve as a baseline demand for which LCTs profiles can then be assigned and modelled. Further to this, the bootstrap method can also be expanded to incorporate LCT profiles as another group from which pre-specified uptakes can be sampled. This way both the behavior uncertainty and the LCT diversity are simultaneously generated [36, 37]. Load flow analysis can then be performed, see [37] for examples.

# Appendix

## Customer Groups

The customers are split into groups according to their profile class, council tax band and the presence of photovoltaic equipment. There are eight generic Elexon profile classes representative of large populations of similar customers. Two classes correspond to domestic customers and distinguish between two tariffs, "Standard" and "Economy 7". The latter provides cheaper rates overnight at the expense of increased day-time charges. Council Tax is a local taxation system used in Great Britain essentially based on property value. Each property is assigned one of eight bands (A–H) in ascending property value.

The grouping used for residential customers is shown in Table 3.1. The commercial customers are not listed, but essentially, a separate profile is used for broad classes of customers (e.g., school, hospital, church, etc.). More details on commercial can be found in Sect. 3.2.5.

Note that the main purpose for the grouping is to reduce the computational cost of the buddying algorithms. However, many other characteristics could be used to group the customers, such as MOSAIC socio-demographics classifications, if profile class and/or council tax band are not available. For other countries similar socio-demographic/asset based groupings are also valid.

## Genetic Algorithm

The genetic algorithm mimics the process of natural selection and proceeds by creating updates of several collections of monitored customers according to how well they score according to the fitness function (3.4). The basic steps of the genetic algorithm are outlined below, which are also summarised in Fig. 3.14.

**Table 3.1** Table showing grouping assignment for UK residential customers based on profile class, council tax band and whether the property has known photovoltaics

| Group | Profile class | Council tax band | Photovoltaics? (Y/N) |
|---|---|---|---|
| 0 | 1 | A, B, C | N |
| 1 | 1 | D | N |
| 2 | 1 | E | N |
| 3 | 1 | F, G, H | N |
| 4 | 2 | Any | N |
| 5 | 2 | Any | Y |
| 6 | 1 | Any | Y |
| >7 | Non residential | N/A | N/A |

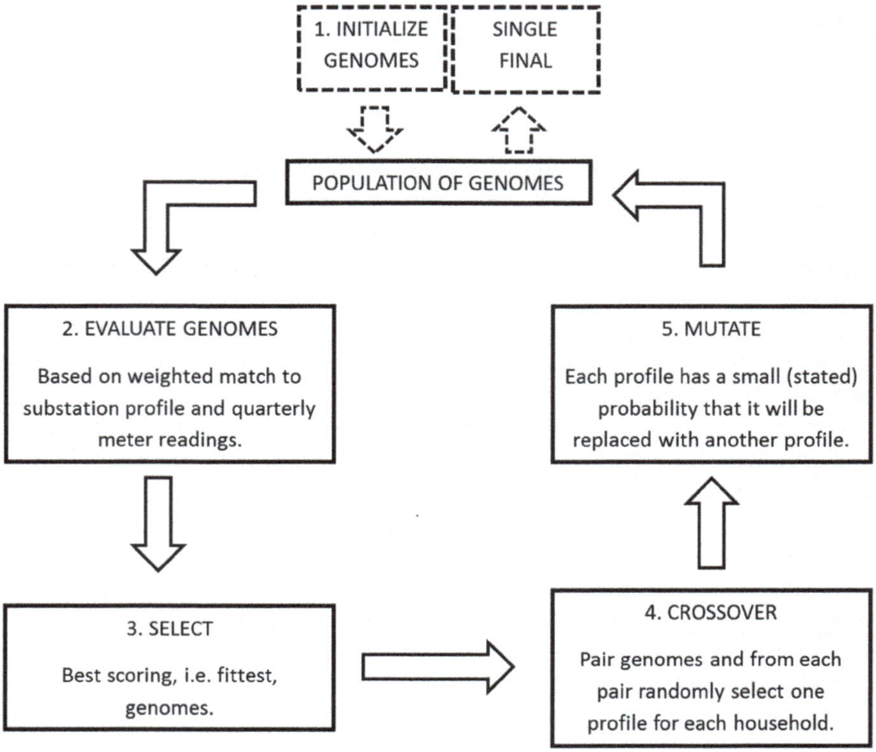


**Fig. 3.14** A graphical presentation of the genetic algorithm

1. Initialise the buddy. Create $G$ genomes each consisting of $M$ randomly selected profiles from $\boldsymbol{P}$ (see Sect. 3.2) for a training period of $H$ half-hours for each customer $c_j, j = 1, \ldots, M$. The selection of the buddies is only restricted so that the buddies belong to the same group as customer $c_j$ (see Sect. "Customer Groups").
2. The fitness of each genome is evaluated using the fitness function (3.4).
3. Select the best-scoring (fittest) genomes.
4. To create each of the $G$ next generation genomes, two of the current best $G' < G$ genomes are randomly selected for crossver. Common profiles are retained while the remaining profiles are selected randomly from one or the other genome.
5. The new genomes are mutated by replacing each profile with a probability $p$ with a new profile (from the same group).
6. Repeat steps 1–3 for 100 generations.

The probability of mutation is free to chose, but is initially set to $p = 0.1$ and slowly decreases as the algorithm progresses. A mutation rate too low and the genomes may lose variation, too high and good solutions may be removed from the population. For step 3, after 40 iterations the genomes are reset, whilst retaining the best genome, to reduce the chances of finding a local minimum.

## References


1. EA Technology, Assessing the impact of low carbon technologies on great britain's power distribution networks (2014) [Online], Available: https://www.ofgem.gov.uk/ofgem-publications/56824/ws3-ph2-report.pdf
2. G. Harrison, A.R. Wallace, Optimal power flow evaluation of distribution network capacity for the connection of distributed generation, in *Proceedings Institute of Electrical Engineering Generation, Transmission and Distribution*, vol. 152 (2005), pp. 115–122
3. P. Lyons, N. Wade, T. Jiang, P. Taylor, F. Hashiesh, M. Michel, D. Miller, Design and analysis of electrical energy storage demonstration projects on UK distribution networks. Appl. Energy **137**, 677–691 (2015)
4. M. Rowe, T. Yunusov, S. Haben, W. Holderbaum, B. Potter, The real-time optimisation of DNO owned storage devices on the LV network for peak reduction. Energies **7**, 3537–3560 (2014)
5. Department of Energy and Climate Change (DECC), Delivering UK Energy Investment: Networks, Jan 2015 [Online], Available: https://www.gov.uk/government/publications/delivering-uk-energy-investment-networks-2014
6. G. Hoogsteen, A. Molderink, J.L. Hurink, G. Smit, F. Schuring, B. Kootstra, Impact of peak electricity demand in distribution grids: A stress test, in *IEEE PowerTech*, (Eindhoven, Netherlands, 2015)
7. G. Hoogsteen, A. Molderink, V. Bakker, G. Smit, Integrating LV network models and load-flow calculations into smart grid planning, in *4th IEEE/PES Innovative Smart Grid Technologies Europe (ISGT EUROPE)* (Lyngby, Denmark, 2013)
8. Ofgem, *Overall criteria for the assessment of Distribution Network Operators' data* (2016)
9. Energy Networks Association, Report on statistical method for calculating demands and voltage regulations on LV radial distributions systems, (Electricity Council, 1989)
10. Elexon, Load profiles and their use in electricity settlement (2013) [Online], Available: https://www.elexon.co.uk/wp-content/uploads/2013/11/load_profiles_v2.0_cgi.pdf
11. C. Muscas, M. Pau, P.A. Pegoraro, S. Sulis, Effects of Measurements and Pseudomeasurements Correlation in Distribution System State Estimation. IEEE Trans. Instrum. Meas. **63**, 2813–2823 (2014)
12. V. Krsman, B. Tesanovic, J. Dojic, Pre-processing of pseudo measurements based on AMI data for distribution system state estimation, in *Mediterranean Conference on Power Generation, Transmission, Distribution and Energy Conversion (MedPower 2016)*
13. A. Poghosyan, D. Greetham, S. Haben, T. Lee, Long term individual load forecast under different electrical vehicles uptake scenarios. Appl. Energy **157**, 699–709 (2015)
14. Ofgem, Low Carbon Network Fund Project Direction: New Thames Valley Vision (2011)
15. G. Giasemidis, S. Haben, T. Lee, C. Singleton, P. Grindrod, A Genetic Algorithm Approach for Modelling Low Voltage Network Demands, Appl. Energy **203**, 463–473 (2017)
16. S. Haben, J. Ward, D. Greetham, C. Singleton, P. Grindrod, A new error measure for forecasts of household-level, high resolution electrical energy consumption. Int. J. Forecast. **30**, 246–256 (2014)
17. M. Neaimeh, R. Wardle, A. Jenkins, J. Yi, G. Hill, P. Lyons, Y. Hubner, P. Blythe, P. Taylor, A probabilistic approach to combining smart meter and electric vehicle charging data to investigate distribution network impacts. Appl. Energy **157**, 688–698 (2015)

18. A. Navarro-Espinosa, L. Ochoa, D. Randles, “Assessing the benefits of meshed operation of LV feeders with low carbon technologies,” in *Proceedings of Innovative Smart Grid Technologies Conference*, (2014)
19. S. Haben, C. Singleton, P. Grindrod, Analysis and clustering of residential customers energy behavioral demand using smart meter data. IEEE Trans. Smart Grid **7**, 136–144 (2016)
20. L. Swan, V. Ugursal, Modeling of end-use energy consumption in the residential sector: a review of modeling techniques. Renew. Sustain. Energy Rev. **13**, 1819–1835 (2009)
21. C. Flath, D. Nicolay, T. Conte, C. van Dinther, L. Filipova-Neumann, Cluster analysis of smart metering data—an implementation in practice. Bus. Syst. Eng. **4**, 31–39 (2012)
22. D. Frame, K. K. Bell, S. McArthur, in A review and synthesis of the outcomes from low carbon networks fund projects, (BUK Energy Research Centre, 2016)
23. S. Haben, M. Rowe, D. Greetham, P. Grindrod, W. Holderbaum, B. Potter, C. Singleton, Mathematical solutions for electricity networks in a low carbon future, in *Proceedings of 22nd International Conference and Exhibition on Electricity Distirbution*, (Stockholm, 2013)
24. F. McLoughlin, A. Duffy, M. Conlon, Charactering domestic electricity consumption patterns by dwelling and occupant socio-economic variables: an Irish case study. Energy Build. **48**, 240–248 (2012)
25. J. Morley, M. Hazas, The significance of difference: understanding variation in household energy consumption, in *Proceedings ECEEE 2011 Summer School, Energy Efficiency First: The Foundation of a Low-Carbon Society* (2011), pp. 2037–2046
26. I. Richardson, M. Thomson, D. Ineld, C. Cliord, Domestic electricity use: a high-resolution energy demand model. Energy Build. **42**, 1878–1887 (2010)
27. J. Widen, E. Wackelgard, A high-resolution stochastic model of domestic activity patterns and electricity demand. Appl. Energy **87**, 1880–1892 (2010)
28. M. Muratori, M.C. Roberts, R. Sioshansi, V. Marano, G. Rizzoni, A highly resolved modeling technique to simulate residential power demand. Appl. Energy **107**, 465–473 (2013)
29. M. Castro, D. Yellen, D. Hollingworth, R. Mukherjee, C. Barteczko-Hibbert, R. Wardle, C. Dent, R. Way, Review of the Distribution Network Planning and Design Standards for the Future Low Carbon Electricity System, (Northern Powergrid, 2014)
30. T. Hong, S. Fan, Probabilistic electric load forecasting: a tutorial review. Int. J. Forecasting **32**, 914–938 (2016)
31. S. Haben, G. Giasemidis, A hybrid model of kernel density estimation and quantile regression for GEFCom2014 probabilistic load forecasting. Int. J. Forecast. **32**, 1017–1022 (2016)
32. R. Koenker, K. Hallock, Quantile regression. J. Econ. Perspect. **15**, 143–156 (2011)
33. B. Efron, R. Tibshirani, *An Introduction to the Bootstrap* (Chapman and Hall, 1993)
34. T. Tilmann Gneiting, A. Raftery, Strictly proper scoring rules, prediction, and estimation. J. Am. Stat. Assoc. **102**, 359–378 (2007)
35. CYME, Secondary Grid Network Analysis [Online]. Available: http://www.cyme.com/software/cymdistsna/. Accessed June 2017
36. L. Hattam, D.V. Greetham, S. Haben, D. Roberts, Electric vehicles and low voltage grid: impact of uncontrolled demand side response, in *24th International Conference & Exhibition on Electricity Distribution (CIRED)*, (Glasgow, 2017)
37. Scottish and Southern Electricity Networks, “SDRC 9.8c Part 1 Smart Analytic and Forecasting Evaluation (2017) [Online], Available: http://www.thamesvalleyvision.co.uk/library/sdrc-9-8c-part-1-smart-analytic-and-forecasting-evaluation/

# Chapter 4
# Coordinated Voltage Control in Active Distribution Networks

**Azah Mohamed and Tengku Juhana Tengku Hashim**

**Abstract** With increasing connection of distributed generations (DGs), the power flow in electric distribution network is no longer unidirectional and the network has become active distribution network. Thus, the connection of DGs in electric distribution networks has created a challenge for distribution network operators due to bidirectional power flow. One of the main technical challenges of an active distribution network is to maintain an acceptable voltage level in the system. This has initiated efforts in using various voltage control strategies to regulate the network voltage profile so that the voltage is maintained at its allowable voltage limits. A number of voltage control methods have been applied to solve voltage control problems associated with the connection of DGs in a distribution system. These voltage control strategies may be classified as decentralized or distributed and centralized or coordinated voltage control. From the literature, a number of coordinated voltage control strategies have been implemented to provide better and faster control to the system. This chapter presents an improved coordinated voltage control method in active electric distribution network by coordinating the three voltage control methods, namely, power factor control, on load tap changer control and generation curtailment control. These voltage control methods are coordinated using fuzzy logic by considering the load bus voltages and DG power as inputs and the voltage control actions as the outputs. The fuzzy logic if-then control rules which are generated to build the fuzzy logic control system are based on the simulations results as well as from the previous works. Results obtained using the fuzzy logic based coordinated voltage control has shown that the voltages are able to be kept within its permissible limits.

A. Mohamed (✉)
Department of Electrical, Electronic and Systems Engineering,
University Kebangsaan Malaysia, Bangi, Selangor, Malaysia
e-mail: azah_mohamed@ukm.edu.my

T. J. Tengku Hashim
Department of Electrical Power Engineering, University Tenaga Nasional,
Kajang, Selangor, Malaysia
e-mail: juhana@uniten.edu.my



A. Arefi et al. (eds.), *Electric Distribution Network Management and Control*,
Power Systems, https://doi.org/10.1007/978-981-10-7001-3_4

**Keywords** Voltage control · Coordinated control · Active distribution network
Distributed generation · Fuzzy logic

## 4.1 Introduction

The increasing rate of distributed generation (DG) penetration in distribution systems has resulted in the transition from the conventional passive power network to the current active power network. The bidirectional power flow in a distribution network due to the presence of DGs has resulted in a more dynamic and active system which is known as active distribution networks. In [1], an active distribution network is defined as a system that adopts integration of control and communication technologies such that the distribution network operators can manage and accommodate the new distribution network. Active distribution network is also defined as a distribution network with systems in place to control a combination of distributed energy resources comprising of generators and storage [2]. There are three main reasons identified as the driving force for the increase in attention to the implementation of active distribution networks [3]. The first reason is due to the fact that customers are looking for a more reliable power delivery and high quality of power supply. Secondly, it is due to the need in exploiting local renewable energy by facilitating the connection of small DG units into the medium and low voltage systems. The third reason is the strong desire in having a better management of assets from the view point of asset utilization, deferral of reinforcement and strategic replacement of aging assets by the distribution network operators.

The fast growing advancement in new technologies as well as deregulation in the electric power system sector has resulted in the application of DGs which are small generation units installed close to load centers with its size ranging from 1 to 10 MW [4]. Due to the connection of DGs in a distribution system, voltage rise issue has become important due to the fact that excessive voltage in the system will affect the system's equipment; increase the system's losses and poor quality of voltage delivered [5]. In [6], voltage rise is the main technical issue which needs to be dealt with when connecting DGs to the distribution network. Hence, for DGs to be connected to the network, it has become the distribution network operators' obligation to ensure that the steady state voltage are kept within permissible limits with limited voltage fluctuations as well as maintaining voltage quality in the system.

Voltage control in active distribution networks is thus considered important and needs to be addressed by both customers and electric providers. Generally, when a DG is connected to a distribution system, the worst case operating scenario is normally considered so as to ensure that the system and customers are not severely affected. Currently, there are a few voltage control methods applied in an active distribution network and these methods may be categorised as coordinated with centralized control and semi-coordinated with decentralized control [7]. Coordinated control determines control actions based on the information from the entire network, which requires data transfer and communication between the network components and devices in the system. Meanwhile, decentralized control performs its control actions with limited communication with other network devices, thereby limiting costs. A number of decentralized voltage control methods are considered, such as power factor control (PFC), on-load tap changer (OLTC) control, generation curtailment and reactive power compensation.

In line with the development in technology in power systems, more focus is given to coordinated voltage control. A number of studies related to coordinated voltage control include the development of centralized distribution management system controller for voltage control in a distribution network with DGs [8]. Centralized coordinated voltage control usually considers coordination of distribution system devices such as OLTCs, voltage regulators, and switched capacitors. For example, a coordinated voltage control method was proposed by coordinating different devices such as the load ratio control transformer, step voltage regulator, shunt capacitor, shunt reactor and static Var compensator [9]. This book chapter presents the technique of coordinating the three different decentralized voltage control methods, namely, PFC, OLTC and generation curtailment using fuzzy logic to provide voltage control to the distribution system with DGs. The rule based fuzzy logic coordinated voltage control is useful in coordinating the control methods because it is able to deal with different ranges of inputs and at the same time giving the desired output control.

## 4.2 Voltage Control Methods in a System with Distributed Generation

Due to the arising voltage rise issue associated with the integration of DGs in a distribution system, there is a need to opt for a systematic control management scheme to help manage and provide voltage coordination in the system. This management scheme which is of interest is known as active network management which is defined as the use of real-time control and communication systems to provide a means to better integrate renewable energy based distributed generators [10]. In particular, the voltage control management scheme may be categorized as centralized or also known as coordinated control, semi-coordinated and decentralized control strategies. Centralized or coordinated control strategy provides voltage control from the substation to the rest of the system, which requires using communication system to coordinate different devices in the system such as the tap changer, voltage regulator and other voltage compensation devices. From the review, the centralized voltage control may be also referred to as the distribution management system which coordinates the distribution system's components and using intelligent technique in coordinating the control. On the other hand, the semi-coordinated and decentralized or distributed control strategies must be able to control the DG unit locally in an active manner while coordinating it with a limited

**Table 4.1** Comparison between centralized and decentralized voltage control methods

| Centralized/Coordinated method | Decentralized/Distributed method |
|---|---|
| Wide coordination, requires communication | No coordination, limited communication |
| High cost | Cost saving |
| Extensive control | Local control |

number of network devices. Table 4.1 illustrates the comparison between centralized and decentralized voltage control methods where the advantages and disadvantages of both methods are presented. From the table, it is clearly shown and substantiated that centralized voltage control method would require extensive and wide coordination as well as communication, hence implying higher costs in implementing such method. On the other hand, the decentralized voltage control manages the voltage rise problem in the system locally with limited communication. It also gives more cost saving as less equipment are involved. However, this local control is less efficient in managing the voltage rise in a distribution system as it does not coordinate or considers the condition of the entire system. Nonetheless, both centralized and decentralized voltage control methods are applied based on the requirement of the system and both methods are able to provide voltage support according to the situation and demand of the system. To date, a number of voltage control methods have been suggested in the literature to manage voltage fluctuation in a distribution system with the presence of DGs [11, 12].

### *4.2.1 Decentralized Voltage Control Methods*

Different decentralized or distributed voltage control schemes have been studied to allow more DG capacity to be connected. The commonly used decentralized voltage control methods are the PFC, OLTC and generation curtailment control described as follows:

A. **Power Factor Control**

Traditionally, it is an obligation for distribution network operators to ensure that all DGs which are connected to the distribution system must operate using the PFC method [13]. PFC method is carried out by keeping the ratio of the real power against the reactive power constant, which is by ensuring that the reactive power follows the variation of the real power. The main advantage of PFC is that it is less disruptive compared to using other network devices, such as the OLTC transformer. However, a disadvantage of the PFC method is that there is only a certain limit of generation that can be connected to the system, whereby, a further increase in generation will result in voltage rise. The method of PFC is sometimes combined with voltage control, hence better known as the Power Factor Control-Voltage Control (PFC-VC) method [14]. This method incorporates the behavior of the generator's operation in two modes, namely, constant power factor and voltage control. At normal condition, where the measured voltage is within the statutory upper and lower limits, the generator will operate in constant PFC mode. However, at times when the voltage deviates above or below its statutory limits, the generator will adopt the VC mode, that is, by varying the excitation of the automatic voltage regulator. In the VC mode, the automatic voltage controller is activated to vary excitation and move the operating point within the bus voltage limit. This method which combines the advantages of automatic voltage regulator and PFC is also termed as automatic voltage/power factor control method. In line with the regulatory rules, there are three different modes of power factor operation, namely, unitary, capacitive or inductive power factor adopted by generators to tackle the voltage fluctuation problem [15]. The PFC control method is implemented by increasing the input generation to the distribution system while maintaining a fixed unity power factor operation to control voltage in the system.

B. **On Load Tap Changer Control**

The OLTC transformers are used to regulate and maintain the voltage which is supplied to consumers within its statutory limits. The OLTC mechanism is a transformer component controlled automatically by a relay to increase or decrease voltage by altering the tap position of a transformer [16]. When the secondary voltage detected is no longer within its permitted dead-band, the relay issues a command to the tap changer mechanism to alter its tap position in order to restore the required voltage level. The OLTC transformer, coupled with its automatic voltage control relay, regulates the transformer output voltage to keep the voltage magnitude within limits. One major disadvantage of this scheme is that the operation of the tap changer is limited to its tapping limits and capability.

With the presence of DG in a distribution network, the automatic voltage control relay performance may be affected, thus resulting in voltage regulation problems due to the interference. The DG integration changes the power flow and sometimes results in reverse power flow as well as voltage increase occurring at the point of connection. The measured voltage is shifted upwards or downwards depending on the power factor of transformer current and direction of power flow to the DG and load [17]. A new voltage control method which controls the voltage relays in OLTC transformer has also been presented [18]. This method deals with the problems associated with the connection of DGs such as inaccurate line drop compensation, increase in voltage level at the point of generator's connection and impaired voltage control for paralleled transformers. An advanced automatic voltage control relay called as the transformer automatic paralleling package schemes is able to control the tap changer locally [19]. This method was proven to be effective under varying power factor and load current, hence maintaining the transformers on a suitable tap position. An automatic voltage reference setting (AVRS) technique which changes the voltage reference for the existing automatic voltage control relays was presented in [20]. The AVRS works by measuring two or more essential voltages along the multiple feeders. From the obtained minimum and maximum voltages, the new voltage reference for automatic voltage control relay is then determined.

C. **Generation Curtailment Control**

Voltage rise can also be mitigated by reducing the active power output of DG. The main drawback of this method is that when a voltage limit is exceeded, only rarely a DG owner might find it beneficial to curtail some of its generation. This is due to the fact that generation curtailment will lead to losses in revenue [8]. The simplest method to implement generation curtailment is to disconnect the required number of generating units when the voltage exceeds its limits. It was suggested that when all the usual means of voltage control have been exhausted, generation curtailment can be used [21]. In [10], generation curtailment method is also only implemented to tackle the voltage rise problem as a last resort if the PFC-VC control mode is not successful. This scheme will reduce a given percentage of the power output when the voltage at the connection bus exceeds its statutory limits. Droop based active power curtailment scheme for managing overvoltage issues has also been presented [22, 23]. Here, the droop control technique is utilized to manage the operation and power sharing among generators, so as to achieve equal sharing of output power losses among inverters.

### *4.2.2 Centralized/Coordinated Voltage Control Methods*

Centralized or coordinated voltage control methods determine their control actions based on information about the whole distribution network and therefore data transfer between network nodes is required. A number of coordinated voltage control methods in distribution systems have been developed with different levels of complexity, effectiveness, communication requirement and investment cost. Examples of coordinated voltage management for distribution systems that have been developed include centralized Distribution Management System (DMS) control and coordinated control of distribution network components such as OLTC and switched capacitors [24]. The use of artificial intelligent techniques for coordinated voltage control has been fast growing as these techniques are more precise and provide automation hence ensuring better voltage control management as a whole [25].

A. **Centralized Distribution Management System Control**

Distribution management system is an active management system which can make control decisions and it can be classified as basic and advanced DMSs. In basic DMS, simple decisions are made by disconnecting distributed energy resources in case of severe network conditions. A basic DMS which considers two new centralized control functions, the volt/Var control and the optimal feeder reconfiguration was presented in [26]. In [8], advanced DMS is considered by controlling all components capable of voltage control through data transfer between network nodes. The substation voltage and reactive power of DG and also other components capable of voltage control are regulated in a coordinated voltage control system. Another method of DMS based voltage control is by using coordinated automatic voltage regulator and OLTC of inter-bus transformers [27]. A DMS coordinated controller that coordinates the OLTC action with the regulation of reactive exchanges between DG plants and distribution feeders was presented in [28]. A power DMS which uses state estimation algorithm coordinated with suitable voltage control equipment was presented in [29]. An advanced DMS incorporating the use of advanced control system which requires inputs such as status of a network, technical constraints and also market information on energy trades has been developed in [30]. The system provides desirable outputs such as amount of generation curtailment and load shedding, ancillary services from DGs, and network configuration. DMSs have also been developed by considering optimization and applied for real time applications. The classical optimal power flow is used for finding the optimal combination of operation options with the aim of minimizing the operation costs due to energy loss, generation curtailment, reactive power and ancillary services, load shedding and energy storage while complying with the technical constraints. In [31], a meta-heuristic optimization technique was developed for optimal coordinated voltage control using voltage and active/reactive power information obtained from the distribution state estimation.

### B. Coordination of Distribution Systems Components

The simplest method of coordinated voltage management is by controlling the substation voltage based on the measured or estimated maximum and minimum voltages in a distribution network. The substation voltage is controlled by changing the set point of the automatic voltage regulator relay which controls the tap changer of the main transformer. In [32], a network voltage controller based on statistical state estimation algorithm is used to control the target voltage of the automatic voltage control relays at the primary substations. A coordinated voltage regulation scheme was presented by combining the contribution of generator and the OLTC in providing voltage control to a distribution system [33]. Coordination between step voltage regulator and DG operations was also developed for voltage profile improvement [34]. Control action coordination between OLTC and DG can be performed by utilizing the priority level of each regulating device through communication [35]. The use of control devices such as step voltage regulator and static Var compensator (SVC), to keep the voltage in a distribution system at its permissible level was presented in [36]. In a related work, the use of voltage regulation method in the presence of DGs was suggested by implementing proper coordination among the OLTC, substation switched capacitors and feeder switched capacitors [37]. In [38], a coordinated control method can be achieved by coordinating different devices such as the load ratio control transformer, step voltage regulator; shunt capacitor, shunt reactor and SVC. The use of a segment controller was suggested for generator automatic voltage control relay utilizing OLTC and collecting local measurements of feeder loads, and remote measurements of voltage and load, to form inputs to the state estimator [39]. A state estimation technique is used to determine the network voltage profile and to adjust the voltage target for the automatic voltage control relay. The generator automatic voltage control relay is considered as an innovative technique for improving voltage control and increase penetration of DGs. An approach for coordinating voltage control of STATCOM and the under-load tap changer (ULTC) transformer was presented in [40]. The ULTC transformer steps are controlled so as to maximize the STATCOM capacity margin, hence increasing the dynamic margin during contingency situations as well as minimizing the number of tap changes. In [41], a coordinated control using OLTC and distribution STATCOM was presented for voltage control in a long radial distribution system with a DG located at the end of the line.

### C. **Coordinated Voltage Control Using Intelligent Techniques**

Coordinated voltage control using intelligent techniques is considered as the most current method in achieving optimal and accurate voltage control in distribution systems with DGs. A number of intelligent techniques have been widely used and tested for decentralized voltage control. Genetic algorithm (GA) was applied to determine the optimal dispatch schedules for OLTC settings at substations and all shunt capacitor switching based on the day-ahead load forecast [42]. The reactive Tabu search was applied in determining the coordinated allocation and control of step voltage regulators and SVCs [43]. An artificial neural network (ANN) based control scheme for the management of ULTC transformer and STATCOM was presented in [44]. In another work, using the combined ANN and fuzzy logic techniques, a coordinated control for managing the main ULTC transformer and the SVC reactive power outputs was developed in [45]. In [46], particle swarm optimization (PSO) technique was applied for solving optimal coordination of ULTC transformer and capacitors on a modified 29-bus distribution system of the Thailand's Provincial Electricity Authority. The findings of this case study showed that optimal coordination between the ULTC transformer and the substation and feeder capacitors for volt/Var control to minimize the DG owner's daily energy payment and energy loss cost can be achieved. A fuzzy logic based voltage controller in coordinated voltage control scheme was implemented by considering the average customer's voltage as input and the preferred tap changer setting as output [47]. GA was applied for coordinated voltage control in a three phase unbalanced distribution system with multiple regulators. Under different load conditions, the optimal settings of the voltage regulators such as the load ratio transformer, SVC and DG reactive powers are determined while maintaining the voltage profile within its specified limits [48]. A non-dominated sorting GA was used for coordinated control of on-load tap changers and reactive power compensator solving a multi-objective optimization problem to achieve optimal voltage control [49].

In this chapter, for intelligent control of voltage in an active distribution system, coordinated control of power factor, generation curtailment of DGs and tap changer setting is proposed using fuzzy logic. The power factor of the DGs at different ranges, tap changer of different settings and amount of DG curtailment are combined and coordinated in the fuzzy logic system. The DG input powers and the voltages at the load buses are considered as input to the fuzzy logic system and the controlled variables are the outputs of the fuzzy system. The proposed fuzzy logic based voltage control method is different from the above-mentioned methods based on the fact that three different voltage control parameters are coordinated compared to the previous methods which coordinate only two control parameters.

## 4.3 Fuzzy Logic for Coordinated Voltage Control in Active Distribution Systems

In this chapter, fuzzy logic technique is applied for coordinating the three voltage control methods which have been compared earlier. Fuzzy logic is one of the most popular technique applied for voltage control in distribution networks with DGs. In the previous works on voltage control, only a single or two methods of voltage control using fuzzy logic was implemented. However, in this chapter, the fuzzy logic is applied to coordinate three different voltage control methods of power factor control, OLTC control and generation curtailment control in an active distribution network. Rule-based system based on experience and off-line studies on the system is the key to implementing fuzzy logic control.

### *4.3.1 Fuzzy Logic and Fuzzy Inference System*

Fuzzy logic is determined as a set of mathematical principles for knowledge representation based on the degree of membership rather than on crisp membership of classical binary logic [50]. Unlike the two-element Boolean logic, fuzzy logic is multi-valued. Fuzzy logic uses the continuum of logical values between 0 (completely false) and 1 (completely true) and can also accept things than can be partly true and partly false simultaneously. The basic idea behind the fuzzy set theory is that an element belongs to a fuzzy set with a certain degree of membership. The degree is typically taken as a real number in the interval [0, 1]. Basically, the first step in designing a fuzzy system is determining the membership functions. The membership functions are typically created based on the experience and knowledge of experts about the problems being solved. There are several types of membership functions of the fuzzy set which can be defined as triangular function, trapezoidal function, Gaussian function and also sharp peak function.

At the root of fuzzy set theory lies the idea of linguistic variables, which are also known as fuzzy variables. In 1973, Zadeh outlined a new approach to analyse complex systems by capturing human knowledge in fuzzy rules. A fuzzy rule can be defined as a conditional statement in the form of

$$\text{IF x is A, THEN y is B} \tag{4.1}$$

where x and y are linguistic variables, and A and B are linguistic values determined by fuzzy sets.

Meanwhile, fuzzy reasoning includes two distinct parts; evaluating the rule antecedent (the IF part of the rule) and applying the result to the consequent (the THEN part of the rule). In fuzzy systems, where the antecedent is a fuzzy statement, all antecedents are true to a certain degree of membership such that the consequent is also true to the same degree. A fuzzy system generally incorporates not one but several rules that describe expert knowledge. Fuzzy logic control uses the principles of fuzzy logic-based decision making to arrive at the control actions. By selecting a suitable input-output linguistic variable utilizing rule base, a wide range of desirable control outcomes can be attained. Control strategies are defined by system operators based on their experience and offline studies, which are then translated into rules of hierarchical fuzzy inference system. To obtain a single crisp solution for the output variable, a fuzzy system aggregates all output fuzzy sets into a single output fuzzy set and then defuzzifies the resulting fuzzy set into a single number. This process is known as fuzzy inference and is one of the most popular applications of fuzzy logic and fuzzy set theory [50]. Fuzzy inference is defined as the process of mapping from a given input and an output by using the theory of fuzzy sets [51].

There are two fuzzy inference techniques, namely, the Mamdani and Sugeno methods. The Mamdani method is widely accepted in fuzzy expert systems for its ability to capture expert knowledge in fuzzy rules. However, the Mamdani method causes computational burden. The Sugeno method improves the computational efficiency of the fuzzy inference and works well with adaptive and optimization techniques, which makes it a suitable choice for controlling dynamic non-linear systems. Tuning a fuzzy expert system to suit the problem encountered generally requires more time and effort than determining fuzzy sets and constructing fuzzy rules. Thus, one has to gather precise knowledge on the system before deciding on the control rules and membership functions to solve a problem.

### 4.3.2 Implementation of Fuzzy Logic for Coordinated Voltage Control

Figure 4.1 presents the implementation process of the proposed coordinated fuzzy logic based voltage control in a distribution system with distributed generators. Initially, the voltage is checked at each load bus of the network in which if the voltage limits are violated at any bus, fuzzy logic actions are taken according to the control algorithm. Fuzzy logic decides from which of the three voltage control methods; PFC, OLTC, and generation curtailment is to be activated and applied

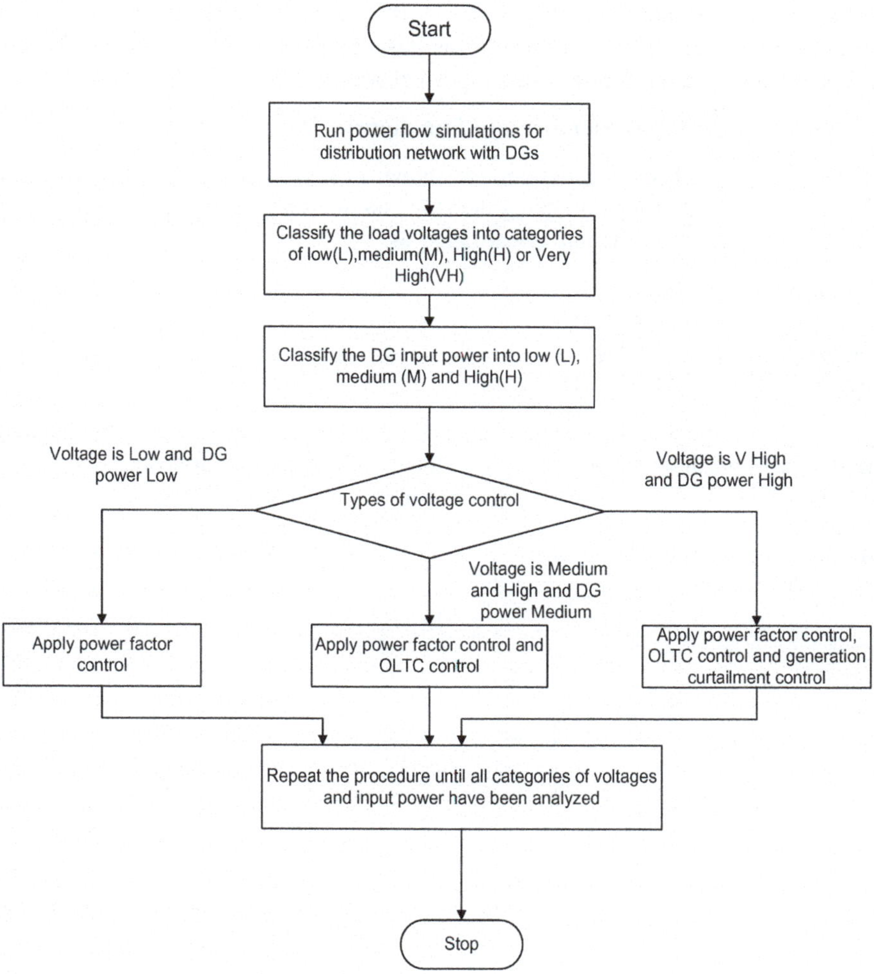


**Fig. 4.1** Flowchart for implementing the fuzzy logic-based coordinated voltage control

when voltage is detected to be beyond its permissible limits. If the chosen control option cannot control the voltage then the fuzzy logic controller decides the next control option to be activated. In this case, the inputs to the system are the load voltages and DG power.

### A. **Selection of Input and Output Parameters**

Two inputs and three outputs are selected as parameters for the implementation of the proposed fuzzy logic inference control system. The two inputs are the amount of active power at each generator and the second input is the voltage recorded at the load buses. On the other hand, the outputs selected are the three control methods of PFC, OTLC and generation curtailment control which must be activated in case the voltage exceed its allowable range of 0.95–1.05 p.u. To be more precise, the control output options are further categorized as described below so as to cater for the different ranges of input power and input voltages.

## B. Fuzzy Logic Input Membership Functions

The input membership functions are shown in Figs. 4.2 and 4.3. The voltages that are considered as input to the system are classified into four different categories and are represented in the numerical range as:

i. Low = [0.9, 0.95]
ii. Medium = [0.925, 1.05]
iii. High = [1.03, 1.072]
iv. Very High = [1.05, 1.1]

The second input to the system is the DG input power, which is also divided and normalised into three categories of low, medium, and high as follows:

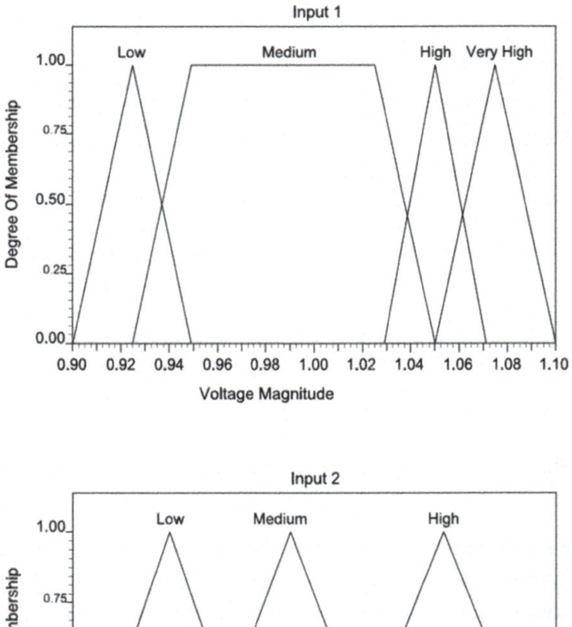


**Fig. 4.2** Input voltage membership functions

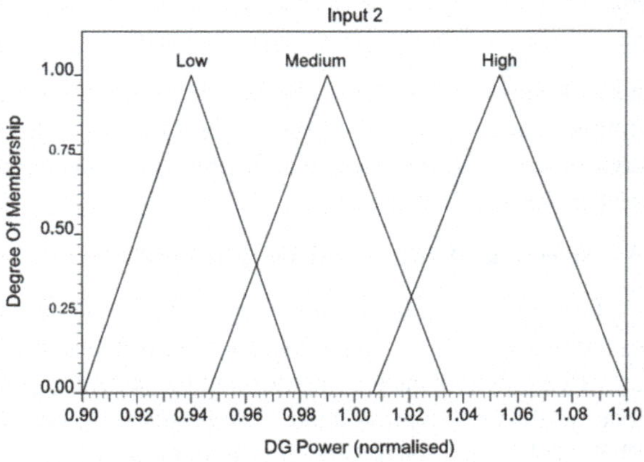


**Fig. 4.3** Input power membership functions

i. Low = [0.9, 0.98]
ii. Medium = [0.946, 1.034]
iii. High = [1.007, 1.1]

C. **Fuzzy Logic Output Membership Functions**

The next step in building a fuzzy logic control system is to build the output membership functions. Here, the output membership functions are the control output actions which are categorized as power factor values leading or lagging, OLTC settings and percentage of curtailed generation. Figures 4.4, 4.5 and 4.6 depict the three fuzzy logic control output membership functions which have been developed. PFC is the first control action that is to be activated in the implementation of the proposed coordinated fuzzy logic control. This is based on the fact that a generator normally operates at a certain power factor and has its own reactive power capability. Power factor indicates the reactive power output of the generating unit maintained in proportion to the real power output such that the power factor

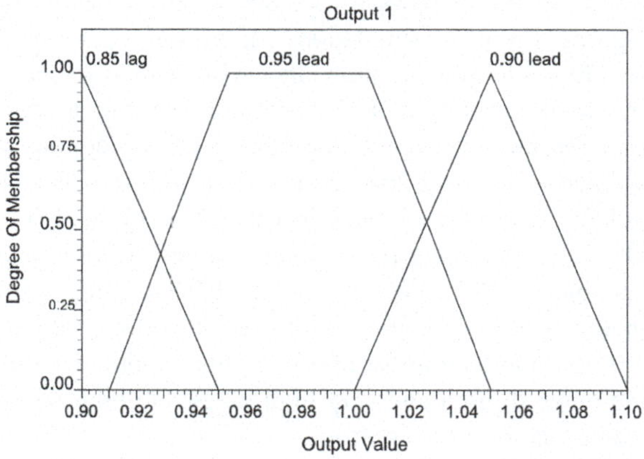


**Fig. 4.4** PFC output membership functions

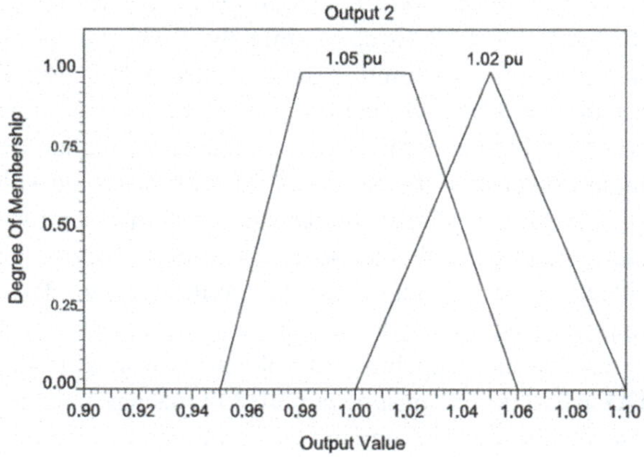


**Fig. 4.5** OLTC output membership functions

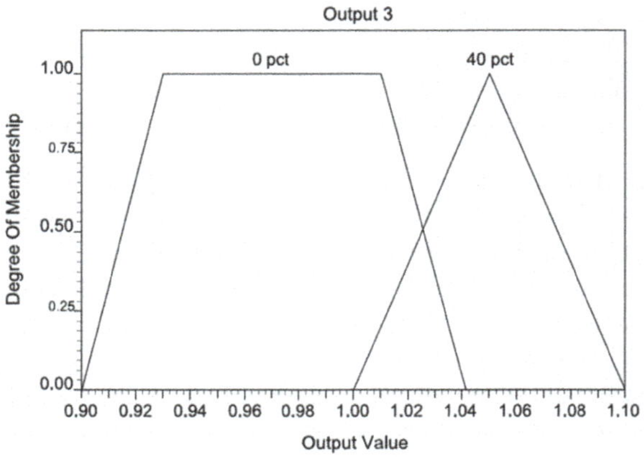


**Fig. 4.6** Generation Curtailment output membership functions

remains constant. For example, Tenaga Nasional Berhad (TNB) which is the Distribution Network Operator (DNO) in Malaysia requires all generators connected to the network to operate at power factors between 0.90 leading and 0.9 lagging. From the simulations carried out, operating a DG at leading power factor was found to mitigate the voltage rise. This is in accordance with the works in [52] in which operating a DG at leading power factor was found to mitigate the voltage rise. On the other hand, operating a DG at lagging power factor will increase the voltage level at the load buses and therefore it is suitable for managing the lower voltage. From the simulations carried out on the IEEE 13-bus test system with two DGs, two operating power factor values of 0.90 and 0.95 leading are found to be suitable for PFC. Power factor of 0.95 leading is suitable for controlling voltage in the medium range of permissible voltage limits of ±5% (between 0.95 and 1.05 p.u.) as well as for managing the middle range of input power. For voltages in the range of high and very high range and high input power, power factor of 0.90 leading is used for control.

The next control option to be activated is the OLTC control. This control action is activated after the limitation of the PFC or reactive power capability has been reached. Several works related to OLTC control have found that OLTC set point in the range of 1.02 to approximately 1.033 p.u. is effective in managing voltage fluctuation and in limiting network losses [53, 54]. Two different OLTC settings; Tapmax = 1.05 and Tapmax = 1.02 are used as shown in Fig. 4.6 where the latter was found to be more effective in managing higher levels of voltage as well as high input power than the setting of 1.05 in managing voltage rise in the system.

The lowest priority or the least preferred option of control is the generation curtailment control since there are various factors that need to be taken into consideration before opting for this control action. This method is most of the time implemented to tackle voltage rise as a last resort when the generators have exhausted their capability of voltage control. In [54], it was suggested that 41% of DG active power must be curtailed to manage voltage rise. In this work, 0 and 40% were chosen as the amount of curtailment to be performed. 0% curtailment was chosen for lower voltage and lower and medium power. On the other hand, the 40% curtailment was chosen for higher voltages and higher input power as shown in Fig. 4.6.

D. **Generation of Fuzzy Logic Control Rules**

Fuzzy logic control rules were developed based on the two input and three output variables. The first input variable (V) is defined in terms of its linguistic variable by using four fuzzy subsets which are symbolized by low (L), medium (M), High (H) and Very High (VH). The membership functions for the first input variables are as shown in Fig. 4.2. The second input variable which is defined as the amount of active power supplied to the DGs is categorized into three fuzzy sets, namely, low (L), medium (M) and high (H) as shown in Fig. 4.3.

The first output variable is defined by its linguistic variable by using three fuzzy sets which are symbolized as 0.85 lag, 0.95 lead, and 0.90 lead which represents the PFC settings as shown in Fig. 4.4. The second output variable is defined by using two fuzzy sets which are symbolized as 1.05 and 1.02 p.u. to represent the OLTC settings as shown in Fig. 4.5. The third output is also classified as two fuzzy sets which is symbolized by 0 and 40 pct to represent the amount of percentage of active power curtailment. The third fuzzy output membership function is depicted in Fig. 4.6.

The fuzzy decision table for the coordinated voltage control using voltages at the load buses and the amount of DG power as inputs and the voltage control methods of PFC, OLTC and amount of generation to be curtailed as outputs is presented in Table 4.2. When the fuzzy decision table is translated into the IF-THEN rules,

**Table 4.2** Fuzzy decision table for coordinated voltage control

| Input 1 | Input 2 | Outputs (voltage control) | | | | | | |
|---|---|---|---|---|---|---|---|---|
| Voltage | DG power | PFC | | | OLTC | | Generation | |
| | | 0.85 lag | 0.95 lead | 0.90 lead | 1.05 p.u. | 1.02 p.u. | 0 pct | 40 pct |
| L | L | ✓ | | | | | | |
| L | M | ✓ | | | | | | |
| L | H | ✓ | | | | | | |
| M | L | | ✓ | | | | | |
| M | M | | ✓ | | | | | |
| M | H | | ✓ | | | | | |
| H | L | | | ✓ | ✓ | | ✓ | |
| H | M | | | ✓ | ✓ | | ✓ | |
| H | H | | | ✓ | ✓ | | ✓ | |
| VH | L | | | ✓ | | ✓ | | ✓ |
| VH | M | | | ✓ | | ✓ | | ✓ |
| VH | H | | | ✓ | | ✓ | | ✓ |

12 rules are generated to construct the fuzzy logic based coordinated voltage control system. The samples of rules generated are as follows:

i. If voltage is low and power is low, then PFC is 0.85 lag.
ii. If the voltage is low and the power is medium, then PFC is 0.85 lag.
iii. If the voltage is low and the power is high, then PFC is 0.85 lag.
iv. If voltage is medium and power is low, then PFC is 0.95 lead.
v. If voltage is medium and power is medium, then PFC is 0.95 lead.
vi. If voltage is medium and power is high, then PFC is 0.95 lead.
vii. If voltage is high and power is low, then PFC is 0.90 lead, OLTC is 1.05 p.u. and generation curtailed is 0%.
viii. If voltage is high and power is medium, then PFC is 0.90 lead, OLTC is 1.05 p.u. and generation curtailed is 0%.
ix. If voltage is high and power is high, then PFC is 0.90 lead, OLTC is 1.05 p.u. and generation curtailed is 0%.
x. If voltage is very high and power is low, then PFC is 0.90 lead, OLTC is 1.02 p.u. and generation curtailed is 40%.
xi. If voltage is very high and power is medium, then PFC is 0.90 lead, OLTC is 1.02 p.u. and generation curtailed is 40%.
xii. If voltage is very high and power is high, then PFC is 0.90 lead, OLTC is 1.02 p.u. and generation curtailed is 40%.

All these rules are generated based on the simulation works carried out on a distribution test system and also based on other works reviewed in the literature to test the settings to be used in the implementation of the fuzzy logic based coordinated voltage control.

## 4.4 Simulation Results of Coordinated Voltage Control Using Fuzzy Logic

Simulations were carried out using the DigSilent Power Factory Version 14 Software on the IEEE 13 bus test system which has been modified to be a balanced system as shown in Fig. 4.7. For the 13 bus test system, the system loads are balanced and considered as spot loads with a total load of 945.3 kW and 873.2 kVar. The data for the 13 bus test system is shown in Appendix 1.A. The system consists of two DGs of synchronous generator type and inverter based DG type placed at bus no 634 and 680, respectively and the OLTC is located between the bus no 633 and 634. The fuzzy logic coordinated voltage control method implemented on the 13 bus test system is to coordinate the three decentralized voltage control methods so that it is able to control the voltage profiles at the load buses within its permissible limits. Figure 4.8 depicts the inputs and outputs captured from the fuzzy logic coordinated voltage control system which utilizes two inputs and three outputs. Figures 4.9, 4.10 and 4.11 show the captured images of

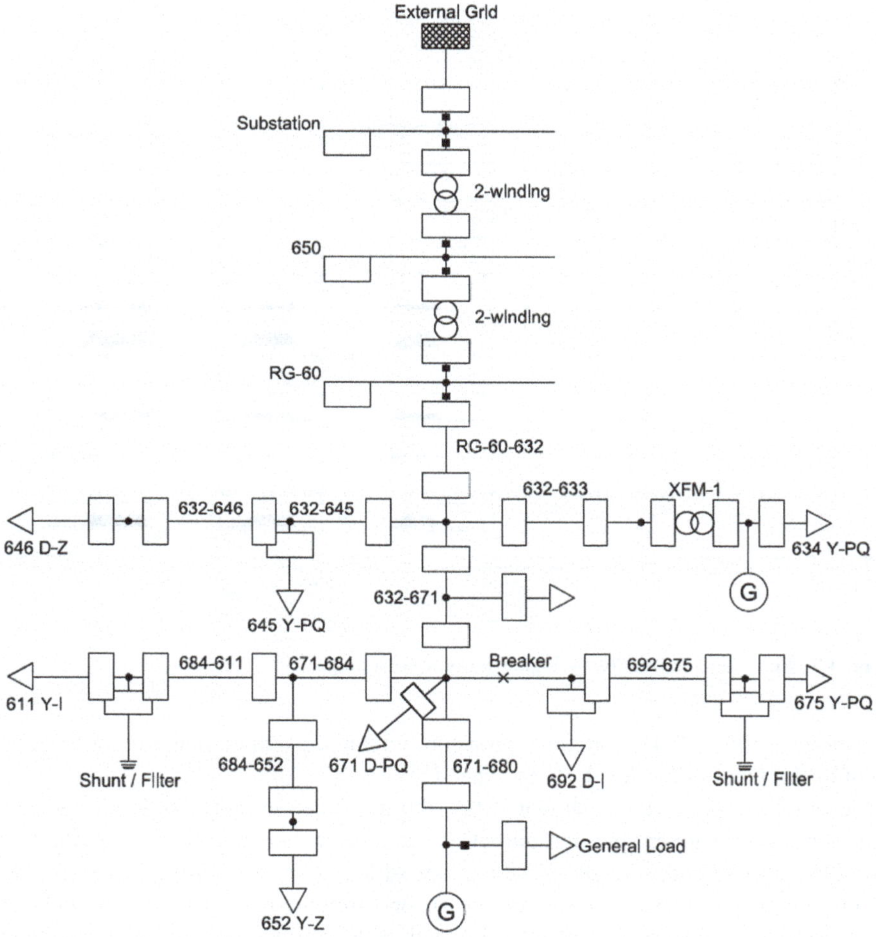


**Fig. 4.7** The IEEE 13 bus test system

the surface viewer of the fuzzy logic control outputs which represents the PFC control output, OLTC control output and the generation curtailment control output. These 3D images represent the two inputs of voltage and power as well as the selected output control method. From Fig. 4.8, the sample of input voltage tested is 1.052 while the input power is 0.9695. Looking at the outputs of the rule viewer, the first output, pfc (power factor setting) is 1.05, second output, oltc (OLTC setting) is 1.01 and the third output, gencurt (generation curtailed) is 0.976. From the implementation and development of the coordinated voltage control using fuzzy logic, the input voltage value of 1.052 falls inside the range of between High and Very High. On the other hand, the input power of 0.9695 falls inside the range of between Medium and High. From the output values recorded for the three control

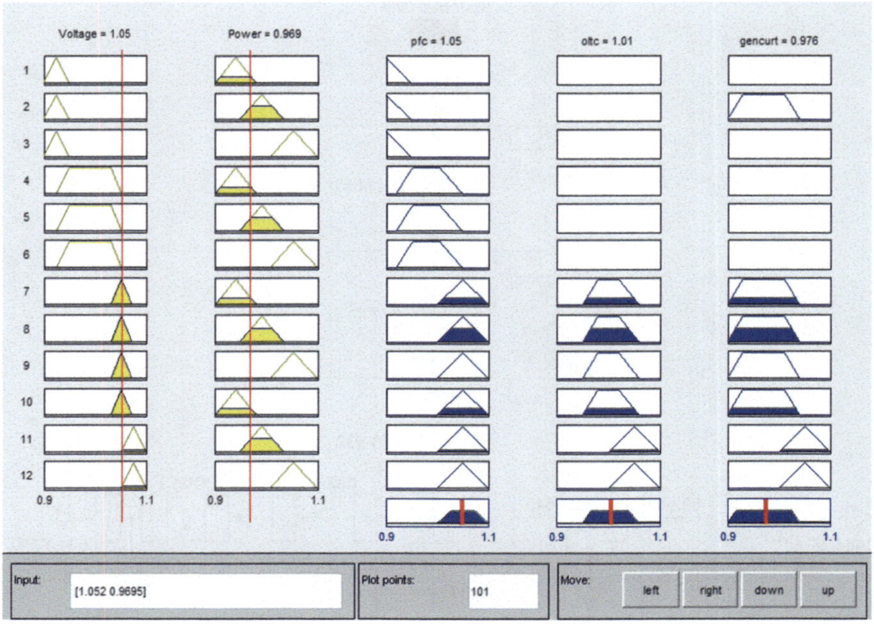


**Fig. 4.8** Rule viewer of the fuzzy logic control system output

methods of PFC, OLTC and generation curtailment, the output values recorded falls within the permissible level of less than 1.05.

From 3D surface viewer shown in Fig. 4.9, for different ranges of input voltages and input power given, the first output of the coordinated control method which is the PFC, still lie within its permissible range of less than 1.05. From Figs. 4.10 and 4.11, the output surface viewer of the second output of OLTC control and the generation curtailment control also lies within the permissible limits of less than 1.05.

To further prove the successful implementation of the coordinated voltage control method based on fuzzy logic, the voltage values at the load buses are compared before and after the implementation of the fuzzy logic technique. First consider a typical daily load data curve shown in Fig. 4.12.

Figures 4.13, 4.14 and 4.15 depict samples of voltage profile results at buses 634, 680 and 675 of the IEEE 13 bus test system before and after implementing the fuzzy logic based voltage coordinated control method. The voltage profile is plotted following the variation in load and the results obtained has proven that the coordinated voltage control using fuzzy logic is able to mitigate the voltage rise and reduce the voltage at the buses within allowable limits. From Fig. 4.13, the voltage profile recorded values at bus 634 ranges from 1.06 p.u. to as high as 1.084 p.u. during the 24 h duration. With the implementation of the fuzzy logic based coordinated voltage control, the voltage values recorded are able to be lowered within the range of 1.03–1.05 p.u., which are the allowable limits of acceptable voltage.

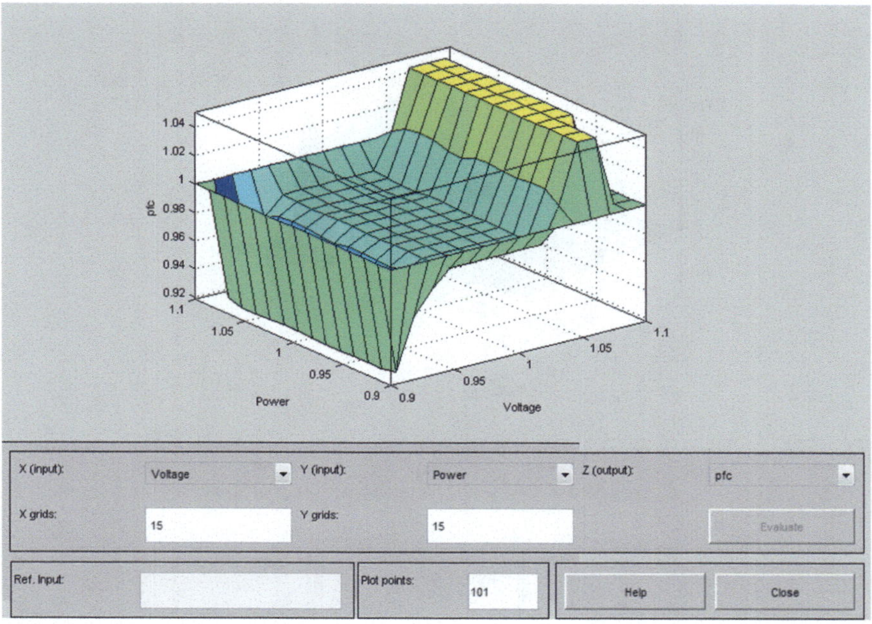


**Fig. 4.9** Surface viewer of the fuzzy logic control output of PFC

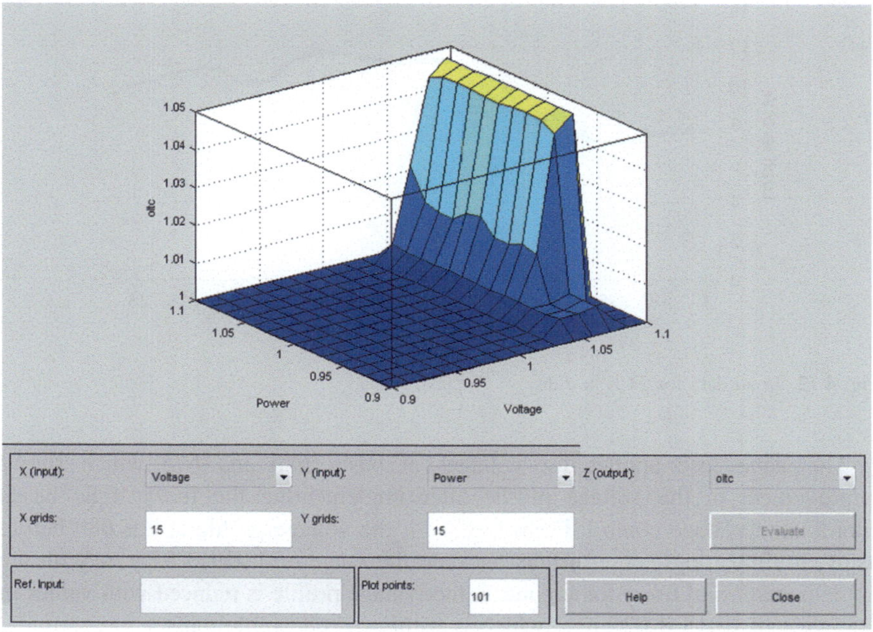


**Fig. 4.10** Surface viewer of the fuzzy logic control output of OLTC control

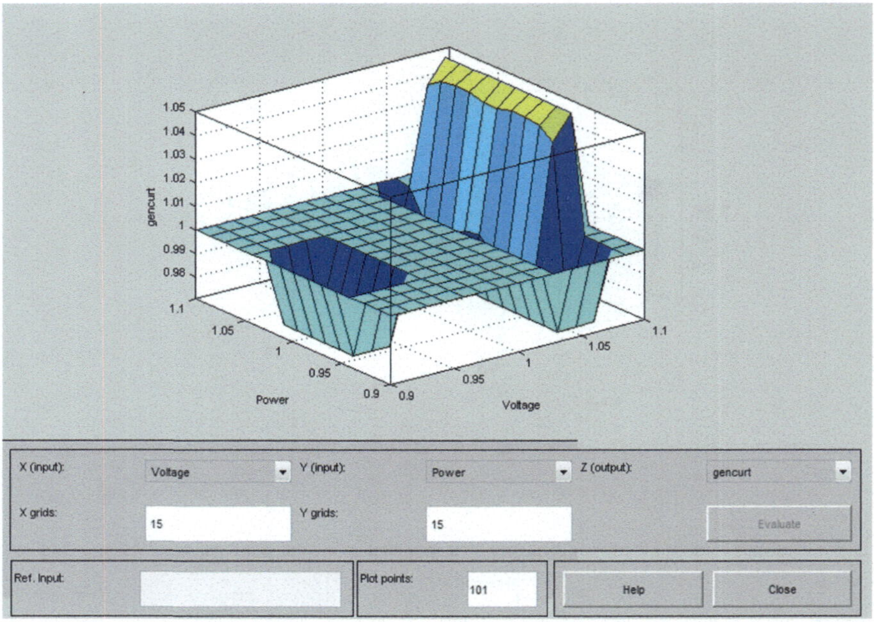


**Fig. 4.11** Surface viewer of the fuzzy logic control output of generation curtailment control

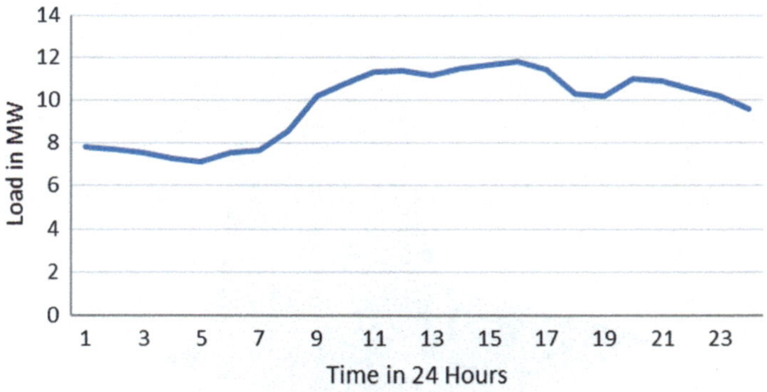


**Fig. 4.12** Load data for 24 h in a day

The same case applies for voltages at buses 680 and 675 which shows improvement in the voltage profile after implementing the fuzzy logic based coordinated voltage control. From Fig. 4.14, the voltage profile at bus 680 before using the fuzzy logic control falls between the range of 1.039–1.07 p.u. With the implementation of fuzzy logic control, the voltage profile is reduced with values in the range of 1.015–1.035 p.u., which is within the allowable range.

From the voltage profile shown in Fig. 4.15, the voltage profile at bus 675 has recorded voltage values which exceeded the permissible voltage value of 1.05 p.u.

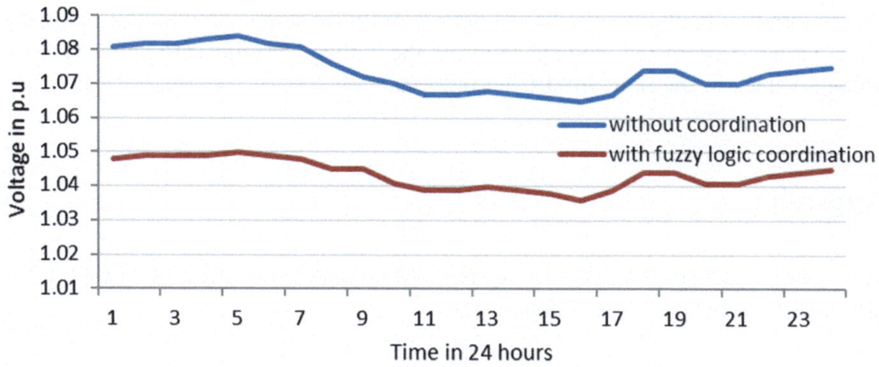


**Fig. 4.13** Voltage profile at bus 634 with and without coordinated voltage control

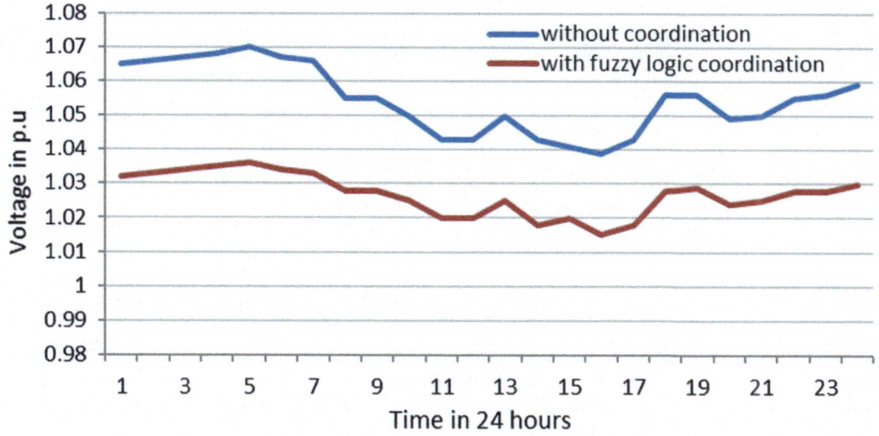


**Fig. 4.14** Voltage profile at bus 680 with and without coordinated voltage control

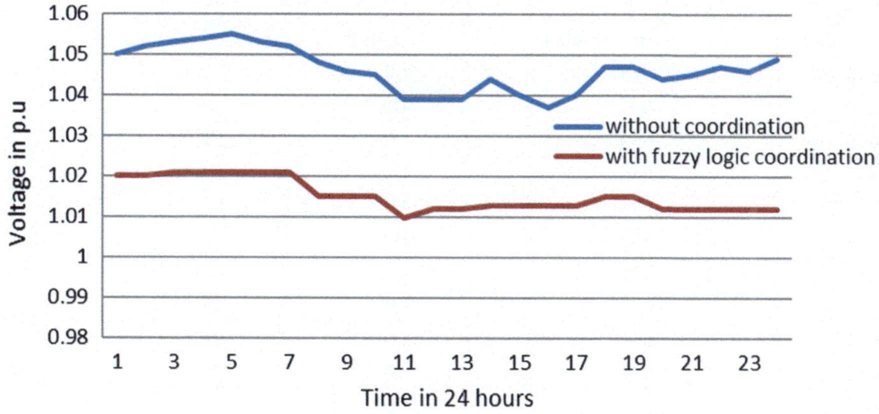


**Fig. 4.15** Voltage profile at bus 675 with and without coordinated voltage control

for the duration between hour 1 until hour 7. With the implementation of the fuzzy logic based coordinated voltage control, the voltage profiles are reduced with voltage values in the range of 1.01–1.022 p.u.

## Appendix

The technical data of the IEEE 13 bus test system under consideration in Fig. 4.2 is provided in Tables 4.3, 4.4, 4.5 and 4.6.

**Table 4.3** Overhead line configuration data of the IEEE 13 bus distribution test system

| Configuration | Phasing | Phase | Neutral | Spacing |
|---|---|---|---|---|
| | | ACSR | ACSR | ID |
| 601 | B A C N | 556,500 26/7 | 4/0 6/1 | 500 |
| 602 | C A B N | 4/0 6/1 | 4/0 6/1 | 500 |
| 603 | C B N | 1/0 | 1/0 | 505 |
| 604 | A C N | 1/0 | 1/0 | 505 |
| 605 | C N | 1/0 | 1/0 | 510 |

**Table 4.4** Underground line configuration data of the IEEE 13 bus distribution test system

| Configuration | Phasing | Cable | Neutral | Spacing ID |
|---|---|---|---|---|
| 606 | A B C N | 250,000 AA, CN | None | 515 |
| 607 | A N | 1/0 AA, TS | 1/0 Cu | 520 |

**Table 4.5** Line segment data of the IEEE 13 bus distribution test system

| Bus number | Bus number | Length(ft.) | Configuration |
|---|---|---|---|
| 632 | 645 | 500 | 603 |
| 632 | 633 | 500 | 602 |
| 633 | 634 | 0 | XFM-1 |
| 645 | 646 | 300 | 603 |
| 650 | 632 | 2000 | 601 |
| 684 | 652 | 800 | 607 |
| 632 | 671 | 2000 | 601 |
| 671 | 684 | 300 | 604 |
| 671 | 680 | 1000 | 601 |
| 671 | 692 | 0 | Switch |
| 684 | 611 | 300 | 605 |
| 692 | 675 | 500 | 606 |

**Table 4.6** Transformer data of the IEEE 13 bus distribution test system

| | kVA | kV-high | kV-low | R (%) | X (%) |
|---|---|---|---|---|---|
| Substation | 5000 | 115-D | 4.16-Gr.Y | 1 | 8 |
| XFM-1 | 500 | 4.16-Gr.W | 0.48-Gr.W | 1.1 | 2 |

# References


1. R. Hidalgo, in *A review of active distribution networks enabling technologies*, IEEE Power and Energy Society Meeting, pp. 1–9 (2010)
2. C. A'damo, S. Jupe, C. Abbey, Global survey on planning and operation of active distribution networks, in *The 20th International Conference and Exhibition on Electricity Distribution—Part 2*, pp. 1–9 (2009)
3. J. McDonald, Adaptive intelligent power systems: active distribution networks. J. Energy Policy **36**(12), 4346–4351 (2008)
4. A.A. Bayod, J. Mur, J.L. Bernal, J.A. Dominguez, Definitions for distributed generation: a revision, in *International Conference on Renewable Energies and Power Qualities*, pp. 1–4 (2005)
5. A. Kulmala, K. Maki, S. Repo, P. Jarventausta, Including active voltage level management in planning of distribution networks with distributed generation. IEEE Bucharest PowerTech, 1–6 (2009)
6. N. Jenkins, R. Allan, P. Crossley, D. Kirschen, G. Strbac, Embedded Generation: The Institution of Electrical Engineers, London (2000)
7. T. Sansawatt, J. O'Donnell, L.F. Ochoa, G.P. Harrison, Decentralised voltage control for active distribution networks, in *Proceedings of the 44th International Power Engineering Conference (UPEC)*, pp. 1–5 (2009)
8. G. Strbac, N. Jenkins, M. Hird, P. Djapic, G. Nicholson, Integration of operation of embedded generation and distribution networks. Final Report Manchester Centre for Electrical Energy (2002)
9. T. Senjyu, Y. Miyazato, A. Yona, N. Urasaki, T. Funabashi, Optimal distribution voltage control and coordination with distributed generation. IEEE Trans. Power Deliv. **23**(2), 1236–1242 (2008)
10. T. Sansawatt, L.F. Ochoa, G.P. Harrison, in *Integrating distributed generation using decentralised voltage regulation*, IEEE Power and Energy Society Meeting, pp. 1–6 (2010)
11. J. Mutale, Benefits of active management of distribution networks with distributed generation, in *IEEE PES Power Systems Conference and Exposition*, pp. 601–606 (2006)
12. T. Xu, P.C. Taylor, Voltage control techniques for electrical distribution networks including distributed generation, in *17th World Congress of the International Federation of Automatic Control*, pp. 11967–11971 (2008)
13. P.N. Vovos, A.E. Kiprakis, A.R. Wallace, G.P. Harrison, Centralized and distributed voltage control: impact on distributed generation penetration. IEEE Trans. Power Syst. **22**(1), 476–483 (2007)
14. A.E. Kiprakis, A.R. Wallace, Maximising energy capture from distributed generators in weak networks. IEE Proc. Gener. Transm. Distrib. **151**(5), 611–618 (2004)
15. W. Freitas, J.C.M. Vieira, A. Morelato, W. Xu, Influence of excitation system control modes on the allowable penetration level of distributed synchronous generators. IEEE Trans. Energy Convers. **20**(2), 474–480 (2005)
16. C. Gao, M.A. Redfern, A review of voltage control techniques of networks with distributed generations using On-Load Tap Changer transformers, in *45th International Universities Power Engineering Conference (UPEC)*, pp. 1–6 (2010)

17. M. Thomson, Automatic voltage control relays and embedded generation. Power Eng. J. **14** (2), 71–76 (2000)
18. J. Hiscock, N. Hiscock, A. Kennedy, Advanced voltage control for networks with distributed generation, in *19th International Conference on Electricity Distribution*, pp. 1–4 (2007)
19. M. Fila, G.A. Taylor, J. Hiscock, M.R. Irving, P. Lang, Flexible voltage control to support distributed generation in distribution networks, in *43rd International Universities Power Engineering Conference*, pp. 1–5 (2008)
20. H.Y. Li, H. Leite, in *Increasing distributed generation using automatic voltage reference setting technique*, IEEE Power and Energy Society General Meeting—Conversion and Delivery of Electrical Energy in the 21st Century, pp. 1–7 (2008)
21. Q. Zhou, J.W. Bialek, Generation curtailment to manage voltage constraints in distribution networks. IET Gener. Transm. Distrib. **1**(3), 492–498 (2007)
22. R. Tonkoski, L.A.C. Lopes, D. Turcotte, Active power curtailment of PV inverters in diesel hybrid mini-grids, in *IEEE Electrical Power & Energy Conference (EPEC)*, pp. 1–6 (2009)
23. R. Tonkoski, L.A.C. Lopes, , T.H.M. El-Fouly, Droop-based active power curtailment for overvoltage prevention in grid connected PV inverters, in *IEEE International Symposium on Industrial Electronics (ISIE)*, pp. 2388–2393 (2010)
24. A. Kulmala, K. Maki, S. Repo, P. Jarventausta, Including active voltage level management in planning of distribution networks with distributed generation. IEEE Bucharest PowerTech, 1–6 (2009)
25. E.M. Davidson, S.D.J. McArthur, M.J. Dolan, J.R. McDonald, in *Exploiting intelligent systems techniques within an autonomous regional active network management system*, IEEE Power & Energy Society General Meeting, pp. 1–8 (2009)
26. I. Roytelman, V. Ganesan, Coordinated local and centralized control in distribution management systems. IEEE Trans. Power Deliv. **15**(2), 718–724 (2000)
27. L. Ran, F. Spinato, P. Taylor, R. Wilson, D. Jackman, Coordinated AVR and tap changing control for an autonomous industrial power system. IEE Proc. Gener. Transm. Distrib. **153**(6), 617–623 (2006)
28. B. Fabio, C. Roberto, P. Valter, Radial MV networks voltage regulation with distribution management system coordinated controller. Electr. Power Syst. Res. **78**(4), 634–645 (2008)
29. T.G. Hazel, N. Hiscock, J. Hiscock, Voltage regulation at sites with distributed generation. IEEE Trans. Ind. Appl. **44**(2), 445–454 (2008)
30. F. Pilo, G. Pisano, G. Soma, Advanced DMS to manage active distribution networks. IEEE Bucharest PowerTech, pp. 1–8 (2009)
31. K. Diwold, W. Yan, L. De Alvaro Garcia, L. Mocnik, M. Braun, Coordinated voltage-control in distribution systems under uncertainty, in *47th International Universities Power Engineering Conference (UPEC)*, pp. 1–6 (2012)
32. C.M. Hird, H. Leite, N. Jenkins, H. Li, Network voltage controller for distributed generation. IEE Proc. Gener. Transm. Distrib. **151**(2), 150–156 (2004)
33. R. Caldon, S. Spelta, V. Prandoni, R. Turri, Coordinated voltage regulation in distribution networks with embedded generation, in *18th International Conference on Electricity Distribution*, pp. 1–4 (2005)
34. L.A. Kojovic, in *Coordination of distributed generation and step voltage regulator operations for improved distribution system voltage regulation*, IEEE Power Engineering Society General Meeting, pp. 1–4 (2006)
35. A.D.T. Le, K.M. Muttaqi, M. Negnevitsky, G. Ledwich, Response coordination of distributed generation and tap changers for voltage support, in *Australasian Universities Power Engineering Conference*, pp. 1–7 (2007)
36. H. Hiroyuki, H. Kobayashi, A study of centralized voltage control method for distribution system with distributed generation, in *19th International Conference on Electricity Distribution* (2007), pp. 1–4
37. F.A. Viawan, D. Karlsson, in *Coordinated voltage and reactive power control in the presence of distributed generation*, IEEE Power and Energy Society General Meeting—Conversion and Delivery of Electrical Energy in the 21st Century (2008), pp. 1–6

38. T. Senjyu, Y. Miyazato, A. Yona, N. Urasaki, T. Funabashi, Optimal distribution voltage control and coordination with distributed generation. IEEE Trans. Power Deliv. **23**(2), 1236–1242 (2008)
39. V. Thornley, J. Hill, P. Lang D. Reid, Active network management of voltage leading to increased generation and improved network utilisation, IET-CIRED Seminar on SmartGrids for Distribution, pp. 1–4 (2008)
40. M.S. El Moursi, B. Bak-Jensen, M.H. Abdel-Rahman, Coordinated voltage control scheme for SEIG-based wind park utilizing substation STATCOM and ULTC transformer. IEEE Trans. Sustain. Energy **2**(3), 246–255 (2011)
41. B.B. Zad, J. Lobry, F. Vallee, Coordinated control of on-load tap changer and D-STATCOM for voltage regulation of radial distribution systems with DG units, in *3rd International Conference on Electric Power and Energy Conversion Systems* (2013), pp. 1–5
42. Z. Hu, X. Wang, H. Chen, G.A. Taylor, Volt/VAr control in distribution systems using a time-interval based approach. IEE Proc. Gener. Transm. Distrib. **150**(5), 548–554 (2003)
43. J. Sugimoto, R. Yokoyama, Y. Fukuyama, V.V.R. Silva, H. Sasaki, Coordinated allocation and control of voltage regulators based on reactive tabu search. IEEE Russia Power Tech. 1–6 (2005)
44. W.K. Gwang, K.Y. Lee, Coordination control of ULTC transformer and STATCOM based on an artificial neural network. IEEE Trans. Power Syst. **20**(2), 580–586 (2005)
45. R.H. Liang, X.Z. Liu, Neuro-fuzzy based coordination control in a distribution system with dispersed generation system, in *International Conference on Intelligent Systems Applications to Power Systems* (2007), pp. 1–6
46. S. Auchariyamet, S. Sirisumrannukul, Optimal daily coordination of volt/var control devices in distribution systems with distributed generators, in *45th International Universities Power Engineering Conference (UPEC)* (2010), pp. 1–6
47. R.A. Shalwala, J.A.M. Blejis, Voltage control scheme using fuzzy logic for residential area networks with PV Generators in Saudi Arabia, in *IEEE PES Conference on Innovative Smart Grid Technologies* (2011), pp. 1–6
48. R. Shivarudrswamy, D.N. Gaonkar, S.K. Nayak, Coordinated voltage control in 3 phase unbalanced distribution system with multiple regulators using genetic algorithm. Energy Procedia, 1199–1206 (2012)
49. S. Abapour, E. Babaei, B.Y. Khanghah, Application of active management on distribution network considering technical issues, in *2nd Iranian Conference on Smart Grids* (2012), pp. 1–6
50. L.A. Zadeh, Fuzzy sets. J. Inf. Control **8**(3), 338–353 (1965)
51. M. Negnevitsky, *Artificial Intelligence, A Guide To Intelligent Systems* (Pearson Education Limited, 2011)
52. A.E. Kiprakis, A.R. Wallace, Maximising energy capture from distributed generators in weak networks. IEE Proc. Gener. Transm. Distrib. **151**(5), 611–618 (2004)
53. F.A. Viawan, D. Karlsson, in *Coordinated voltage and reactive power control in the presence of distributed generation*, IEEE Power and Energy Society General Meeting—Conversion and Delivery of Electrical Energy in the 21st Century (2008), pp. 1–6
54. I. Leibe, Integration of wind power in medium voltage networks—voltage control and losses, Ph.D thesis, Lund University, 2011

# Chapter 5
# Distribution Network Oriented Demand Response

**Mustafa Alparslan Zehir and Mustafa Bagriyanik**

**Abstract** This chapter reviews promising concepts for distribution network oriented demand response. Current demand response (DR) programs are designed for wholesale markets and utility level issues, neglecting local challenges that distribution network operators (DNOs) face in daily operation. Deployment of DR to specific parts of distribution networks can enable additional services and benefits. The literature hosts promising concepts and methods that gain popularity. However, there is a number of conflicting cases that require particular consideration. This chapter presents insight into use of DR in distribution network planning and operation with special focus on promising service opportunities, developing concepts and integration of local DR programs with utility-driven DR programs.

**Keywords** Electric power distribution · Demand response · Distribution networks Incentive-based programs · Price-based programs

## 5.1 Introduction

Power grids with aged infrastructure and conventional management methods are having radical changes. Main issues like continuously growing demand with its peaks, raising concerns on $CO_2$ emissions and increasing volatility of consumption motivate researchers for finding new solutions. The main challenges of today's power system are summarized in Fig. 5.1. Although advances in energy storage technologies and reduction in imported fossil fuels bring a number of advantages, integration of renewables in generation and electric vehicles into networks are expected to breed new issues. In Fig. 5.2, the problems that are expected to become

M. A. Zehir (✉) · M. Bagriyanik
Department of Electrical Engineering, Faculty of Electrical and Electronics Engineering, Istanbul Technical University, Istanbul, Turkey
e-mail: zehirm@itu.edu.tr

M. Bagriyanik
e-mail: bagriy@itu.edu.tr



A. Arefi et al. (eds.), *Electric Distribution Network Management and Control*, Power Systems, https://doi.org/10.1007/978-981-10-7001-3_5

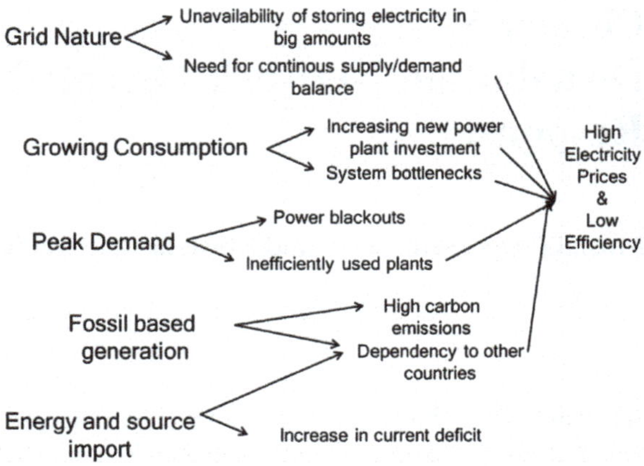


**Fig. 5.1** Main challenges in today's power systems

less critical are colored with green, while possibly growing problems are colored with red. Because utility scale conventional methods are limited to costly capacity investments that take long time and dispatch of high cost/low efficiency peaking plants or periodical avoidance of renewables, electric distribution network operators need to exploit more flexible and cost-effective, fast responding resources. At this stage, demand response is one of the topics that researchers put spotlight on.

Demand Response (DR) focuses on achieving consumption changes at the customer side (either in direct way through remote control by an entity or indirect way via tariffs and notifications to motivate voluntary participation) according to the needs of the grid. It is one of the most customer dependent topics of smart grid. It is mainly located among the applications like microgrids that are enabled through establishment of proper control and communication infrastructures and structured upon the valuable experiences gained from smart meter deployments and substation automation. It also has strong correlations with home energy management systems (HEMS), grid responsive distributed generation and storage management activities. Beyond conceptual development, DR has become an essential part of operation in modernized utilities. According to the sectoral statistics of US in 2014, 9.3 million customers enrolled in DR actions, saving 1.4 million MWh of energy, reducing peak demand by 12.7 GW and receiving $1.2 billion of incentives in turn [1]. It is also noteworthy that, the majority of the participators are residential in number and energy savings, while industrial customers achieved the highest demand reduction and incentives.

The research areas for DR can be categorized under two main groups: technology and non-technology related areas [2]. There are three technology related areas, namely planning (generation capacity planning, transmission planning and forecasting), enabling technologies (end-use device capabilities, communication, measurement, verification, automation and control) and integration (into day-ahead and intraday grid operations, integration of utility-scale renewable energy). Non-technology related areas are forecasting (DR potential and valuation), markets (customer preferences, business models and etc.), methods and policies.

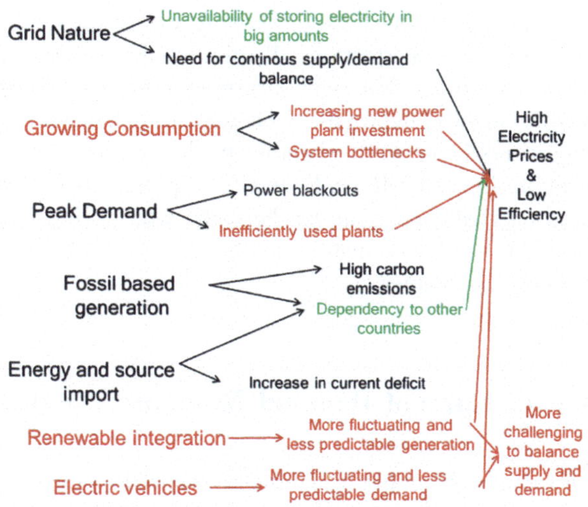


**Fig. 5.2** Expected changes in power system challenges in the near future

DR applications usually follow a bottom up approach from device level to grid level with distinctive requirements at each stage (Fig. 5.3).

Today, DR in all areas of application is based on managing flexible demand to cope with utility scale challenges. On the other hand, there is an increasing interest for using DR to aid distribution network operation.

This chapter reviews promising concepts for distribution network oriented demand response to improve knowledge of interested audiences and foster research

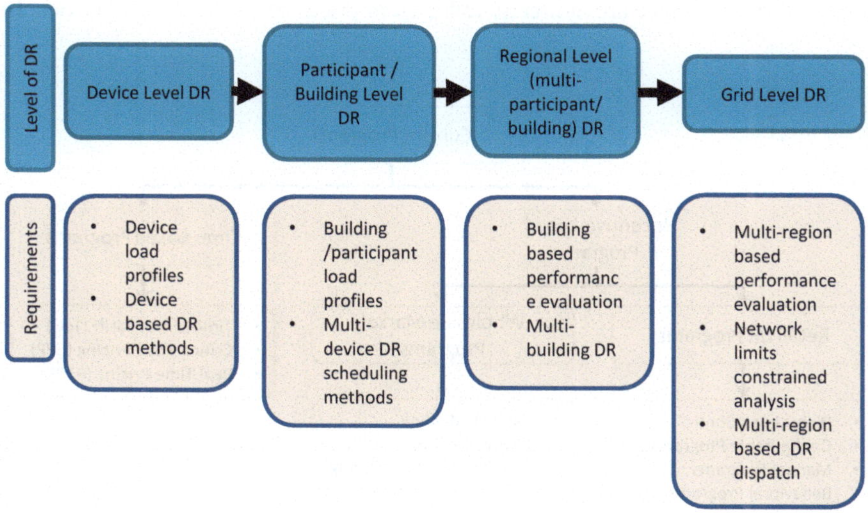


**Fig. 5.3** Different levels of DR with distinctive requirements

activities on the related area of study. Following the discussion of current DR program characteristics and imperfections, distribution network services that can be provided by DR are described. The next section comprises enabling concepts for distribution network oriented deployment of demand response. New approaches including, developments in network planning methods, innovative programs and renewable integration focused DR deployment are presented in this section. The adjacent section is on challenges that can be faced with during concurrent application of utility-driven and electric distribution network oriented DR actions and the last section concludes the chapter.

## 5.2 Overview of Current Demand Response Programs

Understanding of today's DR implementation options is important for development of distribution network oriented concepts and estimation of prospective challenges. The existing DR programs can be mainly categorized as time-based programs and incentive-based programs [1, 3]. There are currently 10 program types, 3 of which are time-based and the remaining 7 are incentive-based (Fig. 5.4).

This section describes the characteristics of each program, provides a general comparison and emphasizes imperfections.

### *5.2.1 Time-Based Programs*

In time-based programs, changes in daily market prices and main trends to maintain supply-demand balance are indirectly reflected to prices of different time periods. In

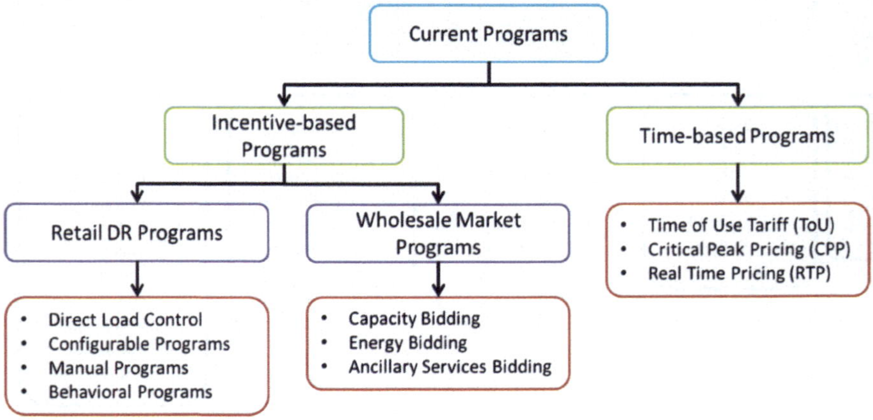


**Fig. 5.4** Current DR programs

response to changes in electricity pricing rates, the consumers are expected to make changes in their consumption and ease supply/demand balancing efforts.

Time-based programs mainly differ in frequency of price changes. Performance of time-based programs depends on how effectively price changes are reflected to customers and how effectively customers can make changes in relation with price of each time period. However, advanced programs in which pricing the periods are small and changes are frequent, require advanced communication, measurement and control infrastructures.

In transforming utilities, grid operators initially prefer basic programs based on small number of pricing periods in a day and manual customer responses. In order to grow benefits, frequency of price changes is increased progressively, while notification and automation systems are improved. The time-based programs are described below from the perspective of complexity in ascending order.

#### 5.2.1.1 Time of Use (ToU) Tariff

In this tariff structure, the day is divided into a number of periods. The most basic ToU tariff include two periods namely peak and off-peak. Electricity prices for each period are announced well in advance and they are rarely updated more than a couple of times annually. There are also some other ToU types with more pricing periods. For instance, in Canada weekdays are divided into three periods, while weekends and holidays are considered as off-peak periods [3]. Additionally, pricing periods of weekdays are different in summer and in winter. Although ToU tariffs are well suitable for manual DR actions, pioneering automated DR activities are available in the field for maximizing achievable benefits.

#### 5.2.1.2 Critical Peak Pricing (CPP) Tariff

CPP is based on application of a high price during critical peak times that occur for short periods in a year. It can be applied either individually or together with a ToU tariff. Unlike ToU, the time periods in which a high price will be applied are not known before they occur. The price for those periods can either be a certain amount that is announced earlier or it can differ according to severity of each peak period.

#### 5.2.1.3 Real Time Pricing (RTP) Tariff

In RTP, hourly wholesale market-clearing prices are proportionally reflected to customers in real time. There are three diversified applications of RTP. In day-ahead RTP, the hourly prices of the next day are determined and announced to the customers; while in intraday RTP, the prices are available just a couple of hours prior to the related hour [4]. In the last type, a demand reference profile is defined for customers and consumptions covering both under and above of that profile are priced expensively. Due to hourly varying dynamic prices, RTP tariffs require advanced communication and automation infrastructures to be effectively deployed.

## 5.2.2 *Incentive-Based Retail Programs*

Time-based programs have limited flexibility, due to their certain operation cycles and predetermined rates. For more effective use of DR in grid operation, a number of incentive-based programs have been developed. These programs usually reward customers using incentives based on the event content (duration, targeted amount of demand management) and participator performance (achieving the requested change in demand for the related period). There are also some programs in which the incentives are paid upon the acceptance of participation or authorization of a remote demand managing entity.

### 5.2.2.1 Direct Load Control (DLC)

DLC is based on remote dispatch of manageable loads by an operator through radio signal, internet and etc. A dispatch action may or may not be announced before the event. It is one of the most effective programs for reducing peak demand in US [5].

### 5.2.2.2 Configurable Programs

Similar to DLC, some loads can be remotely controlled by an operator. The distinctive part of this program is that, device owners can make further changes in control settings to allow or restrict the actions. It is one of the most popular program types.

### 5.2.2.3 Manual Programs

These programs are based on manual control of devices by their owners according to event notifications. Events can be announced using a home energy management device, SMS, e-mail, social media accounts, mobile phone application or web application. Manual programs are not popular except places where the consumers are automatically subscribed. However, they are low cost, easy to deploy programs and usually preferred at the initial step of DR implementations in a utility.

#### 5.2.2.4 Behavioral Programs

These programs aim to socially motivate consumers to make changes in their consumption behavior without providing any monetary incentives. Behavioral programs usually use weekly/monthly reports, real-time feedback using in-home display and e-mail messages to promote energy management [6]. They can also be combined with other program types to increase engagement and operational performance.

### *5.2.3 Incentive-Based Wholesale Market Programs*

The programs that are conducted by the system operator and regulatory organizations are categorized under this title. Achievable benefits generally depend on market-clearing prices and demand response success. The use of DR as a tool in a market mainly reduces the dependency to power plants and other service providers, resulting in lower market-clearing prices [7].

#### 5.2.3.1 Capacity Bidding

It is based on the manageable power that will be used, when power grid operational limits have the risk of being violated. Up on the confirmation to provide this service, a prepayment is settled. If the promised amount of demand reduction is not achieved during an event, the participators are penalized.

#### 5.2.3.2 Energy Bidding

The customers can either directly (mostly industrial customers) or indirectly participate (rather residential and commercial) in the energy market through their bids. The conventional type is day-ahead market, while intraday markets are also established at the further stages of grid modernization. Participators' performances are evaluated according to the performed changes in their estimated base consumption profiles.

#### 5.2.3.3 Ancillary Services Bidding

The reserve bids are in this program. A prepayment is offered after placing a contract. This service type requires frequent dispatch of fast responding devices for short durations of time, offering rather higher incentives compared to other market-driven programs.

### 5.2.4 *Characteristics and Imperfections of Current Retail DR Programs*

The existing programs can be compared considering different characteristics. In [5], a number of important characteristics are provided and the programs are evaluated. Geographic specificity can be defined as the local or zonal applicability of a DR request. A system-wide deployment has low specificity, while deployment at medium voltage distribution scale have medium and low voltage feeder specific deployments represent the highest. Signal variability is another characteristic that represents the change in the content of the signal to describe events or requests in a more detailed way. Signals with only a static value to trigger DR actions have low variability, while a number of different cases can be represented with predetermined several signals in medium variability and system/market states are effectively reflected to DR actions through dynamic signals. Temporal variability, is the ability of a program to trigger DR actions at specific time periods or whenever needed. Availability, represents how frequent DR can be deployed and advanced notice corresponds to time needed for notification before each event. The last characteristic is automation, representing the level of automation needed to deploy a program effectively.

The common imperfection of all the programs is geographic specificity. There is generally medium and high level of signal variability and temporal variability, while availability is limited except RTP programs. Time needed for notification before DR events is rather shorter in incentive-based programs, where automation is not a must except remote control programs.

## 5.3 Prospective DR Services for Distribution Networks

Distribution network operators (DNOs) have a number of challenges in operation, which can be aided by DR actions [5].

One of the most critical challenges is overloading. DNOs should always ensure that electric distribution network equipment and line loadings are inside tolerable operational limits. Beyond maximum demand forecasting and necessary infrastructure investments at planning stage, operational relief mechanisms are deployed in electric distribution network threatening overloading and peak demand cases. Maximum Capacity Relief, is a proactive and planned mechanism to prevent estimated cases in the near future (such as peaks in hot days), while Emergency Load Transfer is a reactive real-time service to cope with unexpected peaks or equipment failures. The main strategy is to reduce loading of the related section. It is achieved through transfer of some loads from one feeder to another, either locally by directed field personnel or remotely by reconfiguration systems. From the perspective of DR, flexible loads at the related area can respond to such cases and aid capacity relief actions. Compared to many other distribution network services, events are infrequent, have up to tens of minutes to prepare and they require duration of response for a couple of hours.

Another essential issue is to maintain voltage levels inside acceptable bounds. In Steady State Voltage Management, voltage magnitude is continuously monitored at substation level and corrective actions are done through on-load tap-changers of transformers, regulators and capacitor banks. DR can be used in several ways to help voltage management. One solution is to manage flexible loads regardless of their type to decrease loading of a feeder, resulting in reduced voltage drop at consumer nodes. An alternative solution is based on specific management of inductive loads and inverter-based loads to decrease reactive power consumption or even to inject reactive power to the electric distribution network. For voltage management, closeness to the affected area is of importance, requiring high geographic specificity.

Power quality is also among the challenges in distribution networks. Transients and harmonics have the risk of affecting devices and processes, causing comfort reduction and additional expenses. It is not quite feasible for DNOs to use advanced monitoring devices with high sampling rates continuously to identify severity and sources of power quality problems. In customer reported cases, specific monitoring and analysis are conducted at the customer facilities followed by establishment of compensation or filtering equipment. Inverter-based loads can be a part of the solution by making changes in the consumption of manageable loads and reducing the severity of transients.

Phase balancing is a typical problem especially in residential areas with high percentage of single phase loads (Fig. 5.5). It may have several impacts on the electric distribution network such as voltage problems, overloading and increased losses. The theoretical solution, transferring some loads from a highly loaded to phase to a lightly loaded phase is not widely deployed by DNOs. DR can offer a fast and significant alternative by reducing or increasing the demand in different phases. One approach can be reducing the demand of two loaded phases, bringing their loads closer to the lightly loaded third phase. However, in some cases there may be a huge gap between the most loaded and the least loaded phases, requiring dispatch of many loads from one phase. An alternative idea is to make changes in the loadings of two extreme phases (the most loaded and the least loaded) and bring them closer to the medium loaded third phase. For the latter option, demand increment is needed for the least loaded phase, which requires consideration of inactive loads and active loads with load increment potential. Steady State Voltage Management, Power Quality and Phase Balancing are frequently occurring events, needing rapid and continuous response.

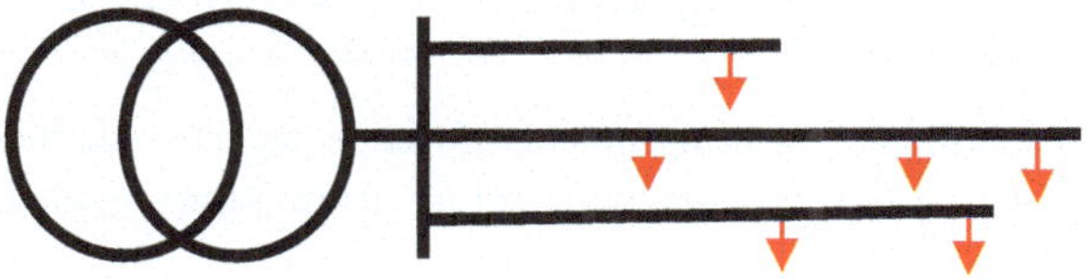

**Fig. 5.5** A representative diagram of a distribution network with phase imbalance due to different single phase loads

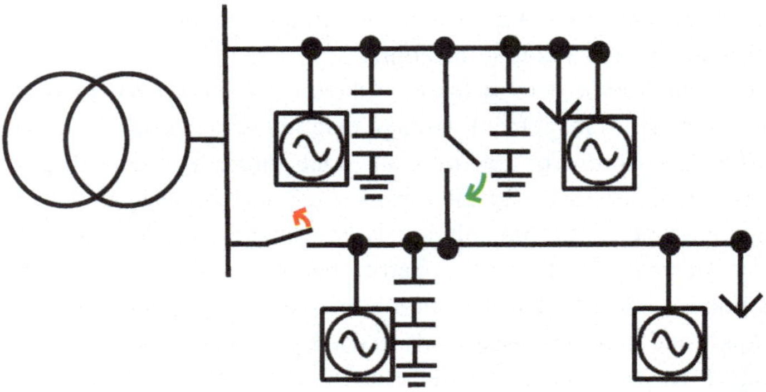

**Fig. 5.6** A representative diagram of reconfiguration in a distribution network comprising local generation and storage devices

When a part of the electric distribution network is subject to outage, recovery times have crucial importance. After a long outage period, thermostatically controlled loads (such as water heaters, air conditioners, refrigerators, deep freezers and many more) may become undiversified and cause extreme loading during reconnection with the rest of the grid. Consequently, outage time and energy not supplied increases. Conventionally, it is coped with using manual reconfiguration (Fig. 5.6) and load recovery by field crew. DR can be used for coordinated and staggered load pickup, reducing recovery time and reconnection issues. The events are infrequent and usually need a response time of less than 1 h. On the other hand, wide deployment of DR actions can affect load diversity and rebound effect can cause unexpected peaks and overloadings.

As explained with details in this section, DR actions can be deployed to provide many different services in distribution networks. These use cases lead to development of enabling concepts comprising effective methods, tariffs and programs.

## 5.4 Enabling Concepts for Distribution Network Oriented DR

There is a growing interest towards distribution network oriented DR studies in the literature. This section provides a number of the promising approaches.

### *5.4.1 DR Integrated Active Distribution Network Planning*

Network expansion and reinforcements are planned considering a number of worst-case scenarios with future estimations evaluating reliability and security.

These scenarios include maximum demand during times of minimum supply (from distributed generators) and maximum local supply during minimum consumption throughout the network. Rather than single values that represent extreme cases, the trend is through the use of load profiles, in order to better consider stochastic nature of aggregated demand and impacts of demand response in network planning [8]. The load profiles are usually organized to represent different days (such as weekday and weekend) from different seasons. Data resolution may range from 1 min to 1 h. Customers with similar behavior may be clustered.

During planning stage, one method for taking into account the impact of DR actions is to reflect the permanent consequences of DR on consumption patterns. This method requires accurate estimation of DR options on future load profiles. This can be done through observed percentages of DR effectiveness in pilot field applications. Another method is to tolerate overload cases with low probability, assuming certain level of support from DR rarely. It is also important to consider local participation rates. As indicated in [9], dispersed participation provides more benefits in rural areas, while concentrated participation is in favor of urban networks. The main benefits are deferral or avoidance of network reinforcements. Economic benefits are heavily dependent on network topology and congestion level. A study provides an additional perspective by explaining the reduction of social costs (emissions, losses, occurrence and duration of outages) due to deployment of DR in distribution network during emergency cases [10]. However, DR may not always guarantee reduction of energy losses in long term. This is explained as reinforcements are delayed, some equipment will be loaded more causing higher energy losses. On the other hand, there are financial benefits for participating customers as incentives or savings from their energy bills and for the DNOs as reduced investment need. In another study [11], researchers suggested use of probabilistic analysis with uncertainty rather than deterministic to better analyze the promising impact of DR in planning. It is noted in the same study that, DR payments should be included as financial costs for more effective analysis and better evaluation of available solutions.

Probabilistic analysis with numerous DR options that can be preferred solely or in combination with electric distribution network reinforcements, adds complexity to planning processes. In [12], probability density functions are used together with Monte Carlo simulations, representing a multi-objective optimization problem. It is beneficial for planners to make use of iterative or recursive algorithms to optimize investment strategies [13].

It is also of importance to include negative impact of utility-driven DR on distribution networks during planning stage. Wide deployment of DR can affect diversification and cause high simultaneity. Since network equipment is sized considering maximum coincidence coefficients, critical loading of components and violation of voltage limits are the major risks. It is stated in [14] that, rural electric distribution network tend to face voltage problems prior to loading issues, while urban networks with high load density and short line pairs are more likely to have loading problems before extension of voltage limits.

### 5.4.2 Innovative Retail Programs

Value of different DR services provided by different customers may have dissimilar values for a distribution network. For this reason, pricing of the services provided by DR participators to DNOs, needs particular consideration. There are several options for packaging and pricing DR services [15]. Granular retail rate, based on discrete pricing of each service is one of the options. The rates for using or providing a specific service are the same, indirectly motivating customers to deploy DR actions for achieving savings. The services like peak reduction, power factor control and voltage support can be rewarded using different rates and monthly revenues can be summarized to the participators via addition of some details to their energy bills. Determination of prices according to location and time (such as use of higher rates in an electric distribution network that requires update in the near future) is a topic open for debate. An alternative model is buy/sell arrangement based on different pricing of used and provided services. Payments for provided services can be in the form of bill credits or direct payments. Similar to Direct Load Control (DLC) programs, customers can be rewarded for accepting to participate and additionally for their response performance in each event. As mentioned in granular retail rate explanations, rates can be location-specific. An innovative model is procurement, in which third-party aggregators have direct business relationship with DR participators with competitive pricing. In this model, several aggregators submit bids to meet procurement needs announced by the network operator. Aggregators with winning bids have to coordinate their customer portfolio to achieve successful response. It is mainly done through contracts between the third parties and customers, including several incentives without strict constraints of regulations. Another option is to price different devices that provide DR individually, considering their distinctive response characteristics. This approach can also be useful to promote grid-responsive products. For instance, customers with controllable electric water heaters can be charged less (or rewarded more) than others.

For design and implementation of new pricing options, a number of issues exist. The first is design considerations. It is a challenge to keep a balance between effective feedbacks and simplicity in monthly bills and reports. On the other hand, customers with advanced monitoring and control options become more interested in energy management than past. Time-specific and location-specific pricing is also an important design element. Pricing structure can be one or a combination of fixed charge, energy based charge, demand based charge including comparisons with observed peaks or installed capacity. Furthermore, hourly data measurements can be used to identify permanent high capacity users and rarely peak consumers. The second is deployment options. DR participators can be considered as a separate class and benefit from special rates. Moreover, pricing can be mandatory, opt-out (customers are automatically enrolled with freedom to change their tariff) or opt-in (customers should apply voluntarily for joining to the new tariff). The third issue is interactions with existing policies. There may be possible conflicting cases that need further modifications of pricing programs at the design stage. There are also some other implementation issues like feasibility of the new methods, market structure with competition and long term changes in the value of provided services. During design stage of new pricing models, evaluation criterions like economic efficiency, fairness, customer satisfaction, utility revenue stability and customer price stability should be taken into account.

### 5.4.3 Distribution Level Energy Market and Locational Marginal Price

Deregulation of electricity markets and privatization of electricity generation pave the way for transparent and competitive environments at transmission level. In modernized day-ahead and intraday markets, utility operator announces needed amount of supply for specific upcoming time slots and many different market players place some bids that can fully or partially meet the request. The bids are sorted according to their prices and a market clearing price occurs when the total value of bids meet the requested amount. A company with an accepted bid have to fulfill its goal at the time of event. Otherwise, it should find another supplier with relatively higher cost of service to fill its gap, or the market operator finds another supplier and makes the company with accepted bid to pay the expenses. Because of the lack of proper communication and automation infrastructures, dynamic changes are not directly reflected to end users, limiting electric distribution network flexibility and demand elasticity.

A promising method is to reflect spot market prices proportionally to the consumers at distribution level. This is expected to facilitate DR in an effective way for retailer-spot market relations [16]. The second option is to deploy a real time pricing program based on distribution locational marginal prices (DLMP). In [17] it is found that DR can reduce peaks and congestions through the use of locational marginal prices. Another method is a distribution level market, where prices reflect the state of the local network considering energy costs, losses and congestions [18]. It is assumed to be a real time intraday market, with short time slots. Since it is totally based on the locational state of distribution network, it can even be used in islanded microgrid operation cases. Such a market will require automated transactive controllers at customer sites for flexible loads. These controllers are expected to consider current state of the electric distribution network, customer preferences, flexible device operational settings and price forecasts to bid their manageable demand. Aggregators are in contact with several houses and bid their aggregated demand to the local market. The market operator is considered as DNO that will select suitable bids, announce market clearing price and determine prices of each bus of the local network. An additional benefit of using locational marginal prices is observation of the highly loaded and stressed parts of a network for further reinforcements and investments [19].

### 5.4.4 DR for Maximizing Distributed Generation Hosting Capacity of Electric Distribution Networks

The concept of distributed generation (DG) grows in popularity throughout the world. There are numerous technical, financial and environmental advantages that foster its development [20, 21]. Contrarily, high penetration of renewable-based DG into electric distribution networks can breed new issues. The prominent challenges are increasing volatility in daily net load profiles (which is named as duck curve) [22], extreme surplus supply from DGs [23] and bidirectional power flows [24]. These stimulate the use of high cost/low efficient peaking plants, cause periodical avoidance of renewables and limit the penetration of DGs in generation mix [25, 26].

Low voltage (LV) radial distribution networks with high R/X ratios are more severely affected by DG penetration than the distribution and transmission networks with higher voltage levels. Conventionally, secondary distribution networks have been designed for one-way power flow from substation to consumers using a radial topology (Fig. 5.7).

If a distributed generator is established at one of the busses, it serves as a sending bus for some neighboring busses (Fig. 5.8).

Whilst bidirectional power flows in a radial network may cause overvoltage and line overloading problems, they can also increase the electric distribution network losses. In addition to the point of connection, neighboring busses are also being affected [24]. Considering penetration levels, distribution networks are affected even at early stages of DG deployments [27].

Conventionally, distribution network operators consider the most threatening operating conditions to scale the capacity of new DG installations. However,

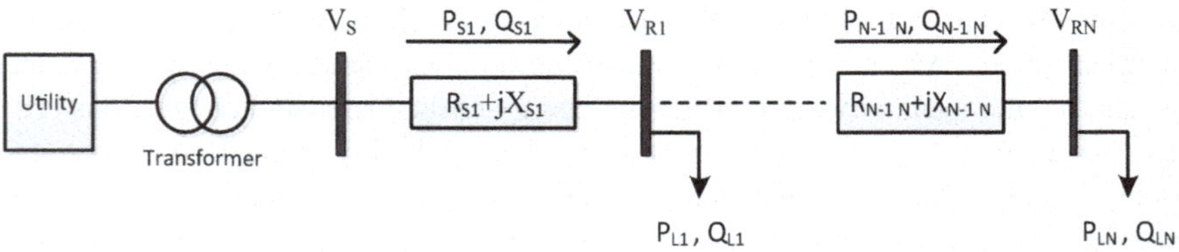


**Fig. 5.7** Unidirectional power flow in a radial LV distribution feeder with multiple busses

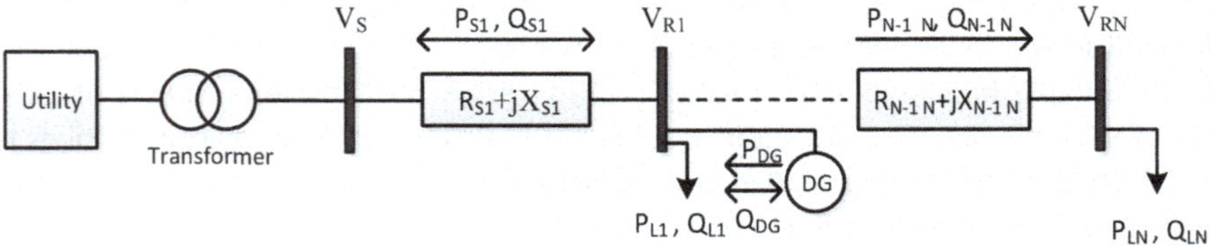


**Fig. 5.8** Bidirectional power flow in a radial LV distribution feeder with DG

determination of tolerable DG hosting capacity of electric distribution networks depending solely on the worst-case scenarios is very constraining for the utilities [28]. Wind turbines generally operate at a level less than their rated capacities, because of variation in wind speed. The highest output from wind turbines is generally provided for a couple of hours after midnight, while photovoltaic (PV) panels reach their peaks during midday just for a short period of time. Therefore, peak outputs of wind generation and photovoltaic generation rarely coincides. Moreover, developments in distributed storage facilities can provide more uniform generating profiles and more effective utilization of renewable generation. Besides, common load factors for buildings are approximately 40%, reducing the occurrence possibility of extreme (minimum generation-maximum loading) conditions [29].

DR can be used as an additional tool to cope with rarely occurring overloading and under/over voltage issues due to high integration of renewables in distribution networks. As stated in [30], impact of DG on a network is dependent on customer load at the same feeder. Closeness to DGs make DR one of the most effective tools for mitigating the negative impacts on the network. The main idea is to stimulate self-consumption or nearby consumption, shortening the path of inverse power flow on the network, reducing losses and voltage issues. Reference [31] introduced contribution of DR to wind turbine hosting capacity and to reduction of losses in a distribution network. Another study shows that load shaping is among the solutions for solar intermittency problems and flexible loads in a residential house can be scheduled to operate in synchronization with high generation from PV [32].

### 5.4.5 Optimal Power Flow Considering DR

Demand response is majorly based on using flexible devices located at distribution level to serve the need of the transmission level. As DR becomes widespread and reaches high number of participation rates, it may have unintended impacts on distribution networks. The possible impacts are reduced diversity of loads, rebound effect (similar to cold load pickup) after a long DR period, phase imbalances, uncoordinated voltage regulation related response performance reduction and increase of energy losses in transformer and capacitors [33]. Therefore, DNOs should consider impacts of wide DR actions in their analysis to prevent possible violations. A study that uses optimal power flow at the planning stage of DR is proposed in [34], demonstrating rebound and location effects.

### 5.4.6 Demand Response in DC Distribution Networks

DC distribution grid is one of the research areas that gain interest in recent years. Especially in undeveloped countries, villages far from urban areas and without proper infrastructure nearby are electrified through islanded DC microgrids. These electric distribution networks usually use PV panels, batteries and residential loads which are connected through a DC bus. Use of DC in small microgrids can overcome many disadvantages of AC electric distribution networks such as reactive power flow, synchronization need, phase angle and many more [35]. There are two pioneer companies that have field applications. BBOXX provides monitoring, remote control of devices together with home appliances and have activities in Kenya, Rwanda and Uganda [36]. Another company, SOLshare has modular devices that allow peer-to-peer electricity trading for rural households [37]. They have a pilot in Bangladesh. A study investigates DR methods in DC networks, to keep loading of DC power sources in operational limits and to deliver power fairly to different locations in a DC network [38]. The researchers develop their approach for more effective use of energy generated from PV panels in another study [39]. DC microgrids is a promising solution especially for rural and islanded communities with physical and economical grid connection barriers in Brazil [40].

## 5.5 Challenges in Concurrent Application of Utility-Driven and Distribution Network Oriented DR

Current demand response (DR) programs have the risk of triggering local or regional problems in distribution networks while providing services to wholesale markets for the favor of the transmission grid. As an example, responding loads to a utility scale demand increment action (to balance surplus generation from renewables for a short time) may cause overloading of some critically loaded distribution lines and transformers. Similarly, a demand reduction to mitigate insufficient supply may cause overvoltage in some parts of distribution networks with high amount of generation from distributed generators. Furthermore, simultaneous dispatch of high numbers of loads in a distribution area can affect their operational diversity and cause unexpected peaks. Moreover the equipment can be subject to many switchings and transients fastening aging and reducing lifetime.

It may be possible to categorize flexible loads in a local network and assign different objectives. The flexible loads of a distribution feeder out of the zone of a prospective local undervoltage issue (Fig. 5.9)—a highly loaded zone, consisting of a group of loads at the end of the feeder with a group of distributed generators near it—can be managed to increase their consumption, because of a utility driven DR case that requires load increment. At the same time, the flexible loads in the zone of prospective local undervoltage issue can be managed to reduce their consumption to mitigate the changes in the specific bus voltages and reduce the load of the highly loaded line pairs. In the case of a local overvoltage (Fig. 5.10) and load reduction request from utility, a similar combined approach can be preferred. Combined DR

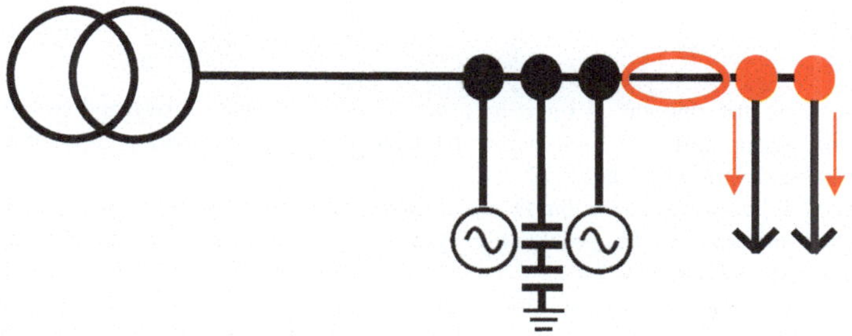

**Fig. 5.9** A general representation of local undervoltage case for a distribution feeder with DG and/or storage

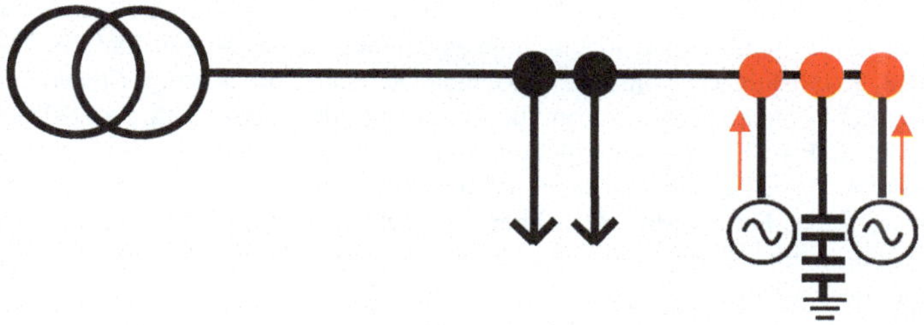

**Fig. 5.10** A general representation of local overvoltage case for a distribution feeder with DG and/or storage

applications based on specified load reduction for the utility and specified load increment for local overvoltage can be implemented.

For responding loads, transition from utility-driven DR to local DR may be needed in the case that local problems become more severe. For instance, the flexible loads of a distribution feeder out of the zone of a prospective local overvoltage issue—a group of distributed generators with high amount of supply, located at the end of the feeder, followed by a group of loads closer to the substation—can be managed to reduce their consumption, because of a utility driven DR case that requires load reduction. At the same time, the flexible loads in the zone with prospective local overvoltage issue can be managed to increase their consumption to mitigate the changes in the specific bus voltages; but it may not be sufficient as a solution. In this case, the loads used for utility response can be used for mitigating the local overvoltage issue by transition to an opposite DR action (from demand reduction to demand increment).

## References

1. U.S. Energy Information Administration, Today in Energy—Demand response saves electricity during times of high demand (2016) [Online], Available: https://www.eia.gov/todayinenergy/detail.php?id=24872
2. Bonneville Power Administration, Demand Response Technology Roadmap (2015) [Online], Available: https://www.bpa.gov/Doing%20Business/TechnologyInnovation/Documents/DR-Technology-Roadmap.pdf
3. Ontario Energy Board, Time-of-Use Electricity Prices (2013) [Online], Available: https://www.oeb.ca/oeb/_Documents/Consumer%20Brochures/brochure_02_Time-of-use%20prices%20for%20electricity.pdf
4. ComEd, ComEd's Hourly Pricing Program—Pricing Dashboard (2016) [Online] Available: https://hourlypricing.comed.com/live-prices/
5. P. Cappers, J. MacDonald, J. Page, J. Potter, E. Stewart, Future Opportunities and Challenges with Using Demand Response as a Resource in Distribution System Operation and Planning Activities (2016) [Online], Available: https://emp.lbl.gov/sites/all/files/lbnl-1003951.pdf
6. T. Mukai, K. Nishio, H. Komatsu, T. Uchida, K. Ishida, Evaluating a behavioral demand response trial in Japan: evidence from the summer of 2013. Energ. Effi. **9**(4), 911–924 (2016)
7. M.H. Albadi, E.F. El-Saaddany, A summary of demand response in electricity markets. Electr. Power Syst. Res. **78**(11), 1989–1996 (2008)
8. F. Pilo, S. Jupe, F. Silvestro, C. Abbey, A. Baitch, B. Bak-Jensen et al., Planning and optimization methods for active distribution systems, in *CIGRÉ C6 Study Committee (Distribution Systems and Dispersed Generation)*, (2014)
9. M. Vallés, J. Reneses, P. Frías, C. Mateo, Economic benefits of integrating active demand in distribution network planning: a Spanish case study. Electr. Power Syst. Res. **136**, 331–340 (2016)
10. E.A.M. Ceseña, P. Mancarella, Distribution network reinforcement planning considering demand response support, in *Power Systems Computation Conference* (2014), pp. 1–7
11. J.A. Schachter, P. Mancarella, J. Moriarity, R. Shaw, Flexible investment under uncertainty in smart distribution networks with demand side response: assessment framework and practical implementation. Energy Policy **97**, 439–449 (2016)
12. G. Mokryani, Active distribution networks planning with integration of demand response. Sol. Energy **122**, 1362–1370 (2015)
13. E.A.M. Ceseña, P. Mancarella, Practical recursive algorithms and flexible open-source applications for planning of smart distribution networks with demand response. Sustain. Energy Grids Netw. **7**, 104–116 (2016)
14. B. Gwisdorf, S. Stepanescu, C. Rehtanz, Effects of demand side management on the planning and operation of distribution grids, in *IEEE PES Innovative Smart Grid Technologies Conference Europe (ISGT Europe)* (2010), pp. 1–5
15. R. Hledik, J. Lazar, L. Schwartz, Distribution System Pricing with Distributed Energy Resources (2016) [Online], Available: https://emp.lbl.gov/sites/all/files/feur_4_20160518_fin-links2.pdf
16. S. Sekizaki, I. Nishizaki, T. Hayashida, Electricity retail market model with flexible price settings and elastic price-based demand responses by consumers in distribution network. Electr. Power Energy Syst. **81**, 371–386 (2016)
17. F. Sahriatzadeh, P. Nirbhavane, A.K. Srivastava, Locational marginal price for distribution system considering demand response, in *North American Power Symposium* (2012), pp. 1–5
18. P. Siano, D. Sarno, Assessing the benefits of residential demand response in a real time distribution energy market. Appl. Energy **161**, 533–551 (2016)
19. R. Palma-Behnke, J.L. Cerda, L.S. Vargas, A. Jofré, A distribution company energy acquisition market model with integration of distributed generation and load curtailment options. IEEE Trans. Power Syst. **20**(4), 1718–1727 (2005)

20. A. Colmenar-Santos, C. Reino-Rio, D. Borge-Diez, E. Collado-Fernández, Distributed generation: a review of factors that can contribute most to achieve a scenario of DG units embedded in the new distribution networks. Renew. Sustain. Energy Rev. **59**, 1130–1148 (2016)
21. M. Stadler, G. Carodo, S. Mashayekh, T. Forget, N. DeForest, A. Agarwal et al., Value streams in microgrids: a literature review. Appl. Energy **162**, 980–989 (2016)
22. M.A. Zehir, A. Batman, M. Bagriyanik, Review and comparison of demand response options for more effective use of renewable energy at consumer level. Renew. Sustain. Energy Rev. **56**, 631–642 (2016)
23. P.S. Moura, A.T. Almeida, The role of demand-side management in the grid integration of wind power. Appl. Energy **87**(8), 2581–2588 (2010)
24. M.A. Mahmud, M.J. Hossain, H.R. Pota, Voltage variation of distribution networks with distributed generation: worst case scenario. IEEE Syst. J. **8**(4), 1096–1103 (2014)
25. T. Ackermann, E.M. Carlini, B. Ernst, F. Groome, A. Orths, J. O'Sullivan et al., Integrating variable renewables in Europe. IEEE Power Energy Mag. **13**(6), 67–77 (2015)
26. I. Stadler, Power grid balancing of energy systems with high renewable energy penetration by demand response. Utilities Policy **16**(2), 90–98 (2008)
27. International Energy Agency, High Penetration of PV in Local Distribution Grids Subtask 2: Case-Study Collection (2014) [Online], Available: https://nachhaltigwirtschaften.at/resources/iea_pdf/reports/iea_pvps_ task14_report_2014_high_penetration_of_pv_in_local_distribution_grids.pdf
28. M.A. Zehir, A. Batman, M.A. Sonmez, A. Font, D. Tsiamitros, D. Stimoniaris et. al., Impacts of microgrids with renewables on secondary distribution networks. Applied Energy, in press, corrected proof, (2017)
29. A.D. Peacock, M. Newborough, Controlling micro-CHP systems to modulate electrical load profiles. Energy **32**(7), 1093–1103 (2007)
30. N. Mahmud, A. Zahedi, Review of control strategies for voltage regulation of the smart distribution network with high penetration of renewable distributed generation. Renew. Sustain. Energy Rev. **64**, 582–595 (2016)
31. A. Soroudi, A. Rabieem, A. Keane, Distribution networks' energy losses versus hosting capacity of wind power in the presence of demand flexibility. Renew. Energy **102**(B), 316–325 (2017)
32. R. Perez, K.R. Rábago, M. Trahan, L. Rawling, B. Norris, T. Hoff et al., Achieving very high PV penetration—the need for an effective electricity remuneration framework and a central role for grid operators. Energy Policy **96**, 27–35 (2016)
33. J. Medina, N. Muller, I. Roytelman, Demand response and distribution grid operations: opportunities and challenges. IEEE Trans. Smart Grid **1**(2), 193–198 (2010)
34. W. Shi, N. Li, X. Xie, C. Chu, R. Gadh, Optimal residential demand response in distribution networks. IEEE J. Sel. Areas Commun. **32**(7), 1441–1450 (2014)
35. R.S. Balog, W. Weaver, P.T. Krein, Low-voltage dc distribution system for commercial power systems with sensitive electronic loads. IEEE Trans. Power Deliv. **22**(3), 1620–1627 (2007)
36. BBOXX, Impact (2017) [Online], Available: http://www.bboxx.co.uk/customers/
37. SOLshare, In the News (2017) [Online], Available: https://www.me-solshare.com/in-the-news/
38. H. Mohsenian-Rad, A. Davoudi, Demand Response in DC Distribution Networks, in *IEEE SmartGridComm Symposium* (2013), pp. 564–569
39. H. Mohsenian-Rad, A. Davoudi, Towards building an optimal demand response framework for DC distribution networks. IEEE Trans. Smart Grid **5**(5), 2626–2634 (2014)
40. T.R. de Oliveira, P.F. Donoso-Garcia, Perspectives for DC distribution adoption in Brazil, in *IEEE First International Conference on DC Microgrids* (2015), pp. 359–364

# Chapter 6
# Achieving Efficiency and Fairness in Dynamic Demand Response

**Zhechao Li and Xuejun Zheng**

**Abstract** This chapter discusses the feasibility of using customer coupon demand response in meshed secondary networks. Customers are rewarded by coupons to achieve the objective of optimal operation cost during peak periods. The interdependence of the locational marginal price and the demand is modeled by an artificial neural network. The effect of multiple load aggregators participating in customer coupon demand response is also investigated. Because load aggregators satisfy different proportions of the objective, a fairness function is defined that guarantees that aggregators are rewarded in correspondence with their participation towards the objective. Energy loss is also considered in the objective as it is an essential part of the electric distribution networks. A dynamic coupon mechanism is designed to cope with the changing nature of the demand. To validate the effectiveness of the method, simulations of the presented method have been performed on a real heavily-meshed distribution network in this chapter. The results show that customer coupon demand response significantly contributes to shaving the peak, therefore, bringing considerable economic savings and reduction of loss.

**Keywords** Customer coupon demand response · Fairness · Load aggregator Locational marginal price · Meshed secondary network

## 6.1 Introduction

In the context of the smart grid, demand response has received great attention in recent years as it can shave the peak using financial incentives [1–3]. Demand response programs provide opportunities to balance the supply and demand during

---

Z. Li (✉) · X. Zheng
State Key Laboratory of Advanced Electromagnetic Engineering and Technology, Huazhong University of Science and Technology, Wuhan, China
e-mail: zcl1237@gmail.com

X. Zheng
e-mail: xuejun_zheng@hust.edu.cn



A. Arefi et al. (eds.), *Electric Distribution Network Management and Control*, Power Systems, https://doi.org/10.1007/978-981-10-7001-3_6

the peak period [4–7]. Most Load Serving Entities (LSEs) are companies that purchase electricity at real-time Locational Marginal Price (LMP) from the whole-sale market and supply electricity at a flat rate to the customers. The risk for an LSE mainly comes from the uncertainty with the wholesale LMP. Under the scheme of demand response, the risk of the fluctuating wholesale LMP can be largely transferred from the LSE to the customers.

Several kinds of time-based rate demand response pro-grams have been investigated, such as Time-of-Use (TOU) pricing, Real-Time Pricing (RTP), Critical Peak Pricing (CPP), and Critical Peak Rebates (CPR) [8–13]. CPR has been implemented in several pilot experiments. However, the rebates paid to the customers are pre-determined fixed value, which cannot satisfy different operating conditions. Therefore, there are some papers proposing another demand response program called Customer Coupon Demand Response (CCDR), in which the coupon value can be an optimization variable [14, 15]. Under CCDR, LSEs broadcast the coupon value to the customers. Customers are rewarded by the reduction of their demand with a coupon. Note that LSEs have flexibility issuing a dynamic coupon at the peak period. Customers participating in CCDR pro-gram still pay the residual demand at a flat rate to be compatible with the existing electricity bill design.

As with other demand response techniques, LSEs would implement the CCDR program only when the marginal price exceeds the flat rate. LMP is the price of electricity at different locations and consists of three components: energy, congestion, and loss [16]. The LMP represents the marginal cost of electric demand at different locations, accounting for the patterns of demand [17, 18]. When implementing CCDR, the demand will be reduced which may have influence on the LMP. Therefore, the interdependence between the LMP and the demand cannot be neglected in the CCDR program.

References [14] and [15] calculate the LMP using dc optimal power flow. They provide a straightforward way to model the relationship between LMP and demand based on the known network topology. However, in some areas, the network topology cannot be directly accessed and merely demand bid data is published by the Independent System Operator (ISO). Therefore, machine learning algorithms have been developed to model the LMP and demand with-out knowing the network topology. Among them, Neural Networks (NNs) have received great attention because of their good resolution to model complex nonlinear relation-ships [19, 20]. It is known that LSE bids and LMP are mutually dependent [21, 22]. There exists a nonlinear relationship between LSE bids and LMP. Hence, an Artificial Neural Network (ANN), as a powerful machine learning method, is applied to estimate the LMP based on the demand data because of its improved accuracy and good performance. Most of the studies estimate the LMP based on the demand of the specific region and neglect that the LMP of the specific region can be influenced by other regions. The above issues are discussed in this chapter using an ANN to estimate the LMP based on the demand bid from several LSEs.

Due to the weak ability of individual customers to re-duce demand, customers are encouraged to apply CCDR through aggregators unless they are able to reduce 50 kW or more [23]. Therefore, load aggregators are necessary participants in

CCDR programs. Today, there is no research studying CCDR with multiple load aggregators. Additionally, the aggregators take different proportions of the operation cost of the LSE because of the location diversity. Therefore, the coupon for each aggregator should be differentiated. It would be fair for the aggregators to be rewarded in accordance to their contribution to the reduction of the operational cost. In this chapter, a fairness function is defined to determine the coupon values for multiple load aggregators.

Previous CCDR publications [14, 15] have not paid attention to the energy loss of the electric distribution networks. Since the CCDR is implemented in an electric distribution networks, the energy loss should be taken in account because of the relatively high resistance to inductance ratio of low voltage electric distribution networks. Most of the demand response pro-grams are implemented in metropolitan areas which consist of numerous residential and commercial customers. Meshed networks are commonly used in metropolitan are as in North America because of the necessary high level of reliability [25, 26]. Therefore, it is essential to investigate the impact of the demand response program on meshed networks. In this chapter, the energy loss has been considered in a real meshed secondary network.

The main contributions of the chapter are: (1) to consider the case when multiple load aggregators participate in the CCDR program. A fairness function is defined to determine the coupon values for different aggregators; (2) to perform simulations of the presented method on a real heavily-meshed distribution network, in which the energy loss is also considered. Existing papers on CCDR have not taken the energy loss into account; (3) to design a dynamic coupon mechanism considering the changing behavior of the demand within one day. Existing publications on CCDR have not discussed the coupon value according to the variation of the demand. Therefore, none of the methods in exiting publications can determine the dynamic value of the coupon for an entire day.

Numerical results show that the presented method greatly contributes to shaving of the peak, which brings significant saving to the operation cost.

## 6.2 Problem Formulation

In the real-time wholesale market, the price of electricity is determined by the ISO in a clearing auction. Suppliers provide the exact amount of electricity for a given price. LSEs choose how much electricity they want to purchase based on their own load forecast. Then the ISO selects the suppliers with least cost to meet the hourly load demand with requirements of reliability and efficiency. For the LSEs, the ISO establishes the price at the specific locations, the so-called Locational Marginal Price (LMP). LMP can be used to reflect the value of electricity at different locations considering the cost of loss and congestion under the operating circumstances.

As shown in Fig. 6.1, the ISO collects the demand bids from different LSEs and determines the LMP for each LSE. Normally, power plants have their own bilateral contracts with fuel companies to hedge the risk of price variation. Thus, it is

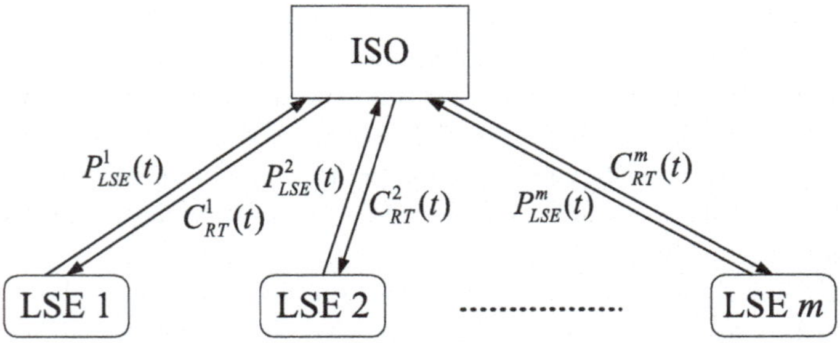


**Fig. 6.1** Operation procedure of the electricity market

assumed that the coefficients for the generation cost are flat in the short time forecast. It is assumed that all the LSEs in the specific region have been included and considered. Under this assumption, the LMP of the specific LSE is mainly correlated with its own demand and the demand bid from other LSEs which can be formulated as

$$C^m_{RT}(t) = f\left(P^1_{LSE}(t), P^2_{LSE}(t), \ldots, P^m_{LSE}(t)\right) \tag{6.1}$$

where $C^m_{RT}(t)$ and $P^m_{LSE}(t)$ are the real-time LMP for LSE $m$ at time $t$ and demand bid for LSE $m$ at time $t$.

Note that, although LSE may have a minor impact on the LMP, for completeness, the model of the marketing clearing process of the ISO is provided.

## 6.3 Data Classification

It is a fact that the CCDR program can only be implemented when real-time LMP is greater than the flat rate during the peak period [14]. Therefore, it is necessary to distinguish between the peak period and off-peak period before the LMP estimation. Due to the different characteristics between peak period and off-peak period, the peak period can be estimated based on the load data. In this chapter, Support Vector Machine (SVM) is used to estimate the peak period based on the load characteristic. SVM is a supervised learning model which is widely applied for classification analysis [27].

## 6.4 LMP-Demand Model

Customers participating in the CCDR program would reduce the system demand depending on the coupon value which may decrease the LMP as a consequence. As a result, the relationship between the demand and LMP should be taken into

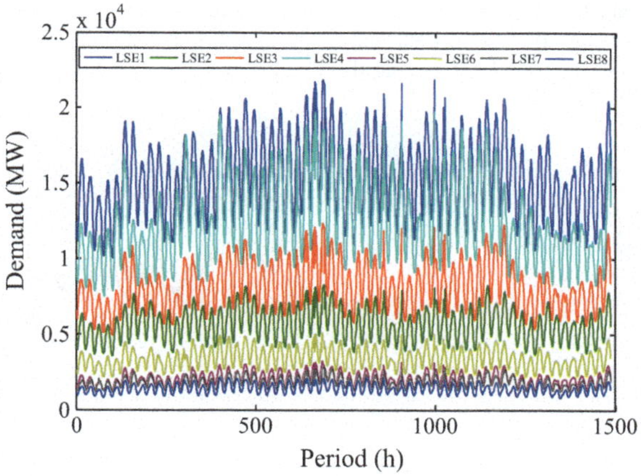


**Fig. 6.2** Actual demand of eight LSEs served by one sub-region of PJM from July 1st to August 31st, 2015 [28]. The top line corresponds to $LSE_1$, the second from the top to $LSE_2$, and so forth

consideration in the coupon optimization problem. LMP is mostly related to the network topology of electric distribution networks, load bid data, and generator bid data. However, some operators do not have direct access to the information of the network topology. They have to depend on data published by the ISO. Therefore, machine learning algorithms have been developed to model the LMP and demand without knowing the network topology.

To calculate LMP after CCDR, first the LMP-Demand model is built. In this chapter, an ANN is trained to build the LMP-Demand model rather than a market simulation. Hourly demand bid data of different LSEs are available online from PJM. The training data is collected from one of the sub-regions served by PJM Interconnection LLC (PJM) [28]. This sub-region of PJM includes eight LSEs (LSE 1 to LSE 8). For the LMP estimation, the input is the demand from several LSEs at time $t$ and output is the real-time LMP for LSE $m$ at time $t$ based on (6.1). Figure 6.2 shows the recorded demand data of these LSEs from July 1st to August 31st of 2015. For simplicity, it is assumed that CCDR is implemented in only one of the eight LSEs (LSE $m$). After applying CCDR, the demand bid from LSE $m$ will change when compared to the recorded data. Other LSEs are still operating independently and bidding as per recorded data. Therefore, the ANN method provides the updated LMP after applying CCDR for LSE $m$.

As discussed previously, the CCDR program is operated only during the peak period. After data classification, 74 among the 1488 h can be considered as peak period which becomes the training set. Figure 6.3 provides estimated results of the LMP for LSE 5 during the peak period. It proves that a well-trained ANN model is accurate enough to model the relationship between demand and LMP.

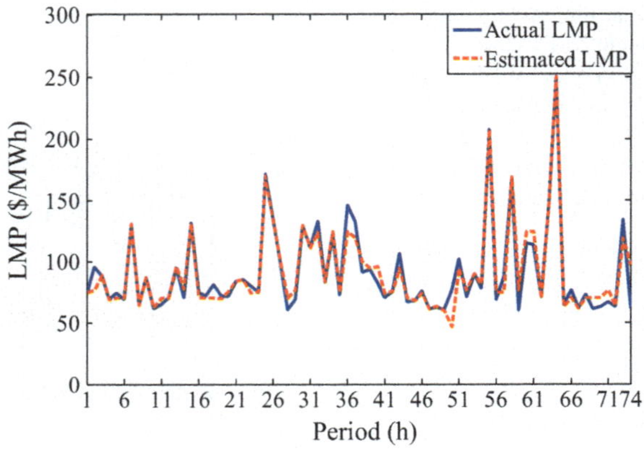


**Fig. 6.3** Comparison between the actual LMP and the estimated LMP during the peak period

## 6.5 Minimization of the LSE Net Loss

The LSE sells electricity at a flat rate to customers and pays for the time-varying LMP purchasing price from the wholesale market. Therefore, the purchasing cost has more fluctuation than the selling revenue due to the variation of the wholesale LMP. During the peak period, the wholesale LMP exceeds the flat rate, which leads to a loss for the LSE. The LSE net loss equals the purchasing cost minus the selling revenue. LSE net cost minimization can be formulated as

$$\min_{C_C^k(t)} \sum_{t=1}^{T} \left[ C_{RT}^m(t) \left( \sum_{k=1}^{K} (P_0^k(t) - \Delta P^k(t)) + P_{NP}(t) + P_{loss}(t) \right) \right. \\ \left. - C_{FR}^m \left( \sum_{k=1}^{K} (P_0^k(t) - \Delta P^k(t)) + P_{NP}(t) \right) + \sum_{k=1}^{K} C_C^k(t) \Delta P^k(t) \right] \tag{6.2}$$

where $P_0^k(t)$, $\Delta P^k(t)$, and $C_C^k(t)$ are the original demand, demand reduction based on coupons, and coupon value for load aggregator $k$ at time $t$. $P_{NP}(t)$, $C_{FR}^m$, $K$, and $T$ represent the time-varying demand which does not participate in the CCDR program, flat rate for LSE $m$, total number of aggregators, and total amount of time periods.

In this chapter, power flow results are obtained using OpenDSS, which is a comprehensive simulation tool for electric utilities. The electric distribution networks is fully modeled and analyzed with OpenDSS. For simplicity, the power loss in the electric distribution networks at time $t$ is denoted as a nonlinear function $G$, where the residual demands after CCDR are taken as input and power flow results are provided as the output. The updated G function is provided as

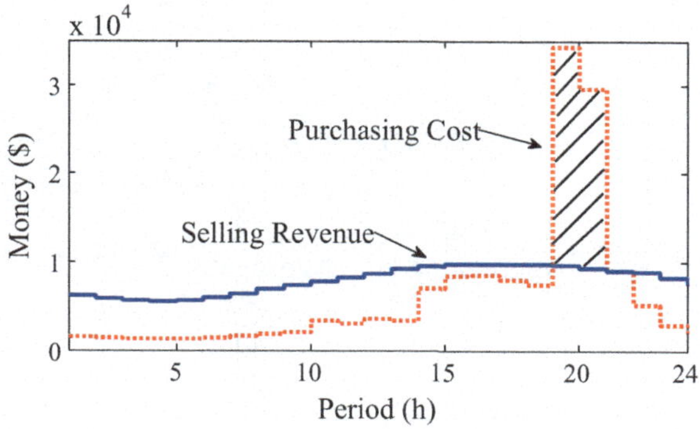


**Fig. 6.4** Illustration of the purchasing cost and selling revenue for LSE; data from [28]

$$P_{loss}(t) = G\left[\left(\boldsymbol{P}_0^1(t) - \Delta\boldsymbol{P}^1(t)\right), \ldots, \left(\boldsymbol{P}_0^k(t) - \Delta\boldsymbol{P}^k(t)\right), \boldsymbol{P}_{NP}(t)\right] \tag{6.3}$$

where $\boldsymbol{P}_0^k(t)$ is a vector that denotes the original demand of the aggregator $k$, which includes the original demand of every end-user within the aggregator $k$. $\Delta\boldsymbol{P}^k(t)$ is a vector that denotes the demand reduction of the aggregator $k$, which includes the demand reduction of every end-user within the aggregator $k$. $\boldsymbol{P}_{NP}(t)$ is a vector that includes the demand of every end-user not participating in the aggregators.

Figure 6.4 gives the hourly purchasing cost and selling revenue without CCDR for the entire day based on the recorded data of LSE 5 on July 20th in 2015 by PJM [28]. The shaded region in Fig. 6.4 shows the total net loss during the peak period.

## 6.6 Fairness Among Multiple Load Aggregators

Electric load aggregator is an organization which clusters customers together to increase the market power of individual customers. As shown in Fig. 6.5, the LSE $m$ broadcasts the coupons to different load aggregators to minimize its net loss. Customers adjust the load consumption pattern and post the demand reductions to the aggregators. The load aggregators collect the total demand reduction and submit to the LSE. For simplicity, the CCDR program is assumed to be implemented in the area served by one LSE and the competition among multiple LSEs is neglected.

Most research papers related to coupon demand response focus on achieving a system-level optimization objective. However, they neglect the factor of fairness, which is important for policy makers [24]. In other words, the rewards that customers can get are not proportional to their contribution towards the system-level objective. The aggregators take different proportions of the operational cost of the LSE because of the location diversity. Intuitively, the contribution is largely dependent on size and location of the load aggregators. In order to measure the impact in a mathematical way, the Shapley Value is employed in a cooperative

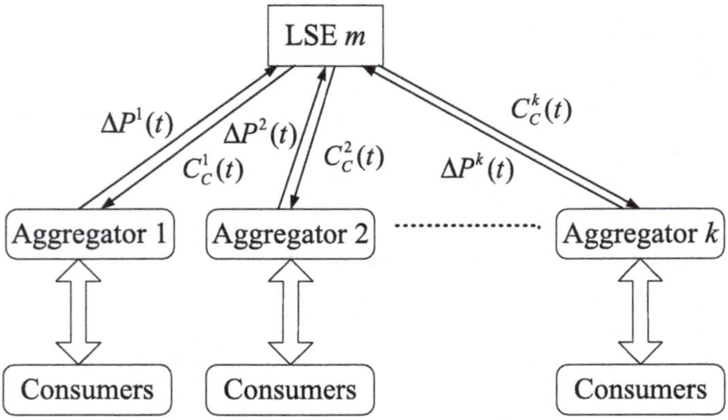


**Fig. 6.5** Interaction between the LSE and load aggregators

game model. Note that, the CCDR is a top-down scheme and aggregators can be seen as cooperative participants from the perspective of the LSE.

Without the CCDR program, the net loss of the LSE at time $t$ is derived from (2) as

$$\begin{aligned} V(t) = C_{RT}^{m}(t)\left(\sum_{k=1}^{K}\left(P_0^k(t) - \Delta P^k(t)\right) + P_{NP}(t) + P_{loss}(t)\right) \\ - C_{FR}^{m}\left(\sum_{k=1}^{K}\left(P_0^k(t) - \Delta P^k(t)\right) + P_{NP}(t)\right) \end{aligned} \tag{6.4}$$

The assuned definition of fairness is that each aggregator gets coupons according to its impact on $V(t)$. The fairness function is defined as

$$C_C^k(t) \propto \frac{\varphi^k(t)}{\sum_{j=1}^{N}\varphi^j(t)} \tag{6.5}$$

where $C_C^k(t)$ is the coupon value for load aggregator $k$, $\varphi^k(t)$ and $\varphi^j(t)$ denote the impact of load aggregators $k$ and $j$ on $V(t)$, and $N$ is the total number of aggregators. When large customers participate independently (not aggregated), they are treated as other aggregators. In this chapter, (6.5) is presented as an axiomatic index to measure fairness. It is assumed that it is fair to distribute the coupon value for each aggregator using (6.5). Under this assumption, the coupons that customers get are proportional to their impact on the system-level objective (6.4).

To calculate the $\varphi^k(t)$, the coupon allocation is modeled as a cooperative game. Therefore, the Shapley value can be used to distribute the net loss to the players in a fair way [24]. In the Shapley method, $\varphi^k(t)$ can be determined from two scenarios.

First, is the case when aggregator $k$ is not part of the grid. Second, is the case when aggregator $k$ is within the grid. Then we can calculate the Shapley value as

$$\varphi^k(t) = \frac{1}{N!} \sum_{S \in N \setminus \{k\}} S!(N - S - 1)! \left[ V_{S \cup \{k\}}(t) - V_S(t) \right] \tag{6.6}$$

where $S$ is a subset of $N$ not containing aggregator $i$. $V_{S \cup \{k\}}(t)$ and $V_S(t)$ are the net loss function of subset $S$ with aggregator and without aggregator $k$, respectively.

## 6.7 Maximizing the Customer Utility Function

In economics, utility function is used to measure welfare or satisfaction of the customers over a set of goods and services. In this chapter, it is assumed that the customers of CCDR choose to reduce the demand with the objective of maximizing their own utility function [14]. The electric load aggregator is an organization that clusters customers together to increase their market power. It is assumed that the customers make commitments with their aggregator. To keep fairness among customers within one aggregator, the amount of demand reduction of customers is assigned in proportion to their size with the same coupon value.

Economists use the demand curve to model the relationship between price and quantity demanded [30]. The quantity demanded is the amount of the good that customers are willing to purchase. Figure 6.6 shows the typical elastic demand curve. $C_w^k$, $P^k(t)$, $P_{min}^k$, $P_x^k(t)$, and $P_0^k(t)$ are the price that customer $k$ is charged, quantity demanded of customer $k$ at time $t$, the minimum demand value of customer $k$ which is the inelastic demand amount, the optimal demand value of customer $k$ at time $t$ under the CCDR program, and the initial demand value of customer $k$ at time $t$ without the CCDR program, respectively. Based on theory of economics [30], the CCDR will shift the demand curve to the left because it gives the customer incentive to reduce the demand. Therefore, the LSE will have less demand while

Fig. 6.6 Impact of CCDR on the price elastic demand curve

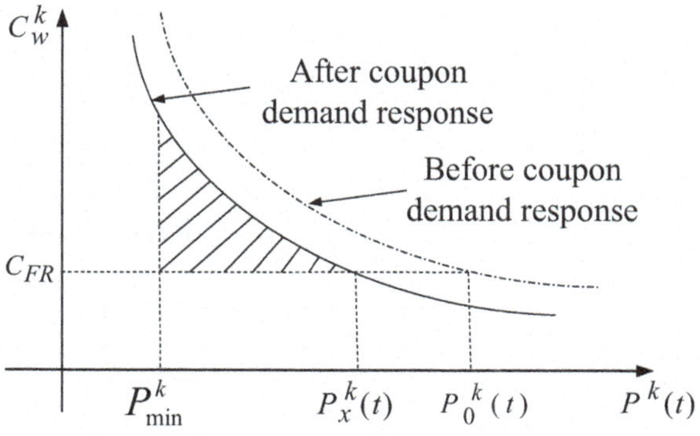

keeping the flat rate. It has been demonstrated that the CCDR is effective in encouraging customers to reduce load voluntarily [14].

Price elasticity of load aggregator $\varepsilon^k$ is defined as the relative change in demand that results from a relative change in the price [31, 32]. In general, measuring price elasticity $\varepsilon^k$ is a complex task and includes large uncertainties. Reference [33] defines short-run and long-run price elasticity. In this chapter, only short-run elasticity is considered which can be expressed as

$$\varepsilon^k = \frac{\Delta P^k / P^k_{ref}}{\Delta C^k / C^k_{ref}} \tag{6.7}$$

where $\Delta P^k$, $\Delta C^k$, $P^k_{ref}$, and $C^k_{ref}$ denote the variation of demanded quantity of aggregator $k$, price variation for aggregator $k$, reference demand of aggregator $k$, and reference price for aggregator $k$, respectively.

From [33], the function of demand curve can be formulated as

$$P^k(t) = a^k \cdot \left(C^k_w\right)^{\varepsilon^k} \tag{6.8}$$

where $a^k$ is the coefficient for the demand curve that can be calculated by putting reference values $P^k_{ref}$ and $C^k_{ref}$ into (6.8).

From the view of an economist [30], the area below the demand curve and above the price measures the customer surplus in a market (see the shaded region in Fig. 6.6). Besides, the customers get the reward from the demand reduction which should be also taken into account. Note that, the coupon should be greater than the specific value (flat rate of electricity), otherwise customers will not give up the comfort for less than the flat rate they consume. In this chapter, the customers' utility function is defined as the sum of the surplus and the coupon reward. Maximizing the utility function can be formulated as

$$\max_{P^k_x(t)} \left[ \int_{P^k_{min}}^{P^k_x(t)} \left( \frac{P^k(t)}{a^k} \right)^{1/\varepsilon^k} dP - C^m_{FR}\left(P^k_x(t) - P^k_{min}\right) + \left(C^k_c(t) - C_{min}\right)\left(P^k_0(t) - P^k_x(t)\right) \right] \tag{6.9}$$

subject to

$$P^k_x(t) = P^k_0(t) - \Delta P^k(t) \tag{6.10}$$

$$P^k_{min} \le P^k_x(t) \le P^k_0(t) \tag{6.11}$$

$$C_{min} \le C^k_c(t) < C_{max}(t) \tag{6.12}$$

where $C_{min}$ and $C_{max}(t)$ are the flat rate of electricity and the wholesale electricity prices at time $t$, respectively. This is because the LSEs would not pay for the demand reduction at a price that is more expensive than the wholesale price.

Due to the convexity of (6.9), (6.11), and (6.12), the global optimal solution can be found by the Karush-Kuhn-Tucher (KKT) condition [34]:

$$P_x^k(t) = a^k \cdot \left(C_{FR}^m + C_c^k(t) - C_{min} + \lambda_1 - \lambda_2\right)^{\varepsilon^k} \tag{6.13}$$

$$\lambda_1 \cdot \left(P_x^k(t) - P_0^k(t)\right) = 0 \tag{6.14}$$

$$\lambda_2 \cdot \left(P_{min}^k - P_x^k(t)\right) = 0 \tag{6.15}$$

$$\lambda_1, \lambda_2 \geq 0 \tag{6.16}$$

where $\lambda_1, \lambda_2$ are Lagrange multipliers.

Hence, the optimal demand value of customer $k$ at time $t$ under the CCDR program is determined as

$$P_x^k(t) = \begin{cases} a^k \cdot \left(C_{FR}^m + C_c^k(t) - C_{min}\right)^{\varepsilon^k} & \lambda_1, \lambda_2 = 0 \\ P_{min}^k & \lambda_1 = 0, \lambda_2 \neq 0 \\ P_0^k(t) & \lambda_1 \neq 0, \lambda_2 = 0 \\ \emptyset & \lambda_1 \neq 0, \lambda_2 \neq 0 \end{cases} \tag{6.17}$$

## 6.8 Overall Procedure

As discussed in previous sections, the objective of the CCDR program is to minimize the net economic loss of the LSE and maximize the utility function of customers. The objectives of the co-optimization (or bi-level optimization) problem can be formulated as (6.2) and (6.9), subject to (6.10), (6.11), and (6.12).

As shown in Fig. 6.5, the LSE broadcasts the coupon value $C_c^k(t)$ to the aggregators. As a result, the consumers within aggregators reduce demand $\Delta P^k(t)$. The demand reduction $\Delta P^k(t)$ in (6.2) is decided by the utility function of customers (6.9), however, the coupon value $C_c^k(t)$ depends on the net economic loss function of LSE (6.2). Therefore, (6.2) and (6.9) form a bi-level optimization problem with correlated variables in both levels. Note that, the real-time LMP for LSE $C_{RT}^m(t)$ depends not only on its own demand bid but also demands on bids from other LSEs, which is illustrated by Fig. 6.1. Therefore, $C_{RT}^m(t)$ in (6.2) could be obtained from a well-trained ANN, as (6.1). The demand bid $P_{LSE}^m(t)$ of LSE $m$ in (6.1) can be calculated by

$$P^m_{LSE}(t) = \sum_{k=1}^{K} \left(P^k_0(t) - \Delta P^k(t)\right) + P_{NP}(t) + P_{loss}(t) \tag{6.18}$$

In this chapter, the bi-level optimization problem is regarded as a mathematical program with equilibrium constraints (MPEC). First, the lower level optimization (6.9) is solved with KKT optimality condition and then an analytic solution is obtained using (6.17). Subsequently, the solution is employed in the upper level optimization (6.2). Therefore, the multi-objective optimization problem is converted into a single-objective optimization problem, as

$$\begin{aligned} \min_{C^k_C(t)} \sum_{t=1}^{T} \Bigg[ & C^m_{RT}(t)\left(\sum_{k=1}^{K}\left(P^k_x(t)\right) + P_{NP}(t) + P_{loss}(t)\right) - C^m_{FR}\left(\sum_{k=1}^{K}\left(P^k_x(t)\right) + P_{NP}(t)\right) \\ & + \sum_{k=1}^{K} C^k_C(t)\left(P^k_0(t) - P^k_x(t)\right)\Bigg] \end{aligned} \tag{6.19}$$

By replacing $P^k_x(t)$ in the objective function (6.2), the coupon value $C^k_C(t)$ has become the only decision variable of the optimization problem (19). To reduce the search space, practical conditions are established: (1) the lower bound is set to the flat rate ($100 /MWh) and the upper bound is set to the wholesale price. This is because the LSEs would not pay for a demand reduction at a price that is cheaper than the flat rate or more expensive than the wholesale price; (2) the coupon value is set to be an integer variable. Therefore, the step size of the coupon $\Delta C^k_C$ is chosen as $1 /MWh. To solve the optimization problem, an exhaustive method is used: Increase coupon value by fixed increment $\Delta C^k_C$ and then calculate the objective (6.19) iteratively until the maximum value is reached. Finally, the optimal value of coupons are defined for the load aggregators that maximize the objective function of the LSE.

Several main steps are involved as follows:

*Step* 1: For a given coupon value of load aggregator $k$, the fair distribution of coupon for the other aggregators can be determined. Initial coupon value is $C_{min}$.
*Step* 2: Customers participating in the CCDR program choose to reduce their demand to the optimal value to maximize their utility function.
*Step* 3: Based on the residual demand, the LSE employs SVM to estimate whether it is still peak period. If it is still peak period, it is possible for LSE to distribute more coupons. Then, proceed to step 4. If not, the LSE would not distribute more coupons and stop iterating. The coupon of this iteration is chosen as optimal coupon value.
*Step* 4: ANN is applied to estimate the LMP based on the characteristics of load. To highlight the effect of CCDR, it is assumed that the program is only implemented by one LSE and other LSEs remain with the original demand profiles.

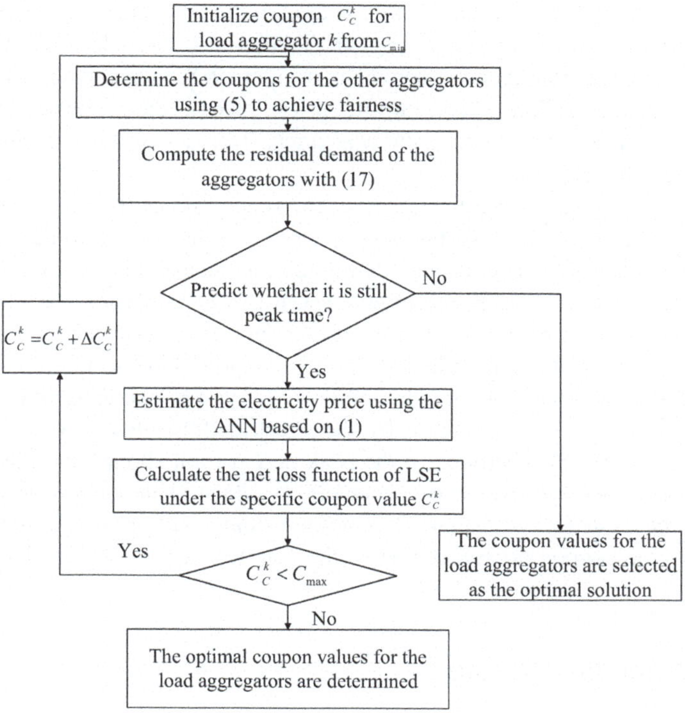


**Fig. 6.7** Flowchart of the presented method

*Step* 5: Calculate the objective function of the LSE under the given coupon value.
*Step* 6: Increase coupon value by fixed increment $\Delta C_C^k$ and then iterate from step 1 to step 6 until reaching the maximum value $C_{max}$. Finally, the optimal value of coupons are defined for the load aggregators that maximize the objective function of the LSE.

The steps above are executed based on one hour of the day. Hourly CCDR can be determined by repeating the above steps. The flowchart of the above steps is illustrated in Fig. 6.7.

## 6.9 Practical Implementation

The two-settlement mechanism, which includes day-ahead market and real-time market, is widely used in North American electricity markets. Most LSEs would lock the energy price in forward or day-ahead markets to hedge the risk of real-time price variation. However, due to the space limitations, this chapter focuses on CCDR under real-time markets. As references [14, 15] point out, the energy cleared in the real-time market is around 2–8%. Although the percentage seems small, it is significant for a demand response program. To diminish the risk of real-time price volatility, the LSEs would encourage the consumers to reduce demand by financial incentives such as CCDR when real-time price spikes occur. With the smart grid technology such as smart meters, the consumers' response to the CCDR could be realized close to real-time.

Figure 6.8 gives a better illustration on the time scale of the presented coupon based operation. First, the LSE prepares to broadcast the coupon to the consumers when a wholesale price spike occurs according to the updated report by ISO. Due to the huge data processing pressure in practice, it is impossible for LSEs to interact with consumers to determine the optimal coupon value iteratively. Hence, LSEs should be aware of the approximated optimal coupon value so that the iterations between LSEs and consumers can be minimized. The pre-operating interval of the real-time markets is set to 60 min by the Electric Reliability Council of Texas (ERCOT) [14, 15, 35]. Therefore, if the peak period is one hour long, there will be approximately one hour for the consumers to adjust their electricity usage. In addition, facilitated by enhanced communication technologies, consumers' response to the coupon price could be realized in near real-time (e.g., 10–15 min).

## 6.10 Simulation Results

Simulations of the presented method have been performed on a real heavily-meshed distribution network that has 1905 buses, 5 substation transformers, and 210 secondary transformers (4 transformer connected with spot loads at 480 V and 206 transformers connected with secondary network at 208 V, see Fig. 6.9. The peak demand is 97.9 MW at 0.91 power factor (lagging). The substation supplies power through several MV radial feeders. The secondary network is fed by the radial feeders through network transformers. Most regular loads are connected to the secondary network at 208 V and a few large customers (so-called spot loads) are supplied at 480 V.

The method is illustrated when three load aggregators are present with percentage of the total load given in Table 6.1. Note that the situation that some customers are not willing to participate in the CCDR is also considered.

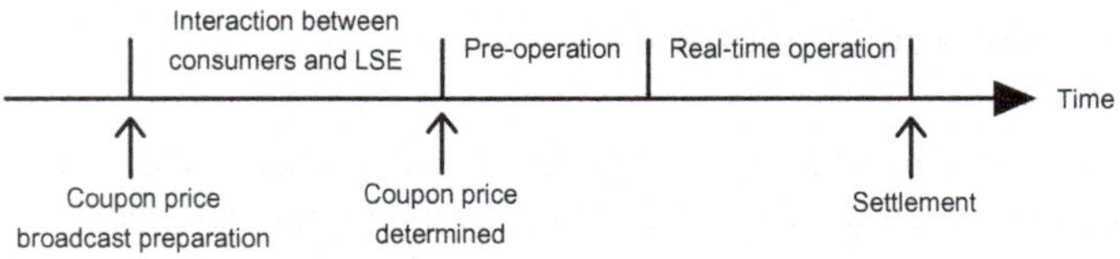


**Fig. 6.8** Timeline of the CCDR implementation

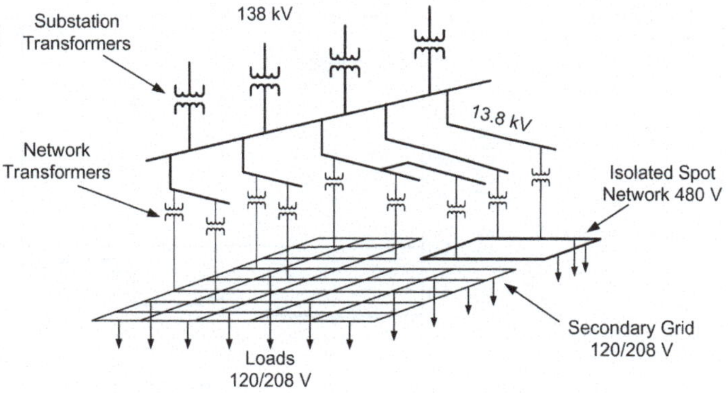


**Fig. 6.9** Structure of the meshed network used in the study

**Table 6.1** Characteristics of the load aggregators

| Load type | Aggregator 1 | Aggregator 2 | Aggregator 3 | Others not in CCDR |
|---|---|---|---|---|
| Percentage (%) | 24.58 | 31.20 | 21.19 | 23.03 |

**Fig. 6.10** Demand profile on July 20th, 2015 [28]

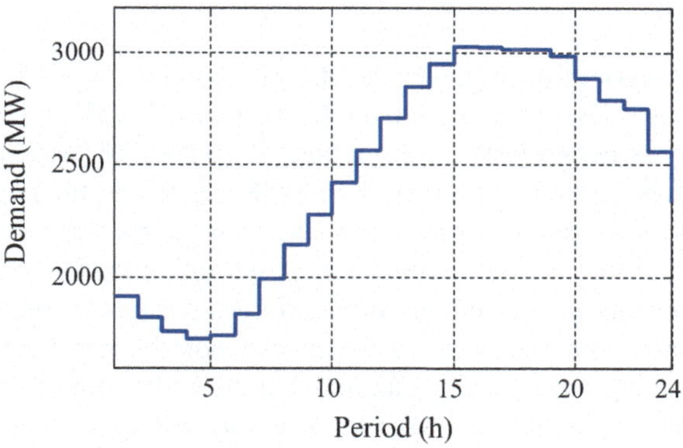


LSE 5 is assumed to be the only one that participates in the CCDR program. However, estimation of the LMP for the LSE 5 needs the demand data from other seven LSEs. To evaluate the effectiveness of the presented method, the load profile and the LMP curve of the LSE served by PJM is investigated [28]. The period of July 20th of 2015 is selected as the study period which has a typical summer load pattern. The load profile and the LMP profile for the LSE 5 on that day are presented in Figs. 6.10 and 6.11. The demand data of the other seven LSEs can be found in Fig. 6.2. As shown, the LMP and the demand behave in a similar manner during the day. The peak of the LMP is almost coincidental with the peak of demand (around 7:00 PM). Since the LMP has direct and positive correlation with demand, the LMP has good possibilities to be reduced when the demand is

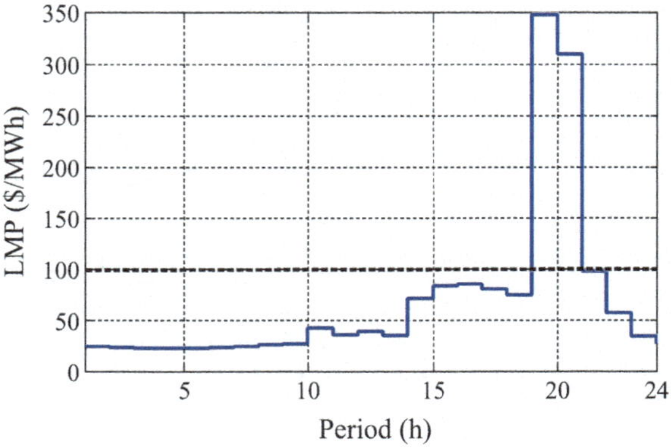


**Fig. 6.11** LMP profile on July 20th, 2015 [28]

stimulated to decrease under the implementation of CCDR. This will largely relieve the burden of purchasing electricity at peak period for the LSE, which gives the CCDR great potential to be implemented.

## *6.10.1 Base Case*

A system operating under flat rate without the CCDR program is chosen as base case. The electricity retail rate is set to $100 /MWh [29]. As shown in Fig. 6.11, the peak period lasts 2 h from hour 19 to hour 20. As expected, the peak of demand is coincidental with the peak of LMP. Therefore, the LSE will suffer a great economic loss by charging customers less than the purchasing cost.

Table 6.2 gives the hourly system operation cost for the peak period. The demand has a smooth variation during the peak period. However, the LMP has a large spike at hour 19 which expands the gap between the purchasing cost and selling revenue. One can find that the LMP rather than the demand is the key factor to the purchasing cost. Yet, electricity selling revenue has a strong relationship with the demand as the LSE charges the customers a flat rate regardless of peak or off-peak period. The calculations of the operation cost during peak period and the entire day are shown in Tables 6.3 and 6.4. At peak, the average wholesale LMP ($329 /MWh) has exceeded the flat rate ($100 /MWh) which causes a net loss of

**Table 6.2** Operation cost for the base case at each hour of peak period

| Hour | 19 | 20 |
|---|---|---|
| Total demand (MWh) | 96.33 | 93.01 |
| Wholesale LMP ($/MWh) | 348 | 310 |
| Electricity purchasing cost ($) | 34,409.43 | 29,594.78 |
| Electricity selling revenue ($) | 9632.54 | 9300.61 |
| Energy loss ($) | 883.08 | 757.96 |
| Net loss ($) | 24,776.89 | 20,294.17 |

Table 6.3 Comparison of the operation cost between base case and ccdr for peak period

| | Without CCDR | With CCDR | Reduction rate (%) |
|---|---|---|---|
| Total demand (MWh) | 189.33 | 165.52 | 12.58 |
| Average wholesale LMP ($/MWh) | 329.00 | 327.00 | 0.61 |
| Electricity purchasing Cost ($) | 64,004.21 | 55,557.25 | 13.20 |
| Electricity selling revenue ($) | 18,933.15 | 16,552.22 | 12.58 |
| Energy loss ($) | 1641.04 | 1422.51 | 13.32 |
| Net loss ($) | 45,071.06 | 42,870.07 | 4.88 |

Table 6.4 Comparison of the operation cost between base case and ccdr for the entire day

| | Without CCDR | With CCDR | Reduction rate (%) |
|---|---|---|---|
| Total demand (MWh) | 1910.96 | 1887.15 | 1.25 |
| Average wholesale LMP ($/MWh) | 68.80 | 68.64 | 0.24 |
| Electricity purchasing Cost ($) | 150,893.33 | 142,446.36 | 5.60 |
| Electricity selling revenue ($) | 191,096.29 | 188,715.35 | 1.25 |
| Energy loss ($) | 3901.86 | 3683.33 | 5.60 |
| Net loss ($) | -40,202.96 | −42,403.95 | −5.47 |

$45,071.06. Because of the short-lasting peak, the profit made during the off-peak period can fully compensate for the loss during peak, which leads to $40,202.96 of net profit.

### 6.10.2 CCDR

As shown above, the LSE suffers losses because of the gap between the LMP and flat rate at peak period. This leads to the CCDR program implementation. In this case, hour 19 to 20 are considered as the valid CCDR implementation interval. The coupon value $C_C^1$ for aggregator 1 changes from flat rate ($100 /MWh) to wholesale price with $1 /MWh incremental steps. As discussed previously, coupon value should satisfy (5) to achieve fairness. Thus, the coupon value for aggregators 2 and 3 can be calculated once $C_C^1$ is known. The price elasticity reflects the peoples' willingness to adjust their demand pattern based on the price variation. In general, measuring the elasticity is a complex and uncertain task. Therefore, the price elasticity of load aggregators 1, 2, and 3 are chosen as: −0.35, −0.25, and −0.22 based on experience [33].

Figure 6.12 provides the relationship between the coupon value $C_C^1$ and six factors including demand, coupon payment, energy loss, purchasing cost, selling revenue, and total cost at hour 20. As coupon value increases, the customers are

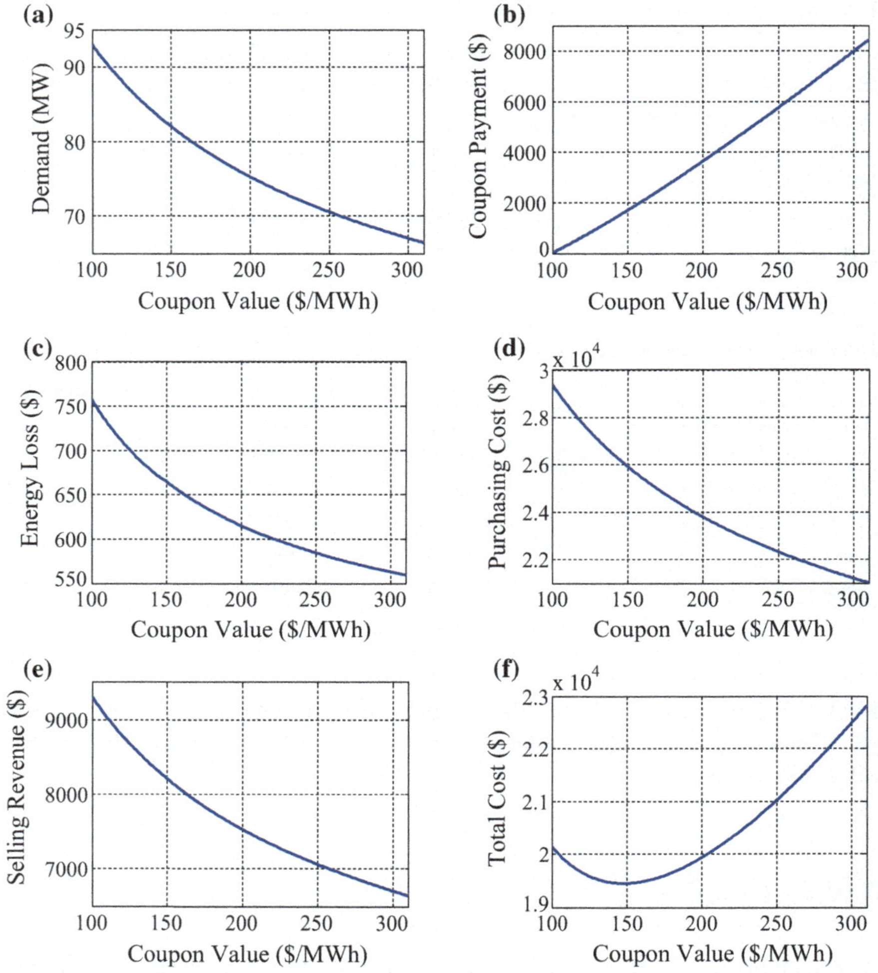


**Fig. 6.12** Six factors of the operation cost versus coupon value; **a** relationship between the demand and coupon value; **b** relationship between the coupon payment and coupon value; **c** relationship between the energy loss and coupon value; **d** relationship between the purchasing cost and coupon value; **e** relationship between the selling revenue and coupon value; **f** relationship between the total cost and coupon value

more willing to reduce the demand which also leads to the growth of coupon payment as shown in Figs. 6.12a, b. This will result in a decrease of energy loss and purchasing cost, see Figs. 6.12c, d. However, it also leads to a reduction of selling revenue as the consequence of losing a part of demand as illustrated in Fig. 6.12e. Therefore, the net loss curve of the LSE will have a minimum. Figure 6.12f shows that the net loss reaches the optimal point, when the coupon value is $147 /MWh.

**Table 6.5** Operation cost of the ccdr case at each hour of peak period

| Hour | 19 | 20 |
|---|---|---|
| Coupon for aggregator 1 ($/MWh) | 162 | 147 |
| Coupon for aggregator 2 ($/MWh) | 176 | 160 |
| Coupon for aggregator 3 ($/MWh) | 151 | 137 |
| Demand reduction within aggregator 1 (MWh) | 5.00 | 3.95 |
| Demand reduction within aggregator 2 (MWh) | 5.44 | 4.43 |
| Demand reduction within aggregator 3 (MWh) | 3.17 | 2.38 |
| Total demand reduction (MWh) | 13.61 | 10.76 |
| Residual demand (MWh) | 83.00 | 82.52 |
| Wholesale LMP ($/MWh) | 346 | 308 |
| Coupon payment ($) | 2249.44 | 1615.60 |
| Electricity purchasing cost ($) | 29,471.70 | 26,085.54 |
| Electricity selling revenue ($) | 8299.77 | 8252.44 |
| Energy loss ($) | 754.50 | 668.01 |
| Net loss ($) | 23,421.37 | 19,448.70 |

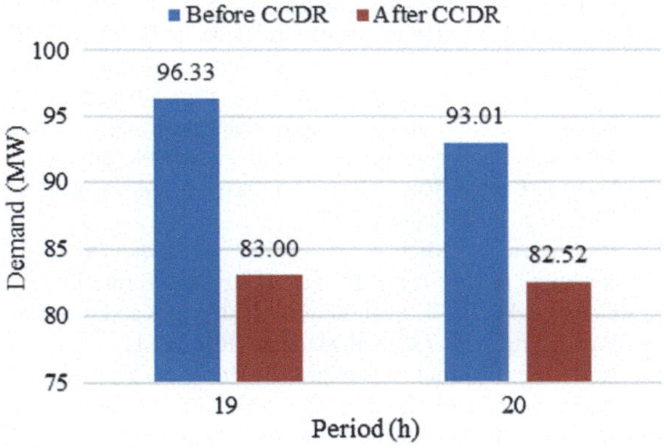


**Fig. 6.13** Impact of CCDR on the demand variation at peak period

The system optimal operation cost for the peak period is shown in Table 6.5. One can find that the optimal coupon value keeps increasing with the increase of LMP. This is because the LSE has the motivation to give out higher coupon values when the wholesale LMP is much higher than the flat rate. The calculations of operation cost during peak period and the entire day are shown in Tables 6.4 and 6.5. At peak period, the demand and average LMP have been reduced by 12.58 and 0.61% compared to the base case, which brings 4.88% reduction of the net loss including 13.32% reduction of energy loss. The comparisons of demand between the base case and the CCDR at peak period are shown in Fig. 6.13. Because the LSE would not implement the CCDR program at off-peak period, the total demand and the average LMP reduction is 1.25 and 0.24% for the entire day. Note that, the

LSE can have a greater profit ($42,403.95) than net profit ($40,202.96) of base case due to the large reduction of losses at peak period (4.88%).

## References


1. H. Mohsenian-Rad, Optimal demand bidding for time-shiftable loads. IEEE Trans. Power Syst. **30**(2), 939–951 (2015)
2. C. Chen, S. Kishore, Z. Wang, M. Alizadeh, A. Scaglione, How will demand response aggregators affect electricity markets?—A cournot game analysis, in Proc. of international symposium ISCCSP, May 2012
3. National Action Plan on Demand Response (FERC, Tech. Rep. Docket AD09-10, 2010 [Online]). Available: http://www.ferc.gov/legal/staffreports/06-17-10-demand-response.pdf
4. C. Vivekananthan, Y. Mishra, G. Ledwich, F. Li, Demand response for residential appliances via customer reward scheme. IEEE Trans. Smart Grid **5**(2), 809–820 (2014)
5. A.H. Mohsenian-Rad, A. Leon-Garcia, Optimal residential load control with price prediction in real-time electricity pricing environments. IEEE Trans. Smart Grid **1**(2), 120–133 (2010)
6. A.H. Mohsenian-Rad, V.W.S. Wong, J. Jatskevich, R. Schober, A. Leon-Garcia, Autonomous demand-side management based on game-theoretic energy consumption scheduling for the future smart grid. IEEE Trans. Smart Grid **1**(3), 320–331 (2010)
7. Q. Hu, F. Li, Hardware design of smart home energy management system with dynamic price response. IEEE Trans. Smart Grid **4**(4), 1878–1887 (2013)
8. Time based rate programs (U. S. Dept. of Energy, Washington, D. C. [Online]). Available: http://www.smartgrid.gov/recovery_act/time_based_rate_programs.html
9. Dynamic pricing and smart grid (Elevate Energy, Chicage, IL [Online]). Available: http://www.elevateenergy.org/home-savings/dynamic-pricing-smart-grid/
10. Federal Energy Regulatory Commission, Assessment of Demand Response & Advanced Metering. Staff Report, 2006
11. B. Severin, J. Michael, R. Arthur, *Dynamic Pricing, Advanced Metering and Demand Response in Electricity Markets* (Univ. California Energy Institute, Berkeley, CA, 2002)
12. C. Kang, W. Jia, Transition of tariff structure and distribution pricing in China, in Proc. IEEE Power Eng. Soc. General Meeting (July 2011)
13. X. Zhang, Optimal scheduling of critical peak pricing considering wind commitment. IEEE Trans. Sustainable Energy **5**(2), 637–645 (2014)
14. H. Zhong, L. Xie, Q. Xia, Coupon incentive-based demand response: theory and case study. IEEE Trans. Power Syst. **28**(2), 1266–1276 (2013)
15. X. Fang, Q. Hu, F. Li, B. Wang, Y. Li, Coupon-based demand response considering wind power uncertainty: a strategic bidding model for load serving entities. IEEE Trans. Power Syst. **PP**(99), 1–13 (2015)
16. Locational marginal pricing (ISO New England, Holyoke, MA [Online]). Available: http://www.iso-ne.com/participate/support/faq/lmp
17. H. Yuan, F. Li, Y. Wei, J. Zhu, Novel linearized power flow and linearized OPF models for active distribution networks with application in distribution LMP. IEEE Trans. Smart Grid (2016). https://doi.org/10.1109/TSG.2016.2594814
18. H. Yuan, F. Li, Y. Wei, LMP step pattern detection based on real-time data, in IEEE PES General Meeting, 1–5 (2013)
19. Y. Y. Hong, C. Y. Hsiao, Locational marginal price forecasting in deregulated electricity markets using artificial intelligence. Proc. Inst. Elect. Eng. Gen. Transm. Distrib. **149**(5), 621–626 (2002)
20. B.R. Szkuta, L.A. Sanabria, T.S. Dillon, Electricity price short-term forecasting using artificial neural networks. IEEE Trans. Power Syst. **14**(3), 851–857 (1999)

21. P. Mandal, A.K. Srivastava, J.W. Park, An effort to optimize similar days parameters for ANN-based electricity price forecasting. IEEE Trans. Ind. Appl. **45**(5), 1888–1896 (2009)
22. P. Mandal, T. Senjyu, N. Urasaki, T. Funabashi, A.K. Srivastava, A novel approach to forecast electricity price for PJM using neural network and similar days method. IEEE Trans. Power Syst. **22**(4), 2058–2065 (2007)
23. Demand response programs details (Con Edison, New York, NY [Online]). Available: http://www.coned.com/energyefficiency/demand_response_program_details.asp
24. Z. Baharlouei, M. Hashemi, H. Narimani, H. Mohsenian-Rad, Achieving optimality and fairness in autonomous demand response: benchmarks and billing mechanisms. IEEE Trans. Smart Grid **4**(2), 968–975 (2013)
25. L. Yu, D. Czarkowski, F. de León, Optimal distributed voltage regulation for secondary networks with DGs. IEEE Trans. Smart Grid **3**(2), 959–967 (2012)
26. P. Chen, R. Salcedo, Q. Zhu, F. de León, D. Czarkowski, Z. Jiang, V. Spitsa, Z. Zabar, R.E. Uosef, Analysis of voltage profile problems due to the penetration of distributed generation in low-voltage secondary distribution networks. IEEE Trans. Power Del. **27**(4), 2020–2028 (2012)
27. M. Mohri, A. Rostamizadeh, A. Talwalker, *Foundations of machine learning* (MA, USA, 2012), pp. 63–84
28. Energy market (PJM, Valley Forge, PA [Online]). Available: http://www.pjm.com/markets-and-operations/energy.aspx
29. Electric power monthly (U. S. Energy Info. Admin., Washington, D. C. [Online]). Available: http://www.eia.com.gov/electricity/monthly/epm_table_grapher.cfm?t=epmt_5_6_a
30. N. G. Mankiw, *Principles of economics*, 6th edn. (OH, USA, 2012), pp. 66–154
31. D. S. Kirschen, G. Strbac, P. Cumperayot, D. de Paiva Mendes, Factoring the elasticity of demand in electricity prices. IEEE Trans. Power Syst. **15**(2), 612–617 (2000)
32. M.G. Lijesen, The real-time price elasticity of electricity. Energy Econ. **29**(2), 249–258 (2007)
33. P.R. Thimmapuram, J. Kim, Consumers' price elasticity of demand modeling with economic effects on electricity markets using an agent-based model. IEEE Trans. Smart Grid **4**(1), 390–397 (2013)
34. S. Boyd, L. Vandenberghe, *Convex optimization* (United Kingdom, 2009), pp. 215–273
35. Nodal Protocol of ERCOT. [Online]. Available: http://www.ercot.com/mktrules/nprotocols/current
36. Z. Li, S. Wang, X. Zheng, F. de León, T. Hong, Dynamic Demand Response using Customer Coupons Considering Multiple Load Aggregators to Simultaneously Achieve Efficiency and Fairness. IEEE Trans. Smart Grid. **PP**(99), 1–10 (2016)

# Chapter 7
# Scheduling in Coupled Electric and Gas Distribution Networks

**Jing Qiu, Zhao Yang Dong and Ke Meng**

**Abstract** This chapter presents a transactive approach to the optimal scheduling for prosumers in coupled electric and natural gas distribution networks, to help the integration of various distributed energy resources (DERs). DERs are co-ordinately operated in the form of a virtual power plant (VPP), which actively participates in the day-ahead and real-time electricity markets, as well as the wholesale gas market. In the day-ahead (DA) electricity and wholesale gas markets, a VPP aims to maximize expected profits by determining the unit commitments and hourly scheduling of DERs. In the real-time (RT) balancing market, a VPP adjusts DER schedules to minimize imbalance costs. This chapter addresses the energy conversions between electric power and gas loads and investigates the interacting operations of electric and gas distribution networks. The simulation results show that hierarchical, coordinated power and gas scheduling can identify more accurate operation plans for coupled transactive energy networks.

**Keywords** Optimal scheduling · Distributed energy resources · Transactive energy · Prosumer

J. Qiu (✉)
Energy Flagship, Commonwealth Scientific and Industrial Research Organization (CSIRO), Sydney, Australia
e-mail: qiujing0322@gmail.com

Z. Y. Dong
School of Electrical Engineering and Telecommunications, University of New South Wales, Sydney, Australia
e-mail: zydong@ieee.org

K. Meng
School of Electrical and Information Engineering, University of Sydney, Sydney, Australia
e-mail: kemeng@ieee.org



A. Arefi et al. (eds.), *Electric Distribution Network Management and Control*, Power Systems, https://doi.org/10.1007/978-981-10-7001-3_7

## 7.1 Introduction

Renewable energy sources, such as wind and solar power, play a critical role in the effective transition towards a low-carbon energy economy. Both bulk injection and the penetration of small-scale distributed renewable power generation at the distribution level (e.g., rooftop solar panels) have increased markedly over the past few decades [1, 2]. However, the intermittency of renewable energy can cause severe power supply-demand imbalances and put the stability of the electric distribution networks at great risk, particularly if strategic ex-ante schedules and proper ex-post backups are lacking [3]. As a result, more controllable loads, distributed generators (DG) (e.g., micro gas turbines) and battery energy storage systems (BESS) are used to actively manage distribution networks and balance renewable power fluctuations [4]. Meanwhile, market-based transactive approaches are emerging as contenders for orchestrating coordinated operation of these distributed energy resources (DERs).

To help integrate numerous flexible DERs into the electric distribution networks, the concept of a virtual power plant (VPP) has been proposed. A VPP manages a large set of DERs (including renewable energy sources), the total capacity of which is comparable to that of a conventional power plant [5, 6]. Thus a VPP allows small owners of DERs to participate in the electricity market in a collective way, and to mitigate the unexpected power fluctuations of renewable sources and demands through the coordinated operation. Within smart-grid architecture, transactive approaches to integrating VPPs promote a new vision for electric distribution network operations [7]. Depending on the energy transactions between participants, a grid customer becomes a mix of consumer and producer: i.e., a prosumer. Therefore, coordinated scheduling for prosumers is a topic needing scientific advancement.

The literature contains several studies on the scheduling of VPPs or microgrids. Generally speaking, these studies can be divided into focusing on energy markets only, such as [8, 9], or focusing on joint energy and reserve markets, such as [10, 11]. A number of references are briefly described here. Reference [1] presents an internal operation scheduling model for microgrid operation. Reference [3] presents a model to optimize the day-ahead (DA) thermal and electrical scheduling of a large scale VPP, aiming to maximize daily profits. Reference [4] studies the optimal scheduling of microgrids, aiming to minimize the expected operational cost and power losses. Reference [12] studies the least-cost unit commitment and the dispatch of renewable-powered microgrids. Reference [13] presents two control strategies for the optimal scheduling of distributed solar, wind and diesel generators with BESS. Reference [14] presents a risk-constrained mean-variance model to manage microgrids in the RT balancing market. Reference [15] investigates the aggregation of DERs in a distribution network to participate in joint energy and reserve markets. Reference [16] presents a price-based market mechanism to alleviate possible electric distribution network congestion. Most of the studies mentioned above have demonstrated the benefits of coordinated scheduling of DERs in a market environment, but failed to investigate the role of distributed thermal units (e.g., gas-fired micro turbines) in backing up renewable energy. Reference [17] presents a DA scheduling model for hybrid thermal/electrical grid-connected energy system, including a fuel-cell with combined heat and power (CHP) and BESS. Reference [18] presents the planned scheduling of a CHP-based microgrid, and considers the optimal siting and sizing of DERs. Reference [19] presents an optimal 24-h scheduling of CHP-based microgrids, aiming to maximize the expected profit of the market. Unfortunately, these references have not taken into account the complex gas network models, and hence gas availability is assumed to be unconstrained. Moreover, the two-stage optimization approach is often used for the scheduling of DERs in electricity markets [20]. For instance, Ref. [7] presents a two-stage stochastic mixed integer linear programming model to address the optimal operation of a VPP in DA and RT balancing markets. The scenario based modeling technique is applied to control the risk of profit variability in relation to renewable energy and DA electricity prices. In [21], a two-level microgrid scheduling model is presented. The upper level minimizes the total cost of the microgrid, while the lower level maximizes renewable power use and minimizes power deviations in the DA and real-time (RT) markets. In [22], the power scheduling and bidding problem of a microgrid is formulated as a two-stage stochastic programming problem. Uncertainties are captured using the Monte Carlo simulation approach. The thermal dynamic characteristics of buildings are modelled, and the formulated model can flexibly help to achieve different tradeoffs between economics and thermal comfort. Reference [23] has investigated how total and surplus profits of a VPP are affected by risk aversion using a two-stage stochastic programming approach. Uncertainties of renewable energy outputs, loads, calls for reserve service and prices in DA market, the spinning reserve market and the RT balancing market are taken into account. However, models of DERs such as BESS or distributed renewable generation are not well presented. Reference [24] presents a two-stage stochastic framework for DA scheduling of microgrids under uncertainties of renewable power generation, RT prices and loads. Different scenarios are generated by the autoregressive moving-average method and then are reduced using the fast-forward technique. The network constraints have been neglected and the reliability of the microgrid is not evaluated. Reference [25] presents a two-stage stochastic programming model for the scheduling DERs. Grid-enabled electric vehicles (EVs) are modelled as energy storage devices, but the ramping limits of distributed thermal units are neglected. Reference [26] presents a bi-level optimization model for a VPP, including the upper level VPP profit maximization problem and the lower level independent system operator (ISO) DA market-clearing problem. Reference [27] presents the risk-based profit allocation to DERs that are integrated as a VPP. A two-stage stochastic programming approach is used and the cooperative Game theory model is presented. In most of the references mentioned above, the conditional value-at-risk (CVaR) is a commonly used risk measure to compute the risk aversion levels of DERs. However, none of them have modelled the risk in relation to gas prices, let alone the operating costs of gas-fired micro turbines.

To sum up, four main research gaps are found in the existing literature, as follows. (1) Most of the models use linear programming and only consider the energy balance constraints, thereby neglecting some important parameters of network constraints (e.g., reactive power and gas pressure). (2) The unit commitment issue of DERs appears extremely simplified. (3) The demand conversion between thermal and electric energy is not well addressed. (4) The interacting operations of electricity and natural gas networks are not modelled (e.g., the fuel prices and requirements of gas-fired micro turbines are based on assumptions).

To address these gaps, this chapter presents here a hierarchical optimal scheduling of prosumers to fulfill the transactive integration of various DERs in coupled electric and gas distribution networks. The operational decisions of DERs are made through a VPP, which acts as an agent to participate in the DA and RT markets. In the upper level, the VPP aims to maximize its expected profits considering a risk metric, while in the lower level, the VPP aims to minimize the power deviations, thus reducing imbalance costs in the RT balancing market. The complex electricity and gas network constraints are explicitly taken into account, and the integrated behavior of coupled transactive electric and gas distribution networks has been investigated. Specifically, the fuel price and availability of gas-fired micro turbines can be modelled. Also, the risk of profit variability in relation to electricity and gas prices, thermal and power demands as well as renewable energy outputs can be analyzed. Therefore, more accurate operational decisions can be made through a market-based coordination mechanism (i.e., coordinated energy transactions).

The remaining chapter is organized as follows: Sect. 7.2 describes the problem formulation, followed by the key mathematical models in Sect. 7.3. The detailed two-stage scheduling model for prosumers is presented in Sect. 7.4, followed by the solution algorithm in Sect. 7.5. Case studies are given in Sect. 7.6. Finally, conclusions are given in the last section.

## 7.2 Problem Description

The electricity market structure studied in this chapter is a joint DA and RT market [16]. The gas market is organized in accordance with the Short Term Trading Market (STTM) in Australia [28]. The STTM is a market for the trading of natural gas at the wholesale level, and operates three gas hubs: Sydney, Adelaide and Brisbane. Gas is traded a day ahead of the actual gas day (a 24-h period) and the day-ahead price is applied to all gas that is supplied according to the schedules on the gas day. The difference between the scheduled and actual quantity of gas is called deviation. Any gas deviation will result in a deviation charge or a deviation payment, i.e., a dual pricing scheme for gas imbalances.

The coupling of the electricity and gas networks is illustrated in Fig. 7.1. DERs in the electricity network are composed of renewable sources, BESS, and controllable loads such as electric vehicles (EVs) and demand response (DR) programs.

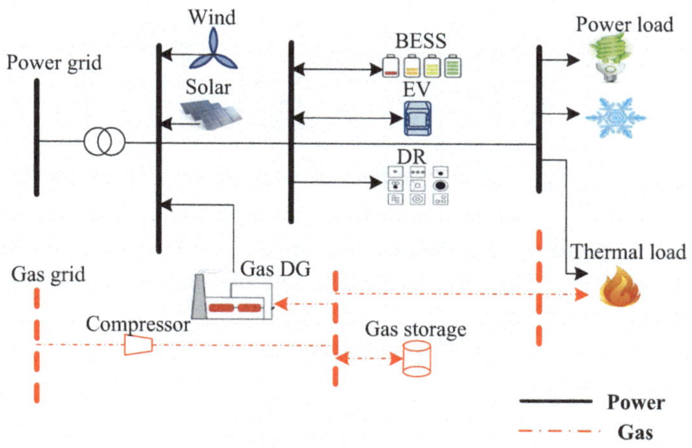


**Fig. 7.1** Paradigm of prosumers in coupled electric and gas distribution networks

Thermal (heating) demands are jointly satisfied by electricity and gas. The fuel availability of micro gas turbines (termed as DG) is also investigated by involving parameters and variables of the two networks in a transactive energy framework. The DERs are coordinated in the form of a VPP, which is responsible for managing, monitoring and controlling the DERs in its portfolio (e.g., unit commitment, energy dispatching and trading decisions). The VPP actively participates in market operations, but is considered as a price-taker.

## 7.3 Mathematical Models

### *7.3.1 Price Model*

In the DA market, the prosumer submits the hourly bid $P_t^{Bid}$ (i.e., quantity of power consumption or production) for the next trading day. If $P_t^{Bid} > 0$, it is a producer; if $P_t^{Bid} < 0$, it is a consumer. Because of the variations in power production and load, there will be a deviation between the bid and the forecasted power. The deviation $\Delta P_t^{\mathrm{Im}b}$ at time $t$ is:

$$\Delta P_t^{\mathrm{Im}b} = P_t^{Fct} - P_t^{Bid} \tag{7.1}$$

The forecasted power $P_t^{Fct}$ can be formulated as

$$P_t^{Fct} = \begin{pmatrix} \sum_{i \in \Omega_{GW}} P_{GWit} + \sum_{i \in \Omega_{BESS}} P_{it}^{BESS} + \sum_{i \in \Omega_{DG}} P_{DGit} \\ -P_{Dt} - P_{Dt}^{Elec,heat} + \sum_{i \in \Omega_D} P_{it}^{IL} \end{pmatrix} \tag{7.2}$$

where $P_{GWit}$, $P_{it}^{BESS}$, $P_{DGit}$ denote outputs of wind power, BESS and DG at time $t$ bus $i$; $P_{Dt}$ and $P_{Dt}^{Elec,heat}$ denote non-thermal power load (e.g., lighting) and electric thermal load, respectively; $P_{it}^{IL}$ denotes interrupted load (IL) (i.e., incentive-based DR); and $\Omega_{GW}$, $\Omega_{BESS}$, $\Omega_{DG}$ and $\Omega_D$ denote sets of wind power, BESS, DG and energy demand respectively.

Thus, the imbalance cost $C_t^{\mathrm{Im}b}$ is calculated as

$$C_t^{\mathrm{Im}b} = \begin{cases} \rho_t^{R+} \cdot \Delta P_t^{\mathrm{Im}b}, & \Delta P_t^{\mathrm{Im}b} < 0 \\ \rho_t^{R-} \cdot \Delta P_t^{\mathrm{Im}b}, & \Delta P_t^{\mathrm{Im}b} \geq 0 \end{cases} \tag{7.3}$$

where $\rho_t^{R+}$, $\rho_t^{R-}$ denote regulation prices for purchasing (up-regulation) and selling (down-regulation) electricity in the RT balancing market at time $t$, respectively.

The regulation price can be calculated as a proportion of the DA market price $\rho_t^{DA}$, since the DA and RT prices are usually correlated [9]:

$$\begin{cases} \rho_t^{R+} = \left(1 + \zeta_t^+\right) \cdot \rho_t^{DA} \\ \rho_t^{R-} = \left(1 - \zeta_t^-\right) \cdot \rho_t^{DA} \end{cases} \tag{7.4}$$

where $\zeta_t^+$ and $\zeta_t^-$ denote the relative differences between the DA price and the up and down-regulation prices.

Therefore, market price uncertainty can be modelled by a random vector, including random DA price and random differences between DA up and down-regulation prices. The distributions of market prices can be obtained by time-series forecasting techniques, such as autoregressive integrated moving average (ARIMA) or artificial neural network (ANN) models, which have been well addressed in [29, 30].

## 7.3.2 Gas System Models

(a) *Linepack model*

Linepack is the pressurized gas stored in pipelines throughout the gas networks [31, 32]. It is proportional to the average pressure in a pipe [33]; increasing the average pressure of a pipe can increase linepack. Gas in pipes can be described by four variables: gas pressure $p$ (kPa), volume $H$ (m$^3$), density $\psi$ (kg/m$^3$) and temperature $\Gamma$(K). According to Boyle's law, gas variables are expressed as [34]:

$$\frac{p_0}{\psi_0\Gamma_0} = \frac{p}{\psi\Gamma} = Z^{Gas}R^{Gas} = \text{Constant} \tag{7.5}$$

$$p^{Aver}H_{ij} = p_0H_0 \tag{7.6}$$

where $R^{Gas}$ and $Z^{Gas}$ are the gas constant and the gas compressibility factor, respectively; and $p_0$, $H_0$, $\psi_0$ and $\Gamma_0$ are gas pressure, volume, density and temperature under normal conditions, respectively. In a pipe, gas volume is equal to the pipe volume capacity: i.e., $H_{ij} = \pi D_{ij}^2 L_{ij}/4$ and $D_{ij}, L_{ij}$ are the internal diameter and length of pipe $i-j$. The steady-state average pressure of a pipe can be expressed as $p_{ij}^{Aver} = \sqrt{1/2\left(p_i^2 + p_j^2\right)}$.

In a steady state, the initial linepack measured in energy (GJ) (denoted by $\Pi_{ij}^{Initial}$) under normal conditions is [34]:

$$\Pi_{ij}^{Initial} = \Lambda \cdot p_{ij}^{Aver}H_{ij}/\psi_0\Gamma_0Z^{Gas}R^{Gas} \tag{7.7}$$

where $\Lambda$ denotes the constant that converts gas volume under the normal condition to energy (GJ/m$^3$).

In dynamic situations, based on the law of conservation of mass, linepack changes with the initial gas stored in pipes, and the net difference between the supplied and consumed gas in a pipe [33]. Gas suppliers include gas producer in gas wells and reserves, liquefied natural gas regasification terminals and gas storages. Gas consumers are compressor loads, non-electric gas loads, DG gas loads and gas storages [35]. Dynamic linepack is described by [34]. Note that other gas storages can be modelled by

$$\Pi_{ij,t+1} = \Pi_{ij,t} + P_{ij,t}^{Gas}\Delta t \tag{7.8}$$

$$P_{ij,t}^{Gas} = P_{Gijt}^{Gas} - P_{Dijt}^{Gas} - P_{ijt}^{Gas,Comp} \tag{7.9}$$

where $P_{Gijt}^{Gas}$, $P_{Dijt}^{Gas}$ and $P_{ijt}^{Gas,Comp}$ denote supplied gas, gas demands, and gas consumed by gas compressors between $i-j$, respectively, and $\Delta t$ denotes a factor that converts power to energy, i.e., time duration.

(b) *Flow Equation*

Gas flows in a steady state can be modeled by Weymouth's formula. Gas flows, $S_{ij}^{gas}$, and directions, $\text{sgn}_{ij}$, are dependent on nodal pressure differences. The steady-state pipeline flow equation for a horizontal pipeline is given as [36]

$$S_{ij}^{Gas} = \text{sgn}_{ij}K\frac{\Gamma_0}{p_0}\sqrt{\text{sgn}_{ij}\frac{\left(p_i^2 - p_j^2\right)D_{ij}^5}{F_{ij}GL_{ij}\Gamma Z^{Gas}}} \tag{7.10}$$

$$K = \sqrt{\pi^2 R^{Air}/64} = 3.2387 \tag{7.11}$$

$$\text{sgn}_{ij} = \begin{cases} +1 & \text{if } p_i - p_j > 0 \\ -1 & \text{if } p_i - p_j < 0 \end{cases} \tag{7.12}$$

where $K$ is an air constant in relation to $R^{Air}$; $G$ denotes the specific gravity ratio as $R_{air}/R_{gas}$ (usually $R^{Air} = 1$; $R^{Gas} = 0.6$) [36]; $p_i, p_j$ is the pressure at nodes $i$ or $j$; and $F_{ij}$ is the dimensionless friction factor depending on $D_{ij}$.

(c) *Compressor Station Equation*

The energy consumed by a compressor depends on the amount of gas flows, and the difference between outlet and inlet gas pressures, as shown in empirical equation of [33]

$$S_{ij}^{Gas,Comp} = \text{sgn}(p_i, p_j) \cdot \frac{HP_{ij}}{\gamma_{1,ij} - \gamma_{2,ij} \cdot \left[\frac{max(p_i, p_j)}{min(p_i, p_j)}\right]^{\gamma_{3,ij}}} \tag{7.13}$$

$$P_i^{Gas,Comp}(HP_{ij}) = a_{3,ij}^{Comp} + a_{2,ij}^{Comp} \cdot HP_{ij} + a_{1,ij}^{Comp} \cdot HP_{ij}^2 \tag{7.14}$$

where $S_{ij}^{Gas,Comp}$ denotes gas flow in compressor between $i - j$; $HP_{ij}$ denotes horse power for gas compressors; $\gamma_{1,ij}$, $\gamma_{2,ij}$ and $\gamma_{3,ij}$ denote coefficients for the compressor between $i - j$; $a_{1,ij}^{Comp}$, $a_{2,ij}^{Comp}$, $a_{3,ij}^{Comp}$ are horse power coefficients of the gas compressor; and $P_i^{gas,Comp}$ is the gas consumption of the compressor. The compressor pressure ratio $CPR_{ij}^{Comp}$ is also bounded as

$$\underline{CPR}_{ij}^{Comp} \le \frac{max(p_i, p_j)}{min(p_i, p_j)} \le \overline{CPR}_{ij}^{Comp} \tag{7.15}$$

### 7.3.3 Models of Other Uncertainties

In this chapter, power and thermal demands, electricity and gas market prices, and wind speed are considered as uncertainties. The historical data in the DA market is used as correlated scenarios; hence, the correlated probability distributions can be estimated based on the statistical correlations among these uncertainties. More detailed information can be found in [9], which comprehensively discusses a variety of time-series-based methods to generate correlated scenarios.

## 7.4 Optimal Scheduling for Prosumers

### *7.4.1 Upper Level Day-Ahead Scheduling*

In the DA market, the prosumer optimizes its bidding strategy (e.g., hourly based) to maximize the total profit over the scheduling horizon. The decisions made in this stage may affect its operational strategies in the second stage. Thus, it is necessary to consider the decision variables pertaining to the RT market while optimizing the bidding strategy in the DA market. However, the second-stage decision variables are not actually implemented in the first stage. The optimal scheduling model in the DA market is formulated as

$$\max f_1 = \left\{ \begin{array}{l} \sum_{t=1}^{T} \rho_t^{DA} P_t^{Bid} \Delta t + \sum_{t=1}^{T} \rho_t^{\mathrm{Retail}} \left( P_{Dt} + P_{Dt}^{Elec,heat} - \sum_{i\in\Omega_D} P_{it}^{IL} \right) \Delta t \\ + \sum_{t=1}^{T} \sum_{i\in\Omega_D} R_{it}^{Heat} - \sum_{t=1}^{T} C_t^{\mathrm{Imb}} - \sum_{t=1}^{T} \sum_{i\in\Omega_D} C_{it}^{IL} - \sum_{t=1}^{T} \sum_{i\in\Omega_{BESS}} C_{it}^{BESS} \\ - \sum_{t=1}^{T} \sum_{i\in\Omega_{DG}} \left( C_{it}^{DG} \chi_{it}^{DG} + SUC_{it}^{DG} \chi_{it}^{SU} + SDC_{it}^{DG} \chi_{it}^{SD} \right) \\ - \sum_{t=1}^{T} \rho_t^{Gas} \cdot P_{Gt}^{Gas} \end{array} \right\} \tag{7.16}$$

where $\rho_t^{\mathrm{Retail}}$ denotes the retail electricity price for customers; $\rho_t^{Gas}$ denotes the gas pool price; $SUC_{it}^{DG}$ and $SDC_{it}^{DG}$ denote start-up (including cold and hot start-up) cost and shut-down cost, respectively; $T$ denotes the total DA scheduling horizon; and $\chi_{it}^{DG}$, $\chi_{it}^{SU}$ and $\chi_{it}^{SD}$ are binary variables denoting commitment status, start-up and shut-down decisions for DG at time $t$ bus $i$, respectively. In (7.16), the first term represents the cost (revenue) by purchasing (selling) electricity from (to) the upper stream networks. The second term represents the revenue by selling electricity to customers, while the third term represents the revenue by selling gas for thermal demand. The remaining terms in (7.16) represent costs of power imbalance, IL, BESS, DG and gas purchase.

The cost of IL $C_{it}^{IL}$ is assumed to be a function of adjusted load $P_{it}^{IL}$ at time $t$ bus $i$, and is modelled by a quadratic polynomial function as [11]

$$C_{it}^{IL} = a_{1i}^{IL} \cdot \left(P_{it}^{IL}\right)^2 + a_{2i}^{IL} \cdot P_{it}^{IL} \tag{7.17}$$

where $a_{1i}^{IL}$ and $a_{2i}^{IL}$ denote cost coefficients of IL. Note that negative IL (i.e., load increase) can be used to balance underestimated renewable energy outputs.

The operational cost of BESS $C_{it}^{BESS}$ at time $t$ bus $i$ generally refers to maintenance cost [37], which can be modelled by a linear function as

$$C_{it}^{BESS} = \beta_i^{BESS} \cdot P_{it}^{BESS} \Delta t + \beta_i^{BESS} E_{it}^{BESS} \eta^L \Delta t \tag{7.18}$$

where $P_{it}^{BESS}$ denote the charged or discharged BESS power at time $t$ bus $i$; $E_{it}^{BESS}$ denotes energy stored in BESS; $\eta^L$ denote leakage loss factor of BESS; and $\beta_i^{BESS}$ is the cost coefficient of the BESS lifetime degradation, which is calculated as [38]:

$$\beta_{it}^{BESS} = \frac{IC^{BESS}}{E_R^{BESS} \cdot (LCN)} \tag{7.19}$$

where $IC^{BESS}$, $E_R^{BESS}$ and $LCN$ denote investment cost, rated energy capacity and total life cycle number of BESS, respectively.

The revenue of selling gas for thermal $R_{it}^{Heat}$ is formulated as

$$R_{it}^{Heat} = \rho_t^{Gas,retail} \cdot P_{Dit}^{Gas,heat} \tag{7.20}$$

where $P_{Dit}^{Gas,heat}$ denotes gas thermal demand (GJ/hour), and $\rho_t^{Gas,retail}$ denotes gas retail price at time $t$($/GJ).

Here DG is only termed as distributed gas-fired units. The DG cost $C_{it}^{DG}$ can be modelled by a quadratic function:

$$C_{it}^{DG} = a_{1i}^{DG} \cdot P_{DGit}^2 + a_{2i}^{DG} \cdot P_{DGit} + a_{3i}^{DG} \tag{7.21}$$

where $a_{1i}^{DG}$, $a_{2i}^{DG}$ and $a_{3i}^{DG}$ denote cost coefficients of DG at bus $i$, which are dependent on gas price and the efficiency of the gas generator; and $P_{DGit}$ denotes active power output at bus $i$ time $t$.

Given the uncertainties in energy markets, (7.16) should be formulated as a probabilistic version. To improve computational efficiency, the initial set of scenarios is reduced to several representative scenarios using an appropriate technique, such as importance sampling, used in [39]. The risk associated with the profit variability is explicitly captured by the model through incorporating the CVaR metric:

$$\max f_2 = \sum_{k \in \Omega_K} \Pr_k \cdot f_{1k} + \varpi \left( VaR - \frac{1}{1-\alpha} \right) \sum_{k \in \Omega_K} \Pr_k \cdot \Phi_k \tag{7.22}$$

where $f_2$ denotes the objective function with the CVaR metric; $\Pr_k$ denotes the probability of scenario $k$; $\Omega_K$ denotes the set of all scenarios; $\varpi \in [0, +\infty)$ is a weighting factor for the incorporation of risk in the objective function [40]; $VaR$ denotes value-at-risk; $\alpha \in (0, 1)$ is the confidence level denoting the risk preference; and $\Phi_k$ denotes an auxiliary, continuous, non-negative variable defined as the maximum between zero and the difference between the VaR minus the profit of each scenario realization [30]. The higher the value of $\varpi$, the more risk averse the scenario.

For each scenario, the complete constraints of (7.22) are given below. To avoid repeated information, the subscript $k$ is removed for variables below.

(1) *Supply-demand balance constraint*

$$\sum_{i\in\Omega_{GW}} P_{GWit} + \sum_{i\in\Omega_{BESS}} P_{it}^{BESS} + \sum_{i\in\Omega_{DG}} P_{DGit} = \left( \begin{array}{l} P_{Dt} + \sum\limits_{i\in\Omega_D} P_{Dit}^{Elec,heat} \\ - \sum\limits_{i\in\Omega_D} P_{it}^{IL} + P_t^{Bid} + P_t^{Loss} \end{array} \right), \forall t \in 1:T \tag{7.23}$$

where $P_t^{Loss}$ denote power loss at time $t$.

(2) *Wind power constraint*

$$\begin{cases} 0 \le P_{GWit} \le \overline{P}_{GWi};\ \forall i \in \Omega_{GW} \\ P_{GWit}/\sqrt{P_{GWit}^2 + Q_{GWit}^2} = \text{constant} \end{cases}, \forall t \in 1:T \tag{7.24}$$

where $\overline{(\bullet)}$ denotes the upper limit, and $Q_{GWit}$ denotes reactive power output of wind at time t bus i.

(3) *DG constraint*

$$\chi_{it}^{DG} \underline{P}_{DGi} \le P_{DGit} \le \chi_{it}^{DG} \overline{P}_{DGi}, \forall t \in 1:T, \forall i \in \Omega_{DG} \tag{7.25}$$

$$\overline{P}_{DGi} \le P_{Dit}^{Gas,DG} \cdot HR^{Gas} \cdot \Delta t \tag{7.26}$$

$$\begin{cases} P_{DGi,t} - P_{DGi,t-1} \le Ramp_{DGi}^{Up}, \text{ if } P_{DGi,t} \ge P_{DGi,t-1} \\ P_{DGi,t-1} - P_{DGi,t} \le Ramp_{DGi}^{Down}, \text{ if } P_{DGi,t-1} \ge P_{DGi,t} \end{cases} \tag{7.27}$$

$$\begin{cases} \left[T_{i,t-1}^{On} - MUT_i\right] \cdot \left[\chi_{i,t-1}^{DG} - \chi_{it}^{DG}\right] \ge 0 \\ \left[T_{i,t-1}^{Off} - MDT_i\right] \cdot \left[\chi_{i,t-1}^{DG} - \chi_{it}^{DG}\right] \ge 0 \end{cases} \tag{7.28}$$

$$\begin{cases} \chi_{i,t}^{DG} - \chi_{i,t-1}^{DG} \le \chi_{it}^{SU} \\ \chi_{i,t-1}^{DG} - \chi_{i,t}^{DG} \le \chi_{it}^{SD} \\ \chi_{i,t}^{DG} - \chi_{i,t-1}^{DG} \le \chi_{it}^{SU} - \chi_{it}^{SD} \end{cases} \tag{7.29}$$

where $Ramp_{DGi}^{Up}$, $Ramp_{DGi}^{Down}$ denote ramping up and ramping down limits of DG at bus $i$; $P_{Dit}^{Gas,DG}$ denotes DG gas demand (GJ/hour); $HR^{Gas}$ denotes gas heat rate

(MW/GJ); $\underline{(\bullet)}$ denotes the lower limit; $T^{On}_{i,t-1}$ and $T^{Off}_{i,t-1}$ denote number of hours for which DG unit has been on or off at time $t$ bus $i$; and $MUT_i$ and $MDT_i$ denote minimum up and down time limits of DG in hours at bus $i$. Equation (7.26) states the outputs of DG are subject to gas availability in the gas network. Note that the energy efficiency of DG has been implicitly taken into account in the heat rate.

(4) *Interruptible load constraint*

$$\left|P^{IL}_{it}\right| \leq \overline{P}^{IL}_{it}, \forall t \in 1:T, \forall i \in \Omega_D \tag{7.30}$$

$$\begin{cases} P_{Dit} - P^{IL}_{it} \geq \underline{P}_{Dit}, & if\ P^{IL}_{it} \geq 0 \\ P_{Dit} - P^{IL}_{it} \leq \overline{P}_{Dit}, & if\ P^{IL}_{it} < 0 \end{cases} \tag{7.31}$$

Equations (7.30) and (7.31) state the maximum amount of IL and ensure that IL is within the upper and lower bounds.

(5) *BESS constraint*

$$E^{BESS}_{i,t+1} = E^{BESS}_{i,t} - P^{BESS}_{it}\Delta t - \left|P^{BESS}_{it}\right|\eta^C \Delta t - E^{BESS}_{i,t}\eta^L \Delta t, \forall t \in 1:T, \forall i \in \Omega_{BESS} \tag{7.32}$$

$$SOC_{it} = E^{BESS}_{it} / E^{BESS}_{R} \tag{7.33}$$

$$\underline{SOC}_{it} \leq SOC_{it} \leq \overline{SOC}_{it} \tag{7.34}$$

$$-\overline{P}^{BESS,Chr}_{i} \leq P^{BESS}_{it} \leq \overline{P}^{BESS,Dis}_{i} \tag{7.35}$$

$$\overline{P}^{BESS,Chr}_{i} \geq 0; \overline{P}^{BESS,Dis}_{i} \geq 0 \tag{7.36}$$

$$E^{BESS}_{i,0} = E^{BESS}_{i,Initial}; E^{BESS}_{i,T} \geq E^{BESS}_{i,End} \tag{7.37}$$

where $\eta^C$ denotes charging/discharging loss factor, and $\overline{P}^{BESS,Dis}_{i}$, $\overline{P}^{BESS,Chr}_{i}$ denote the maximum discharge and charge power limits of BESS. Equation (7.32) states the energy balance of BESS, which includes net energy difference, energy losses during charging or discharging, and leakage loss. Equations (7.33)–(7.36) represent state-of-charge constraints, and maximum charging or discharging power capacity [38]; and (7.37) defines the initial and final energy stored.

(6) *Electricity network constraint*

$$P_{it}(\theta_t, V_t) - P_{Git} + P_{Dit} + P_{Dit}^{Elec,heat} = 0, \forall t \in 1:T, \forall i \in \Omega_N \tag{7.38}$$

$$Q_{it}(\theta_t, V_t) - Q_{Git} + Q_{Dit} = 0, \forall t \in 1:T, \forall i \in \Omega_N \tag{7.39}$$

$$S_{ijt}(\theta_t, V_t) \leq \overline{S}_{ij}, \forall t \in 1:T, \forall i,j \in \Omega_N \tag{7.40}$$

$$\underline{V}_i \leq V_{it} \leq \overline{V}_i, \forall t \in 1:T, \forall i \in \Omega_N \tag{7.41}$$

where $P_{it}(\theta_t, V_t)$, $Q_{it}(\theta_t, V_t)$ denote active and reactive power injection time $t$ bus $i$; $P_{Git}$ and $Q_{Git}$ denote active and reactive power production, while $P_{Dit}$ and $Q_{Dit}$ denote active and reactive power demand; and $S_{ijt}(\theta_t, V_t)$ denotes complex power flow between $i-j$ at time $t$, with voltage angle $\theta_t$ and amplitude $V_t$.

(7) *Gas network constraint*

$$P_{Git}^{Gas} + \sum S_{ijt}^{Gas} = \sum S_{jit}^{Gas} + P_{Dit}^{Gas} + P_{it}^{Gas,Comp}, \forall i \in \Omega_N \tag{7.42}$$

$$P_{Dit}^{Gas} = P_{Dit}^{Gas,DG} + P_{Dit}^{Gas,heat}, \forall t \in 1:T, \forall i \in \Omega_N \tag{7.43}$$

$$P_{Dit}^{Heat} = P_{Dit}^{Gas,heat} + P_{Dit}^{Elec,heat}/HR^{Gas} \tag{7.44}$$

$$P_{Dit}^{Elec,heat}/P_{Dit}^{Gas,heat} \leq \phi, \forall t \in 1:T, \forall i \in \Omega_N \tag{7.45}$$

$$\underline{\Pi}_{ijt} \leq \Pi_{ijt} \leq \overline{\Pi}_{ijt} \tag{7.46}$$

$$\Pi_0 = \sum \Pi_{ij,Initial}; \sum \Pi_{ij,T} \geq \Pi_{End} \tag{7.47}$$

where $P_{Git}^{Gas}$ and $P_{Dit}^{Gas}$ denote supplied gas and gas demand at time $t$ bus $i$. Equation (7.42) states gas nodal balance; (7.43) states gas demand that includes DG gas and gas thermal demands; (7.44) states thermal demands can be satisfied by power and gas; (7.45) states the ratio of electric and gas thermal demand is bounded by the dual fuel ratio [41]; (7.46) states linepack is within the upper and lower limits; and (7.47) states the initial and final linepack of the day.

(8) *Interconnection exchange constraint*

$$\underline{P}^{Exch} \leq P_t^{Bid} \leq \overline{P}^{Exch}, \forall t \in 1:T \tag{7.48}$$

Equation (7.48) states the lower and upper limits of interconnection exchange between the prosumer and the upstream grid.

(9) *Steady security reserve constraint*

$$\begin{aligned}&\sum_{i\in\Omega_{DG}}\overline{P}_{DGi}\chi_{it}^{DG}-\sum_{i\in\Omega_{DG}}P_{DGit}\chi_{it}^{DG}+\sum_{i\in\Omega_{GW}}P_{GWit}\\&+\sum_{i\in\Omega_{BESS}}P_{it}^{BESS}+\sum_{i\in\Omega_{D}}\left(\overline{P}_{it}^{IL}-P_{it}^{IL}\right)\\&+\left(\overline{P}_{t}^{Exch}-P_{t}^{Bid}\right)\geq P_{Dt}+P_{Dt}^{Elec,heat}-\sum_{i\in\Omega_{D}}P_{it}^{IL}+Rsv(t)\end{aligned}\tag{7.49}$$

Equation (7.49) states the system reserve requirement *Rsv(t)* for static security and adequacy.

(10) *Conditional value-at-risk constraint*

$$VaR-f_1\leq\Phi\tag{7.50}$$

$$\Phi\geq0\tag{7.51}$$

Equations (7.50) and (7.51) state the constraint of CVaR at the $\alpha$-confidence level, i.e., $\alpha$ − VaR [40].

### 7.4.2 Lower-Level Real-Time Balancing

RT scheduling maintains the power balance. The frequency of RT clearing is very high (e.g., every 5 min). In this chapter, the time-series forecasting techniques, ARIMA, is used to capture the persistent behavior of uncertainties [30].

The expression of the ARIMA (n, m) model is as [30]

$$\begin{aligned}y_t&=\sum_{i=1}^{n}\ell_{1i}y_{t-1}+\lambda_t-\sum_{j=1}^{m}\ell_{2j}\lambda_{t-j}\\x_t&=\mu_t+\sigma_t y_t\end{aligned}\tag{7.52}$$

where $y_t$ is the time series value at time $t$; $\ell_{1i},\ell_{2j}$ are the auto-regressive and moving average parameters, respectively; $\{\lambda_t\}$ is a normal white noise process with zero mean and variance (a normal distribution); $x_t$ is simulated correlated uncertainties at time $t$, which can be obtained from the mean $\mu_t$ and standard deviation $\sigma_t$ multiplying the time series $y_t$.

The objective is to minimize the imbalance cost.

$$\min f_3 = \sum_{t=1}^{NT} \hat{C}_t^{\mathrm{Imb}} \tag{7.53}$$

where $NT$ denotes total number of scheduling intervals in the RT market, and $\hat{C}_t^{\mathrm{Imb}}$ denotes imbalance cost due to the deviation between the contracted commitment in the upper level and the actual consumption (production) in the lower level.

Note that the symbol $(\hat{\bullet})$ is used to distinguish the variables in the lower level from the variables in the upper level. The detailed calculations of $\hat{C}_t^{\mathrm{Imb}}$ are given in (7.54)–(7.56).

$$\Delta \hat{P}_t^{\mathrm{Imb}} = \hat{P}_t^{RT} - P_t^{Bid} \tag{7.54}$$

$$\hat{P}_t^{RT} = \left( \begin{array}{l} \sum\limits_{i \in \Omega_{GW}} \hat{P}_{GWit} + \sum\limits_{i \in \Omega_{BESS}} \hat{P}_{it}^{BESS} \\ + \sum\limits_{i \in \Omega_{DG}} \hat{P}_{DGit} \end{array} \right) - \left( \begin{array}{l} \hat{P}_{Dt} + \hat{P}_{Dt}^{Elec,heat} \\ + \hat{P}_t^{Loss} - \sum\limits_{i \in \Omega_D} \hat{P}_{it}^{IL} \end{array} \right) \tag{7.55}$$

$$\hat{C}_t^{\mathrm{Imb}} = \begin{cases} \rho_t^{R+} \cdot \Delta \hat{P}_t^{\mathrm{Imb}}, & \Delta P_t^{\mathrm{Imb}} < 0 \\ \rho_t^{R-} \cdot \Delta \hat{P}_t^{\mathrm{Imb}}, & \Delta P_t^{\mathrm{Imb}} \geq 0 \end{cases} \tag{7.56}$$

The complete constraints of (7.53) include (7.23)–(7.27), (7.30)–(7.35) and (7.38)–(7.47).

## 7.5 Solution Algorithms

### *7.5.1 Solution to Day-Ahead Prosumer Scheduling*

The formulated hourly prosumer scheduling in the upper level is a mixed-integer nonlinear programming problem. The model is a here-and-now decision-making process. The decision variables in (7.22) include $P_t^{Bid}$, $P_{it}^{IL}$, $P_{it}^{BESS}$, $P_{Dit}^{Elec,heat}$, $P_{DGit}$, $P_{Gt}^{Gas}$, $\chi_{it}^{DG}$, $\chi_{it}^{SU}$, $\chi_{it}^{SD}$. The enhanced particle swarm optimization (EPSO) algorithm is employed to solve the upper level optimization model [42]. The key operators of EPSO are briefly introduced below.

**Step 0: Input data.**
Input data, such as network, wind speed, load, price.
**Step 1: Initialization.**
Initialize the population within their value limits by employing a uniformly distributed random vector.

**Step 2: Start constraint check and fitness calculation.**

**Step 2.1**: Adjust the elements of the individual to satisfy the operational requirements. The ON/OFF commitment constraint handling algorithm in [43] is applied to adjust the binary states of DG and BESS to satisfy constraints in (7.28)–(7.29) and (7.35)–(7.36). Based on the given values of DG output $P_{DGit}$, BESS output $P_{it}^{BESS}$, electric thermal demand $P_{Dit}^{Elec,heat}$ and IL amount $P_{it}^{IL}$, the power flow function [2] is run to calculate the power losses, the value of exchanged power with the upstream power grid and the gas purchased from the gas pool market. The exchanged power is equal to the bid $P_{t}^{Bid}$ at time $t$. Other constraints, such as the network security and adequacy constraints in (7.38)–(7.41), bid constraint in (7.48) and reserve constraint in (7.49) are also checked, based on running the power flow function. After obtaining the power flow results for all intervals, the risk constraints in (7.50)–(7.51) are checked.
**Step 2.2**: If there is any violation of the constraints, a variable penalty is assigned. The variable penalty is defined as a function of the distance from the feasible area [21]. The summation of absolute distances of violated constraints is scaled by a large penalty factor and is combined with the objective to constitute the fitness value [11].

**Step 3: Individual selection and global best update.**
Find and update the best individual of the current generation. This step involves comparing individuals with each other and selecting the best one whose fitness value is the highest.
**Step 4: Offspring generation.**

**Step 4.1**: Produce the offspring generation according to the evolution rule in EPSO. Typical evolution operations include mutation, recombination and crossover.
**Step 4.2**: The two particles with the lowest and highest fitness values are selected. Using these two particles as the start values, apply the IP method to generate two local optima, and add them to the offspring generation [42].

**Step 5: Check the termination criterion.**
If the termination criterion is satisfied, stop the computation and export the best individual as the final solution. Otherwise, repeat the evolutionary algorithm from Step 2.

## *7.5.2 Solution to Real-Time Prosumer Scheduling*

After solving the optimal bidding model in the upper level, the ON/OFF status of DG and BESS are determined. The prosumer needs to adjust DERs in real time to compensate for the energy deviations. The lower-level model is a nonlinear programming problem with continuous variables. Variables such as $\hat{P}_{it}^{IL}$, $\hat{P}_{it}^{BESS}$ and

$\hat{P}_{DGit}$ are wait-and-see decisions made after the actual realization of the stochastic processes. The commercial solver AMPL/IPOPT 3.8.0 is used to solve the formulated lower-level problem.

## 7.6 Simulation Results

The presented scheduling approach is tested on an 18-bus radial electric and gas distribution network. As seen in Fig. 7.2, the electricity and gas networks are coupled at each bus. There are three wind turbines (WT), three BESS, three gas-fired DG units, one gas compressor (GC) and one gas storage (GS). The total generation capacity is 1500 kW, including 600 kW of wind power and 900 kW of DG. It is assumed that IL is located at every load bus, and up to 10% of the system load can be adjusted if necessary. For simplicity, the resistance and reactance of power lines are 0.01 and 0.04 p.u., respectively, and the distance between each bus is 10 km. For gas pipelines, the diameter is 660 mm and the safety pressure is 140 kPa. The capacity of power interconnection between the VPP and the main grid is set at 800 kW. The power curve of a Vestas V27 wind turbine is used [37]. The rated, cut-in and cut-out wind speeds are 14.0, 3.5 and 25.0 m/s, respectively. Also, the wind power output is calculated using the power-speed curve as

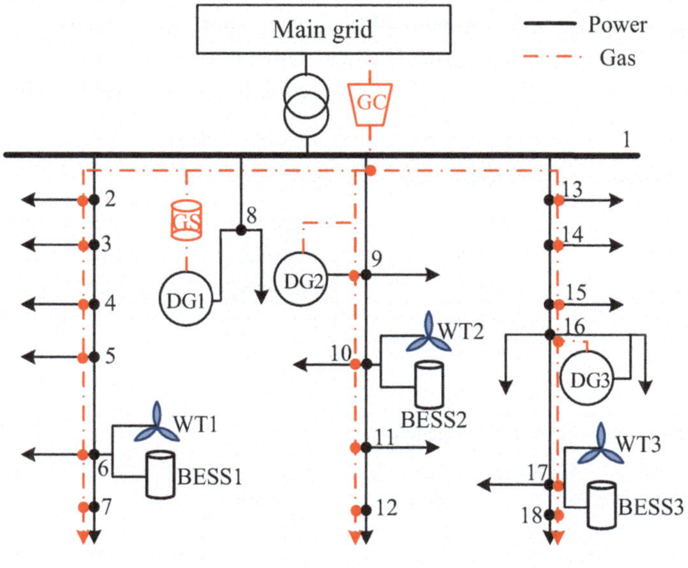


**Fig. 7.2** One-line diagram of 18-bus coupled electric and gas distribution networks

$$P_{GW} = \begin{cases} 0, & \upsilon \leq \upsilon_{In}, \upsilon \geq \upsilon_{Out} \\ \frac{\upsilon - \upsilon_{In}}{\upsilon_R - \upsilon_{In}} \overline{P}_{GW} & \upsilon_{In} \leq \upsilon \leq \upsilon_R \\ \overline{P}_{GW} & \upsilon_R \leq \upsilon \leq \upsilon_{Out} \end{cases} \tag{7.57}$$

where $P_{GW}$, $\upsilon$ denote wind power output and wind speed respectively; $\upsilon_{In}$, $\upsilon_{Out}$, $\upsilon_R$ denote cut-in, cut-out and rated wind speeds; $\overline{P}_{GW}$ denotes rated wind power.

The parameters of each BESS are shown in Table 7.1. The start-up and shut-down costs are assumed to be \$70 and \$20, respectively. Other DG parameters are shown in Table 7.2. The correlated scenarios of energy prices, wind speed, and electricity and gas demands are modeled based on the Australian Energy Market Operator historical data from 2015, which are publically available [28]. The historical wind power production data from Wattle Point wind farm (coordinates 35° 07′ 21″S, 137° 42′ 55″E) are used. The historical spot, up-regulation and down-regulation prices are used to capture the cross-correlation between the market prices. The simulations were completed by a PC with Intel Core i7-6600 CPU @ 2.80 GHZ with 8.00 GB RAM. In total 512 scenarios are generated after using the importance sampling scenario reduction technique.

Three cases are used to verify the presented scheduling approach. (i) **Case 1**: separated gas and electricity scheduling. (ii) **Case 2**: without the lower-level RT balancing problem. (iii) **Case 3**: the presented approach. The confidence level $\alpha$ is set at 0.95 and the weighting factor is set at 0.8. Thus, CVaR denotes the expected value of the 5% scenarios with lowest profit.

As seen in Fig. 7.3, linepack plays a critical role in balancing the deviations between gas supply and demand. When gas supply is greater than demand in off-peaks, linepack is replenished; when demand is greater than gas supply in on-peaks, linepack is depleted. For separated gas and electricity scheduling in

**Table 7.1** Parameters of each battery energy storage system

| | | | |
|---|---|---|---|
| Power capacity | 100 kW | Self-discharge | 3% per month |
| Energy capacity | 400 kWh | Life cycles | 1000 |
| Energy efficiency | 91.4% | Initial energy | 40 kWh |
| Investment cost | 800 $E_R^{BESS}$ | Final energy | 40 kWh |

**Table 7.2** Parameters of distributed generation

| # | Min. (kW) | Max. (kW) | Ramp up (kW/min) | Ramp down (kW/min) | MUT (h) | MDT (h) |
|---|---|---|---|---|---|---|
| 1 | 10 | 200 | 2 | 2.5 | 3 | 3 |
| 2 | 25 | 250 | 2.5 | 3 | 2 | 2 |
| 3 | 35 | 450 | 2 | 2 | 3 | 2.5 |

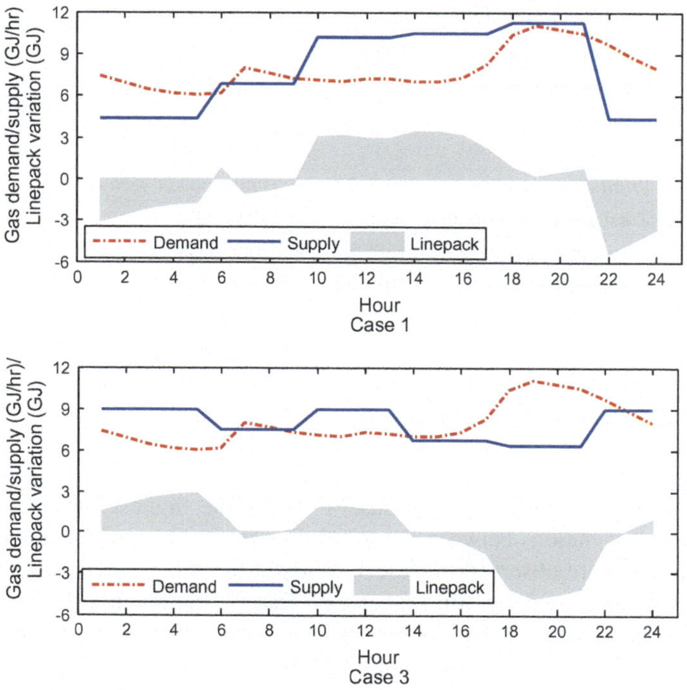


**Fig. 7.3** Gas supply, demand and linepack variations for Cases 1 and 3

Case 1, more gas is purchased at evening peaks (6–10 pm). In contrast, in Case 3, more gas is purchased from 1–5 am and 12–2 pm, when gas prices are relatively lower. In other words, linepack is more efficiently used in Case 3 to reduce gas price volatility.

The detailed financial indicators for Cases 1 and 3 are given in Table 7.3. Compared with Case 1, in Case 3 the revenue (Rev.) from selling gas to customers is higher, while the cost of purchasing gas is lower. Thus the profit to the gas sector for Case 3 is higher ($647.90 for Case 3 and $459.73 for Case 1). For the power sector, the revenue in the DA market is higher for Case 3. However, the revenues from selling electricity are very similar for Cases 1 and 3. The imbalance cost and the costs paid to DERs (e.g., IL, BESS and DG) are lower for Case 3. Therefore, the presented approach in Case 3 can achieve much higher profits in both the gas and power sectors. The total net profits for Case 1 and 3 are $1758.20 and $2565.56, respectively.

Figure 7.4 illustrates the imbalance costs and exchanged power between prosumers and the main grid for Cases 1–3. Since the separated scheduling of power

**Table 7.3** Detailed financial indicators for Cases 1 and 3 ($)

| | | | | |
|---|---|---|---|---|
| Case 1 | Revenue of gas | 1522.42 | Rev. in DA | 571.01 |
| | Gas purchase cost | −1065.69 | Rev. of electricity | 2984.47 |
| | | | Imbalance cost | −555.85 |
| | | | Cost of DERs | −1701.26 |
| | Profit gas sector | 459.73 | Profit power sector | 1298.47 |
| | Total net profit in both sectors | | 1758.20 | |
| Case 3 | Revenue of gas | 1658.82 | Rev. in DA | 895.72 |
| | Gas purchase cost | −1010.92 | Rev. of electricity | 2981.25 |
| | | | Imbalance cost | −290.05 |
| | | | Cost of DERs | −1669.52 |
| | Profit gas sector | 647.90 | Profit power sector | 1917.56 |
| | Total net profit in both sectors | | 2565.56 | |

and gas faces more uncertainties, Case 1 incurs the highest imbalance cost in general. More deviations between the scheduling in the DA and RT markets (solid and dotted lines) are observed in Case 1. Without the RT balancing in Sect. 7.4.2, the deviations in Case 2 are still higher than Case 3, particularly during the afternoon and evening hours. Table 7.4 shows the scheduling results for Cases 1–3. The expected profits in the DA market are $2.3140, $2.5493 and $2.8556 K for Cases 1–3, respectively. Case 3 incurs the lowest imbalance cost and trades the least energy in the RT market ($290.05 and 291.34 kWh), and hence obtains the highest total net profit. Meanwhile, the CVaR for Case 3 is the highest, followed by Case 2 and then Case 1. This implies that Case 3 incurs the lowest risk in the market. The presented approach is effective in hedging against the risk of profit variation. It is noteworthy that the amounts of purchased gas for Cases 1–3 are similar.

The compositions of gas and power demands for Case 3 are shown in Fig. 7.5. The electric thermal demand is flatter than the gas thermal demand, and it accounts for about 20–25% of the total power demand. The majority of purchased gas is used as DG fuel, particularly between 2 and 10 pm. More gas thermal demand is observed during the early morning (1–6 am). The load and energy profiles of DERs for Case 3 are shown in Fig. 7.6. Figure 7.6a shows that BESS is discharged to meet the morning and evening peaks, and hence BESS is almost empty at 9 am and 10 pm. In the early morning, prosumers rely on power imports from the grid (see the green dashed line at 4 am). During peak hours (8–10 am and 6–10 pm) the power surplus is sold back to the grid, since energy prices are relatively high in the

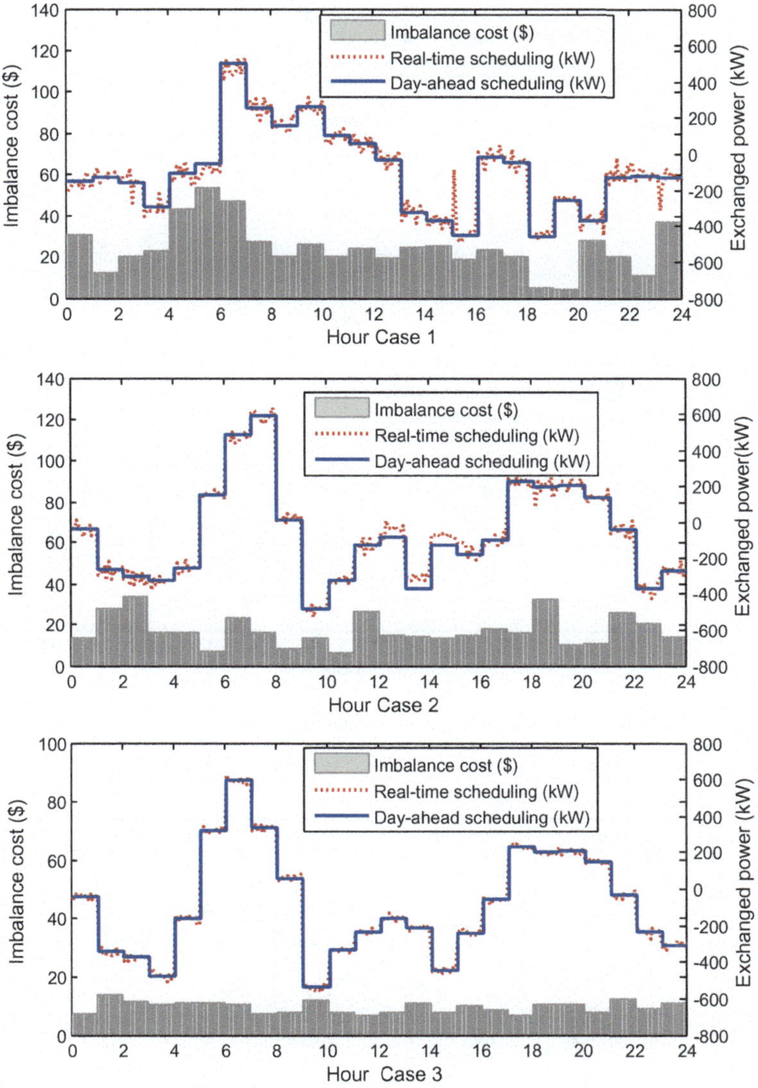


**Fig. 7.4** Imbalance costs and exchanged power for Cases 1–3

**Table 7.4** Results of Scheduling for Cases 1–3

| Case # | Expected profit (k$) | CVaR (k$) | Imbalance cost ($) | Gas purchased (GJ) | Energy traded in RT (kWh) | Net profit ($) |
|---|---|---|---|---|---|---|
| 1 | 2.3140 | 0.86 | –555.85 | 189.38 | 672.11 | 1758.20 |
| 2 | 2.5493 | 1.19 | –422.03 | 187.31 | 608.17 | 2127.28 |
| 3 | 2.8556 | 1.75 | –290.05 | 187.52 | 291.34 | 2565.56 |

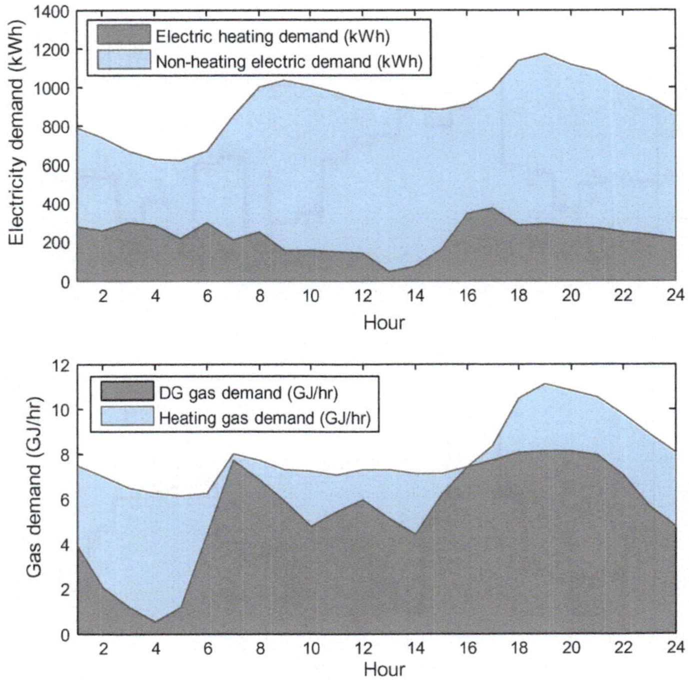


**Fig. 7.5** Composition of electricity and gas consumption

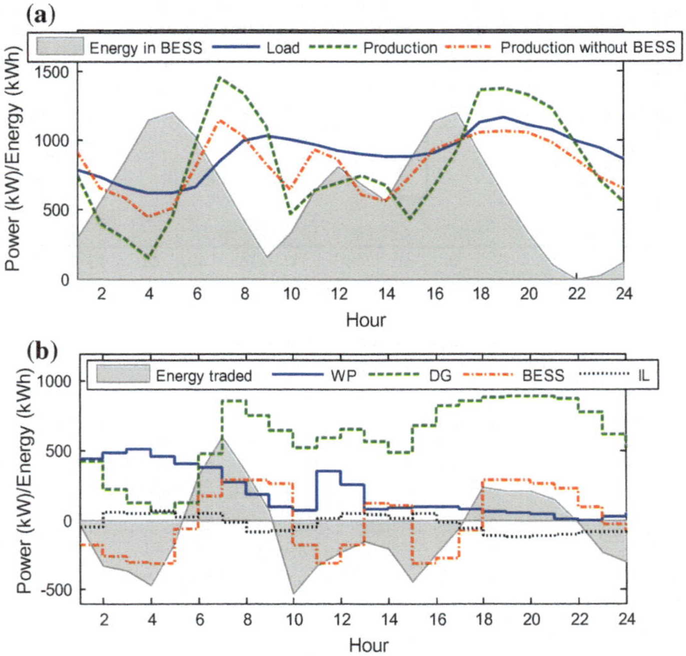


**Fig. 7.6** Load and productions files with and without BESS (**a**); energy traded and scheduling of distributed energy resources (**b**)

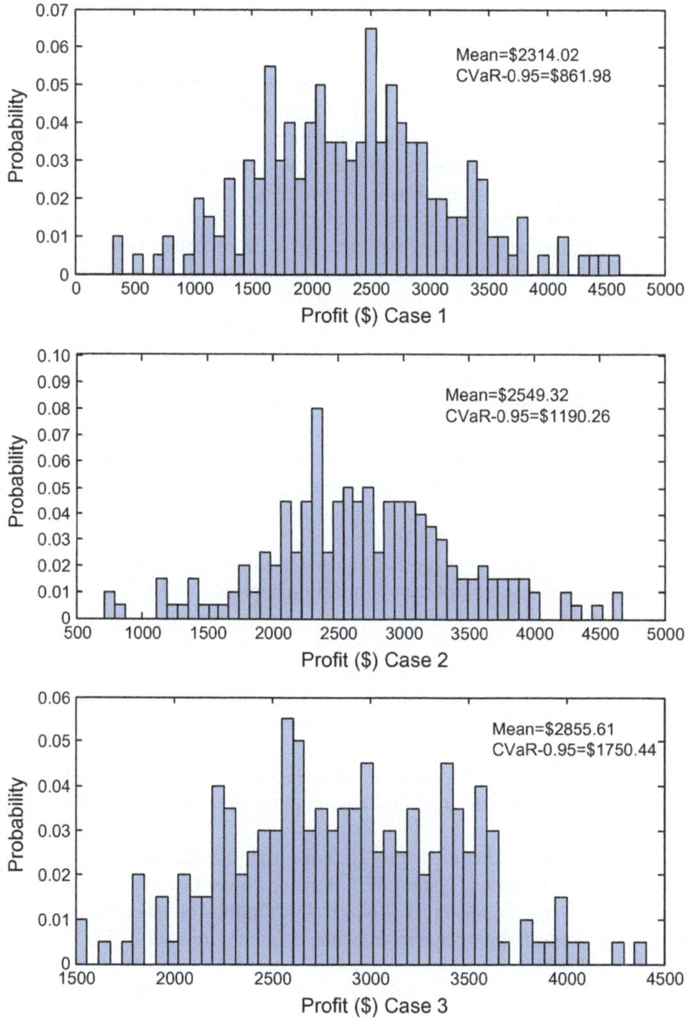


**Fig. 7.7** Probability distributions of profits in DA markets for Cases 1–3

markets and prosumers try to make a higher profit. Figure 7.6b shows that wind power (WP) generates more energy in the morning but not in the evening peaks. DG is used at almost full capacity in peak hours (nearly 900 kW). Load increment (negative IL) balances the sudden increase in WP outputs, while load decrement is mainly used in peak hours. Furthermore, the probability distributions of expected profits in DA markets for Cases 1–3 are shown in Fig. 7.7. It can be seen that Case 3 is superior, since the expected profit is the highest ($2855.61), while the risk is the lowest (CVaR is $1750.44). Note that a higher CVaR value means a lower risk of profit variation due to uncertainties.

# References


1. Z. Ding, W.J. Lee, J. Wang, Stochastic resource planning strategy to improve the efficiency of microgrid operation. IEEE Trans. Ind. Appl. **51**(3), 1978–1986 (2015)
2. S.M. Moghaddas-Tafreshi, E. Mashhour, Distributed generation modeling for power flow studies and a three-phase unbalanced power flow solution for radial distribution systems considering distributed generation. Electr. Power Syst. Res. **79**(4), 680–686 (2009)
3. M. Giuntoli, D. Poli, Optimized thermal and electrical scheduling of a large scale virtual power plant in the presence of energy storages. IEEE Trans. Smart Grid **4**(2), 942–955 (2013)
4. W. Su, J. Wang, J. Roh, Stochastic energy scheduling in microgrids with intermittent renewable energy resources. IEEE Trans. Smart Grid **5**(4), 1876–1883 (2014)
5. P. Asmus, Microgrids, virtual power plants and our distributed energy future. Electr. J. **23** (10), 72–82 (2010)
6. H. Pandzic, J.M. Morales, A.J. Conejo, I. Kuzle, Offering model for a virtual power plant based on stochastic programming. Appl. Energy **105**, 282–292 (2013)
7. M.A. Tajeddini, A. Rahimi-Kian, A. Soroudi, Risk averse optimal operation of a virtual power plant using two-stage stochastic programming. Energy **73**, 958–967 (2014)
8. M. Peik-Herfeh, H. Seifi, M.K. Sheikh-El-Eslami, Decision making of a virtual power plant under uncertainties for bidding in a day-ahead market using point estimate method. Electr. Power Syst. Res. **44**(1), 88–98 (2013)
9. M. Rahimiyan, L. Baringo, Strategic bidding for a virtual power plant in the day-ahead and real-time markets: a price-taker robust optimization approach. IEEE Trans. Power Syst. **31**(4), 2676–2687 (2016)
10. H. Nezamabadi, M.S. Nazar, Arbitrage strategy of virtual power plants in energy, spinning reserve and reactive power markets. IET Gen. Tran. Dis. **10**(3), 750–763 (2016)
11. E. Mashhour, S.M. Moghaddas-Tafreshi, Bidding strategy of virtual power plant for participating in energy and spinning reserve markets-part I: problem formulation. IEEE Trans. Power Syst. **26**(2), 949–956 (2011)
12. B. Zhao, Y. Shi, X. Dong, W. Luan, J. Bornemann, Short-term operation scheduling in renewable-powered microgrids: a duality-based approach. IEEE Trans. Sustain. Energy **5**(1), 209–217 (2014)
13. K. Kusakana, Optimal scheduled power flow for distributed photovoltaic/wind/diesel generators with battery storage system. IET Renew. Power Gen. **9**(8), 916–924 (2015)
14. W. Pei, Y. Du, W. Deng, K. Sheng, H. Xiao, H. Qu, Optimal bidding strategy and intramarket mechanism of microgrid aggregator in real-time balancing market. IEEE Trans. Ind. Inform. **12**(2), 587–596 (2016)
15. P. Karimyan, M. Abedi, S.H. Hosseinian, R. Khatami, Stochastic approach to represent distributed energy resources in the form of a virtual power plant in energy and reserve markets. IET Gen. Tran. Dis. **10**(8), 1792–1804 (2016)
16. W. Liu, Q. Wu, F. Wen, J. Ostergaard, Day-ahead congestion management in distribution systems through household demand response and distribution congestion prices. IEEE Trans. Smart Grid **5**(6), 2739–2747 (2014)
17. H. Karami, M.J. Sanjari, S.H. Hosseinian, G.B. Gharehpetian, An optimal dispatch algorithm for managing residential distributed energy resources. IEEE Trans. Smart Grid **5**(5), 2360–2367 (2014)
18. A.K. Basu, A. Bhattacharya, S. Chowdhury, S.P. Chowdhury, Planned scheduling for economic power sharing in a CHP-based micro-grid. IEEE Trans. Power Syst. **27**(1), 30–38 (2012)
19. M. Alipour, B. Mohammadi-Ivatloo, K. Zare, Stochastic scheduling of renewable and CHP-based microgrids. IEEE Trans. Ind. Inform. **11**(5), 1049–1058 (2015)
20. D.T. Nguyen, L.B. Le, Risk-constrained profit maximization for microgrid aggregators with demand response. IEEE Trans. Smart Grid **6**(1), 135–146 (2015)

21. X. Wu, X. Wang, C. Qu, A hierarchical framework for generation scheduling of microgrids. IEEE Trans. Power Deliv. **29**(6), 2448–2457 (2014)
22. D.T. Nguyen, L.B. Le, Optimal bidding strategy for microgrids considering renewable energy and building thermal dynamics. IEEE Trans. Smart Grid **5**(4), 1608–1620 (2014)
23. S.R. Dabbagh, M.K. Sheihk-EI-Eslami, Risk assessment of virtual power plants offering in energy and reserve markets. IEEE Trans. Power Syst. **31**(5), 3572–3582 (2016)
24. H. Geramifar, M. Shahabi, T. Barforoshi, Coordination of energy storage systems and DR resources for optimal scheduling of microgrids under uncertainties. IET Renew. Power Gener. **11**(2), 378–388 (2017)
25. J. Soares, M.A.F. Ghazvini, N. Borges, Z. Vale, A stochastic model for energy resources management considering demand response in smart grids. Electr. Power Syst. Res. **143**, 599–610 (2017)
26. E.G. Kardakos, C.K. Simoglou, A.G. Bakirtzis, Optimal offering strategy of a virtual power plant: a stochastic bi-level approach. IEEE Trans. Smart Grid **7**(2), 794–806 (2016)
27. S.R. Dabbagh, M.K. Sheihk-EI-Eslami, Risk-based profit allocation to DERs integrated with a virtual power plant using cooperative Game theory. Electr. Power Syst. Res. **121**, 368–378 (2015)
28. Australian Energy Market Operator (AEMO) [Online], Available: http://www.aemo.com.au/
29. A.J. Conejo, J. Contreras, R. Espinola, M.A. Plazas, Forecasting electricity prices for a day-ahead pool-based electric energy market. Int. J. Forecast. **21**(3), 435–462 (2005)
30. A.J. Conejo, M. Carrion, J.M. Morales, *Decision Making Under Uncertainty in Electricity Markets* (Springer, New York, NY, USA, 2010)
31. M. Shahidehpour, Y. Fu, T. Wiedman, Impact of natural gas infrastructure on electric power systems. Proc. IEEE **93**(5), 1042–1056 (2005)
32. L. Wu, M. Shahidehpour, Optimal coordination of stochastic hydro and natural gas supplies in midterm operation of power systems. IET Gen. Trans. Dist. **5**(5), 577–587 (2011)
33. A. Kabirian, M.R. Hemmati, A strategic planning model for natural gas transmission networks. Energy Policy **35**(11), 5656–5670 (2007)
34. M. Chaudry, N. Jenkins, G. Strbac, Multi-time period combined gas and electricity network optimisation. Elec. Power Syst. Res. **78**(7), 1265–1279 (2008)
35. J. Qiu, Z.Y. Dong, J.H. Zhao, K. Meng, Y. Zheng, D.J. Hill, Low carbon oriented expansion planning of integrated gas and power systems. IEEE Trans. Power Syst. **30**(2), 1035–1046 (2015)
36. S. An, Q. Li, T.W. Gedra, Natural gas and electricity optimal power flow, in *Proceedings IEEE PES Transmission and Distribution Conference and Exposition* (Dallas, TX, 2003), pp. 138–143
37. F.J. Luo, Z.Y. Dong, K. Meng, J. Qiu, J.J. Yang, K.P. Wong, Short-term operational planning framework for virtual power plants with high renewable penetrations. IET Renew. Power Gen. **10**(5), 623–633 (2016)
38. Y. Zheng, Z.Y. Dong, F.J. Luo, K. Meng, J. Qiu, K.P. Wong, Optimal allocation of energy storage system for risk mitigation of DISCOs with high renewable penetrations. IEEE Trans. Power Syst. **29**(1), 212–220 (2014)
39. J. Huang, Y. Xue, Z.Y. Dong, K.P. Wong, An efficient probabilistic assessment method for electricity market risk management. IEEE Trans. Power Syst. **PP**(99), 1–1 (2012)
40. K. Leong, R. Siddiqi, Value at risk for power markets, in *Energy Risk Management*, Ed. (New York: McGraw-Hill, 1998), pp. 157–178

41. J. Qiu, H. Yang, Z.Y. Dong, J.H. Zhao, K. Meng, F. Luo, K.P. Wong, A linear programming approach to expansion co-planning in gas and electricity markets. IEEE Trans. Power Syst. **31** (5), 3594–3606 (2016)
42. J.H. Zhao, F.S. Wen, Z.Y. Dong, Y.S. Yue, K.P. Wong, Optimal dispatch of electric vehicles and wind power using enhanced particle swarm optimization. IEEE Trans. Ind. Inform. **8**(4), 889–899 (2012)
43. T. Li, M. Shahidehpour, Z. Li, Risk-constrained bidding strategy with stochastic unit commitment. IEEE Trans. Power Syst. **22**(1), 449–458 (2007)

# Chapter 8
# Switch Deployment in Distribution Networks

**Milad Izadi, Mohammad Farajollahi and Amir Safdarian**

**Abstract** This chapter presents the optimal switch deployment in distribution systems. First, an explanation regarding different types of switches and their functionality is introduced. Then, a fundamental description of fault management procedure in distribution networks is presented. Thereafter, the mathematical formulation of optimal fault management process is described. Optimal switch deployment problem is formulated in the format of mixed integer programming (MIP). The impact of remote controlled switch (RCS) and manual switch (MS) is scrutinized on the interruption cost once they are installed either individually or simultaneously. The concept of switch malfunctions is explained and the influence of this issue on the optimal solution of the problem is discussed. Finally, the effect of uncertain parameters such as failure rate and repair time on the solution of switch deployment problem is investigated. It was shown that the uncertainty imposes a significant risk on distribution companies (DisCos).

**Keywords** Distribution networks · Fault management · Remote controlled switch Manual switch · Network automation · Malfunction · Financial risk

M. Izadi · A. Safdarian (✉)
Electrical Engineering Department, Center of Excellence in Power System Control and Management, Sharif University of Technology, Tehran, Iran
e-mail: safdarian@sharif.edu

M. Izadi
e-mail: izadi_milad@ee.sharif.edu

M. Farajollahi
Department of Electrical and Computer Engineering, University of California, Riverside, CA, USA
e-mail: mfara006@ucr.edu



A. Arefi et al. (eds.), *Electric Distribution Network Management and Control*, Power Systems, https://doi.org/10.1007/978-981-10-7001-3_8

## 8.1 Introduction

Nowadays, the tight dependency of social life to electricity makes the users much further desire to receive electrical services with high level of reliability, appropriate service quality, and enough safety and security. Since a great share of interruptions in power systems is originated from faults in distribution networks, improving the service reliability in distribution level has motivated distribution planners to establish different strategies. To this end, various methods have been proposed, among them deploying monitoring and control devices has caught more attention of distribution companies (DisCos). Sectionalizing switches (SSs), both in remote controlled switch (RCS) and manual switch (MS) types, play a fundamental role in the improvement of service reliability in distribution systems. SSs enable network reconfiguration in both normal and abnormal conditions. In normal conditions, network reconfiguration done through SSs can be applied to enhance network efficiency, while in abnormal conditions, prompt network reconfiguration is conducted to mitigate violations in operational constraints and to restore service to interrupted customers. The principle benefit of the switches corresponds to their ability for the reduction in interruption duration of affected customers. Although both of the switch types are effective in fault management process, RCSs outdo MSs in much faster restoration. Remote switching actions performed by RCSs take few minutes, which is mainly needed for detecting the fault location and making a decision for suitable maneuvers. On the other hand, MSs just can be used for field switching actions, which may take several minutes. From fault management point of view, once a fault occurs in an electric distribution network, field crews can determine the location of the fault by patrolling the suspicious fault zone. Once the fault is located, the customers whose connection point is out of the fault zone are restored by switching RCSs and MSs. The rest of customers should remain interrupted until the fault section is repaired. Although applying SSs brings numerous advantages to DisCos, they impose some costs comprising of investment costs, installation costs, and maintenance costs. In addition, issues such as budget limits prevent the wide deployment of these devices in distribution networks. Also, it is neither necessary nor financially justifiable to fully equip a network with such devices. Hence, cost/benefit analyses are required to determine the optimal number and location of SSs. In the literature, the switch deployment problem was attacked via several optimization approaches including classical optimization methods like mixed integer programming (MIP) and heuristic methods like genetic algorithm, simulated annealing algorithm, particle swarm optimization algorithm, and cooperative agent algorithm.

In addition, there are numerous significant parameters with considerable impacts on the optimal solution of the switch deployment problem such as the interruption cost function of each customer, the switch malfunction probability, and the stochastic nature of contingency events in distribution systems. Customer damage function (CDF) plays an important role in the number of installed SSs such that DisCos are enthusiastic to install more devices when CDF is increased. In case of high value of CDF, the benefits of SSs installation justify the relevant switch costs. The second parameter concerns with switch reliability, assuming the full reliable of SSs is not rational. Hence, considering the impact of SSs malfunction can tamper the cost/benefit analysis of switch deployment problem and leads to change in the final solution of the problem. The last but not least parameter is uncertainties with high volatility in distribution networks. The uncertain behavior of contingencies in practical systems such as stochastic nature of contingency events, uncertain repair time, and failure time impose remarkable financial risk and detract from the worth of SSs installation. Therefore, a risk-averse or a risk-taker behavior of the planners can change the final optimal solution of the problem. To consider this issue, the financial risk evaluation of SS in distribution networks is presented.

## 8.2 Switching Devices and Types

The most common used switching devices in distribution networks include circuit breaker, automatic recloser, sectionalizing switch, and counter sectionalizer. The explanations regarding their characteristics and functionality are described in the following subsections.

### *8.2.1 Circuit Breaker (CB)*

A circuit breaker (CB) is designed to immediately isolate the faulted feeder from the rest of network. CBs are usually installed inside the distribution substation where transmission high voltage is converted to distribution medium voltage. CB may be equipped with various protective relays such as overcurrent and earth fault relays which send signals to CB in order to operate properly. This may cause a considerable interruption in the feeder since customers who are located in downstream of the CB are de-energized.

### *8.2.2 Automatic Recloser (AR)*

Automatic recloser (AR) acts as CB and is also able to distinguish and clear transient faults in addition to permanent faults. According to field observations and experiences, a bulk portion of fault occurrences in distribution systems are related to transient faults which are originated from assorted sources such as temporary tree contact, flashover initiating from lighting strike, conductor clashing, and bird contact, to name just a few. AR is able to interrupt the electric power for a short duration, and then to restore electrical energy. This process can be done for several times and the interruption duration is increased consecutively in order to make sure that the interruption duration is enough to clear the transient fault [1]. For instance, once a fault occurs at the downstream of an AR in an electric distribution network, the AR operates after a short delay in order to check if the fault is transient or permanent. This step is known as the first operation. The AR remains open for a specific duration (i.e., near 0.2 s) and is closed for a predetermined duration again. If the fault current still flows, the AR would disconnect the electric distribution network, which is known as the second open-action (i.e., near 2 s). The described process is iterated for specific number of open-close actions (most of the time at most three iterations are sufficient). In case the AR clears the fault before reaching the maximum number of iterations, the fault is determined as a transient fault, otherwise the fault is permanent and the AR isolates the electric distribution network. Since substantial share of faults in distribution systems are transient, installing AR would play a prominent role in reducing the interruption duration due to transient contingencies and consequently, enhancing the service reliability of distribution networks.

### *8.2.3 Sectionalizing Switch (SS)*

Sectionalizing switches (SSs), both in RCS and MS types, enable network reconfiguration in both normal and abnormal conditions. In normal conditions, network reconfiguration can be applied to enhance network efficiency, while in abnormal conditions, prompt network reconfiguration is conducted to mitigate violations in operational constraints and to restore service to interrupted customers. The principle benefit of SSs corresponds to their ability in isolating healthy zones from faulted section and consequently, shortening interruption duration of affected customers. It is worth mentioning that SSs cannot operate under the excess current like short circuit fault current. Although both types of SS are effective in fault management process, RCSs outdo MSs in much faster restoration. Remote switching actions performed by RCSs take few minutes, which is mainly needed for detecting the fault location and making a decision for suitable maneuvers. On the other hand, MSs just can be used as field switching actions, which may take several minutes.

### *8.2.4 Counter Sectionalizer (CS)*

Counter sectionalizers (CSs) are installed downstream of ARs. They cannot operate under fault or load currents. Rather, they are equipped with a fault counter device in order to count the number of current interruptions made by the upstream AR. These switches are opened once the AR operates and the counter reaches a predefined value. Appropriate coordination of ARs and CSs can limit fault consequences if the fault is permanent and occurs downstream of the CS. When a fault occurs in an electric distribution network, the AR iterates reclosing process and CS counts the number of interruptions. If the fault is permanent, the CS is opened once the counter reaches the preset value. If the fault is located downstream of the CS, the AR does not sense the fault current anymore. So, the customers downstream the CS are interrupted while the upstream customers are isolated from the fault. If the fault is located upstream of the CS, the AR operates and all customers are interrupted.

## 8.3 Fault Management Process

Fault management process is defined as the set of actions conducted in order to alleviate the consequences of an unexpected fault. Fault management includes several sequential processes including fault occurrence notification, locating the faulted section, isolating the fault from the healthy sections, and remedial actions to restore the interrupted customers. In this regard, the flowchart of fault management process is shown in Fig. 8.1.

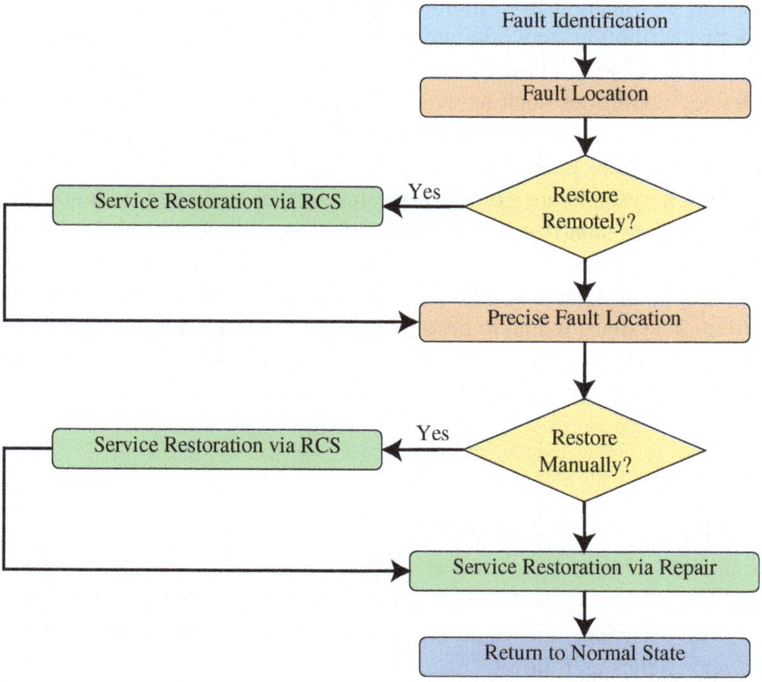


**Fig. 8.1** Fault management process flowchart

### 8.3.1 Step 1: Fault Notification

The first step in fault management process is to notice that a fault occurred in the system. The following indications enable operators to become aware of fault in distribution systems:

- Customers contact
  One of the typical ways to identify fault occurrence is contacts from customers whose services are interrupted. In this regard, a system is established in distribution control center to receive information of interruption events that are provided by customers. The contacts can be in different types such as calls, email services, and web-based event recorders.
- Status change of protection devices
  Distribution operator can check the status of the system-wide protection and control devices through monitoring systems [2–4]. Once the status of a switching device changes, the first point bears in mind is that a fault has occurred somewhere downstream the device.
- Condition change in network operation
  Any significant deviation in network operation can be considered as an indication of fault occurrence [5]. As an example, an abrupt change in loading of a feeder points that a protection device within the feeder is opened following a downstream fault.
- Receiving notification from monitoring devices
  Modern distribution networks are equipped with different monitoring devices such as line sensors. These devices send signals to the control center whenever predefined conditions such as overload are sensed. Fault indicators are among line sensors which are particularly designed for fault identification.
- State-of-art system monitoring
  Distribution systems are going to become further equipped with sophisticated devices such as smart meter. These devices are able to record the interruption and provide prompt notification for the operator. In addition, because of the great benefit of system monitoring through synchronized measurements, the distribution systems intend to become equipped with Phasor Measurement Units (PMUs) [6]. These devices monitor the system in time-series manner and give insight regarding the abnormal event in the system.

### 8.3.2 Step 2: Fault Location

After the operator recognized that a fault is occurred somewhere in the system, it is necessary to determine the fault location suspiciously. There are various fault location approaches as follows:

- Customers contact
  Once an interrupted customer reports an interruption event, the distribution operator can estimate the likely location of the fault. Although this is a typical way to identify the suspicious location of fault, it is not an effective and accurate way.
- Fault distance estimation
  There exist different methods for fault location identification based on fault distance estimation. Broadly speaking, the existing methods to identify fault locations at distribution grid can be categorized into two main groups: impedance based methods and wide-area monitoring. The former class of methods work based on calculating the line impedance between the fault location and sensor location. These methods usually come up with multiple possible locations for the fault. The second group of methods, e.g., the wide-area monitoring, work based on the fact that voltages and currents along the feeder fluctuate following fault events. In this regard, these methods use the pre-event and post-event states of the grid to identify the exact location of the fault [7, 8].
- Fault indicator
  Fault indicator is among line sensors which provide substantial chance to estimate the fault location. Fault indicators which are equipped with communication module, send signals once fault current is sensed. Hence, the operator recognizes that the fault is somewhere downstream the fault indicator whose signals are received. Fault indicators without communication modules are equipped with light bulbs whose blinking lets the field crews understand that the fault is somewhere downstream of the device. This information usually limits suspicious fault area and thus, eases fault location process [9].
- Advanced metering infrastructure (AMI)
  By propagating AMIs in distribution system, DisCos are of great interest to take benefits of these state-of-art infrastructures. One of the most advantages of AMIs is their ability in assisting in fault management process. AMIs installed at the customer level are able to capture the interruptions and immediately report them. Control center collects all the reported interruptions and determine the likely zone of the fault. In particular, an especial type of these devices can provide the voltage sag initiated by a fault. The collected voltage sages can be used in various fault location methods to identify the precise fault location.

Although the above-mentioned approaches provide utmost effect in finding the suspicious location of fault, neither determines the precise location. So, it is always necessary to employ field crews to patrol the suspicious area in order to determine the exact location of fault.

## *8.3.3 Step 3: Service Restoration*

Once the fault is detected and its location is determined, different remedial actions are accomplished to reduce the interruption duration of affected customers. In this situation, the distribution operator should make his/her best decision in order to restore the customers as soon as possible. The interrupted customers can be divided into two main groups. The first group contains the customers whose connection point can be isolated from the faulted area. The second group contains customers who are directly connected to the faulted area. The following steps should be conducted for restoring service to the interrupted customers.

### 8.3.3.1 Step 3-1: Remote Switching Action

After determining the approximate location of the fault, field crews start patrolling the suspicious area. However, since finding the precise location may take considerable time, it makes sense to remotely change the status of available RCSs adjacent to the faulted section. By doing so, some customers would be restored in a short duration. These customers experience interruption duration required for RCS switching action.

### 8.3.3.2 Step 3-2: Precise Fault Location

By restoring some customers via installed RCSs, field crews should find the precise fault location. This process also extends the fault location duration. Finding the precise location depends on several factors such as the geographical location of the electric distribution network (i.e., mountain or residential area, harsh or soft valley, etc.), types of the feeder (i.e., overhead or underground), the number of field crews, and other possible factors which might vary from one distribution system to another.

### 8.3.3.3 Step 3-3: Manual Switching Action

Once the faulted section is located, some customers whose connection points are out of the faulted area can be manually restored through available MSs. To do so, field crews determine boundary MSs and change their status. The customers restored via MSs experience interruption with longer duration that is needed for manual switching action and fault location.

#### 8.3.3.4 Step 3-4: Repair Action

After finding the faulted equipment and restoring service to customers out of the faulted area, repair crews repair the faulted section. The time takes to repair a system element depends on various factors such as the type of the element (e.g., transformer, switch, and line), the number of repair crews, the required tools for repairing, etc. The customers in the faulted area should remain de-energized until the faulted section is repaired. These customers experience the interruption duration associated with the repair time and fault location time.

#### 8.3.3.5 Step 3-5: Returning to the Initial State

Conducting various remedial actions alters the network operation from the optimal condition. So, it is necessary to return the status of switches to their normal condition. To do so, depending on the switches, the customers who were previously restored via appropriate switching actions may be de-energized for a short duration again.

**Example 8.1** The interruption duration of load points fed through a typical feeder following a fault are determined here. The feeder and other required information are shown in Fig. 8.2.

**Assumptions** The feeder is equipped with a CB at the beginning and a tie switch (TS) at the end. Without loss of generality, it is presumed that the time takes to change the status of CB and TS is trivial. The time takes for fault location is assumed to be 20 min. Also, the switching times associated with RCS and MS are considered to be 5 and 60 min, respectively. It takes 180 min to repair the faulted section. Also, it is assumed that operators can recognize the suspicious fault section and consequently, the suspicious fault location duration is neglected. The locations of MS and RCS are determined with circuit and square, respectively.

**CASE I** This case provides information regarding the feeder not equipped with any SS. This case is a comparison benchmark to show the effectiveness of SS deployment.

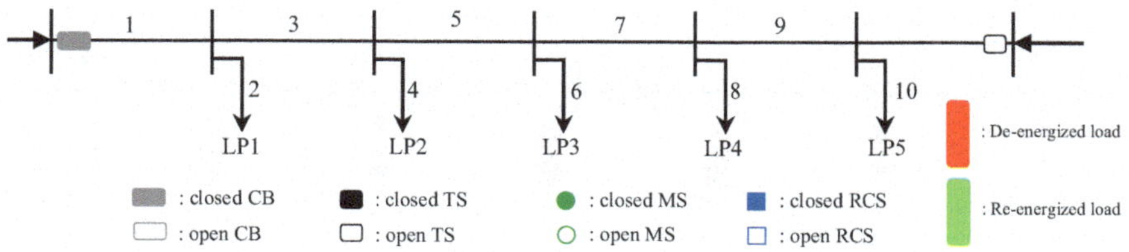


**Fig. 8.2** Representative feeder of Example 8.1

Step I: The operators recognize that a fault has occurred somewhere in the network by the information obtained from either of the above-mentioned ways (e.g., customers contact).

Step II: The operators recognize that the fault has happened in the illustrated feeder. In this way, the approximate fault location is determined by the above-mentioned approaches, e.g., fault indicator, (Fig. 8.3).

Step III: The repair crews patrol the suspicious area in order to identify the faulted section (Fig. 8.4).

Step IV: The repair crews start repairing the faulted section. The repair action takes 180 min (Fig. 8.5).

Step V: After repairing section 3, the CB is closed to restore all customers. Therefore, the customers who are fed through this feeder remain interrupted for 200 (= 20 + 180) min (Fig. 8.6).

The restoration times for the load points are represented in Table 8.1.

**CASE II** In this case, just MS deployment is considered and the allocation of RCSs is ignored. The configuration of the feeder is depicted in Fig. 8.7. The fault management steps are as follows:

**Table 8.1** Restoration time and mode of the load points in CASE I of Example 8.1

| Load point no. | Restoration time (min) | Restoration type |
|---|---|---|
| 1 | 200 | Fault location + repair action |
| 2 | 200 | Fault location + repair action |
| 3 | 200 | Fault location + repair action |
| 4 | 200 | Fault location + repair action |
| 5 | 200 | Fault location + repair action |

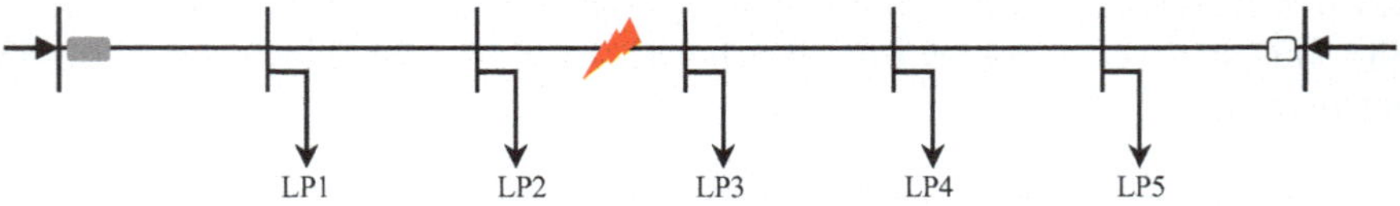


**Fig. 8.3** Representative feeder of Example 8.1 in CASE I

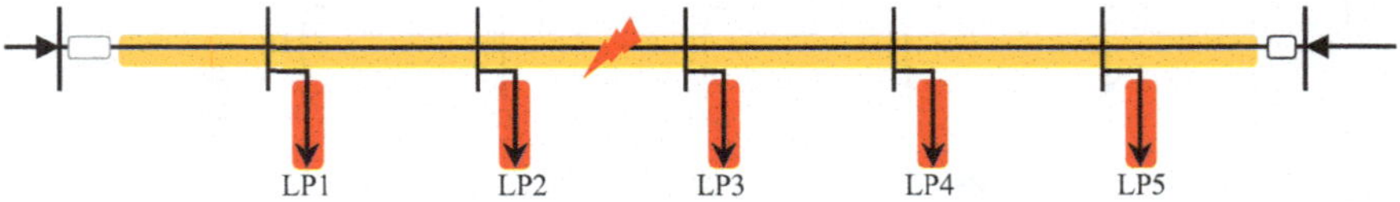


**Fig. 8.4** Representative feeder of Example 8.1 in CASE I-step II

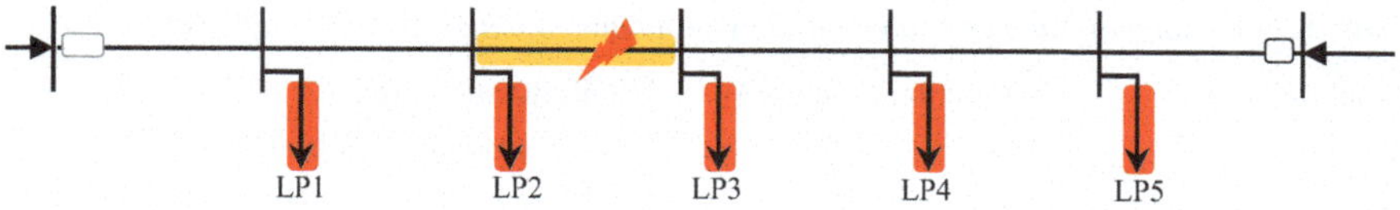


**Fig. 8.5** Representative feeder of Example 8.1 in CASE I-step III

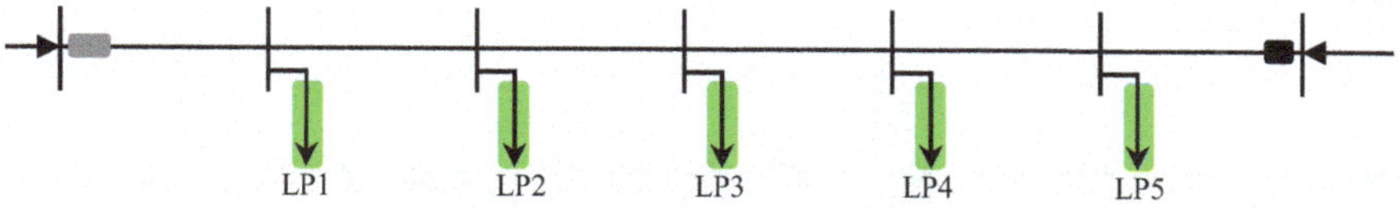


**Fig. 8.6** Representative feeder of Example 8.1 in CASE I-step V

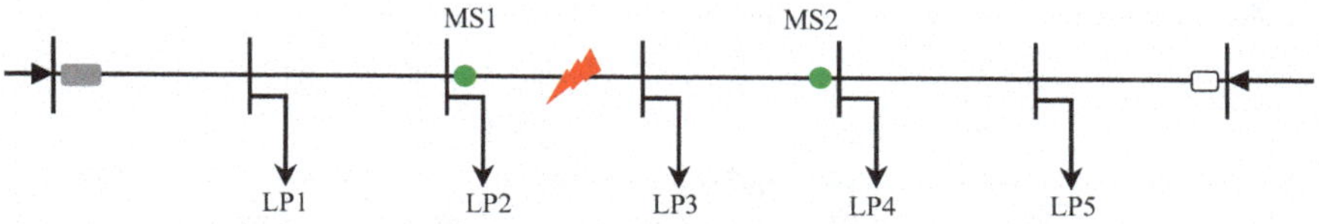


**Fig. 8.7** Representative feeder of Example 8.1 in CASE II

Step I: The operators recognize that a fault has occurred somewhere in the network by the information achieved from the above-mentioned ways (e.g., customers contact).

Step II: They identify that the fault has happened somewhere in the feeder below. In this way, they are able to determine the approximate faulted zone by the above-mentioned approaches, e.g., fault indicator, (Fig. 8.8).

Step III: The repair crews patrol the suspicious area in order to identify the faulted section. By doing so, the crews determine the faulted section (Fig. 8.9).

Step IV: The repair crews manually open MS1 and MS2. Then, CB and TS are closed in order to restore load points 1, 2, 4, and 5. Since the time to arrive the MS location and change its state takes 60 min, the load points remain interrupted for 80 (= 20 + 60) min. Hence, the interruption duration of load points 1, 2, 4, and 5 is equal to 80 min (Fig. 8.10).

Step V: The repair crews commence repairing the faulted equipment. This process takes 180 min.

Step VI: Since load point 3 cannot be restored prior to repair action, it should remain interrupted until section 3 is repaired. So, the customers connected to this load point experience 200 (= 20 + 180) min of interruption (Fig. 8.11).

The restoration times for the load points are represented in Table 8.2.

Table 8.2 Restoration time and mode of the load points in CASE II of Example 8.1

| Load point no. | Restoration time (min) | Restoration type |
|---|---|---|
| 1 | 80 | Fault location + manual switching |
| 2 | 80 | Fault location + manual switching |
| 3 | 200 | Fault location + repair action |
| 4 | 80 | Fault location + manual switching |
| 5 | 80 | Fault location + manual switching |

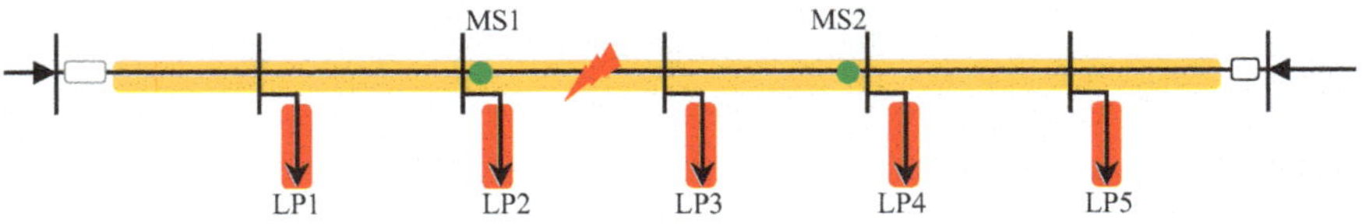


**Fig. 8.8** Representative feeder of Example 8.1 in CASE II-step II

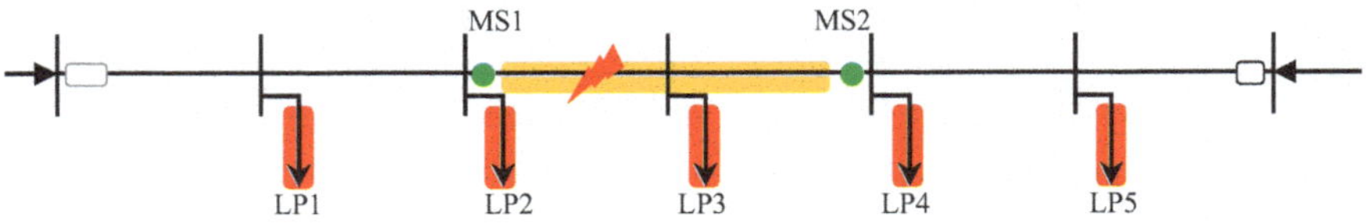


**Fig. 8.9** Representative feeder of Example 8.1 in CASE II-step III

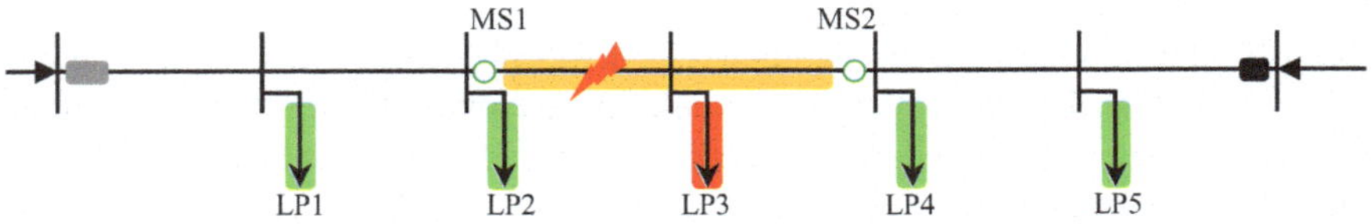


**Fig. 8.10** Representative feeder of Example 8.1 in CASE II-step IV

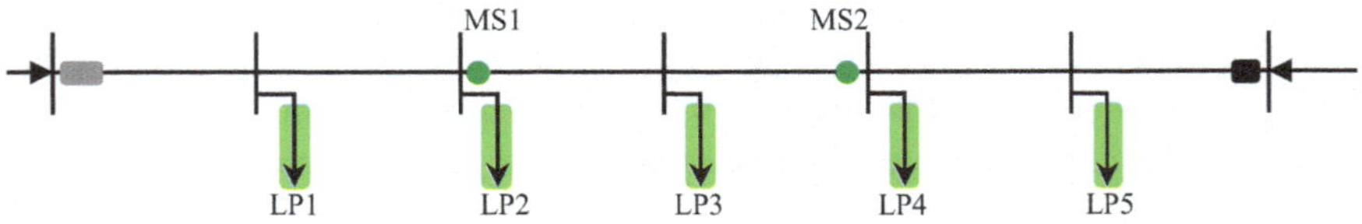


**Fig. 8.11** Representative feeder of Example 8.1 in CASE II-step VI

**CASE III** In this case, just RCS deployment is considered and the allocation of MSs is ignored. The feeder is portrayed in Fig. 8.12. The fault management steps are as follows.

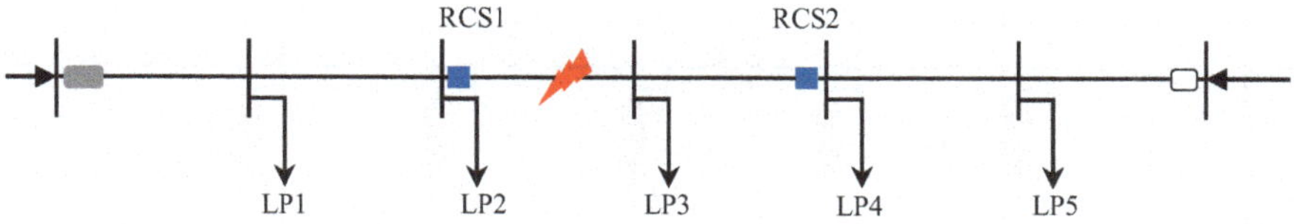


**Fig. 8.12** Representative feeder of Example 8.1 in CASE III

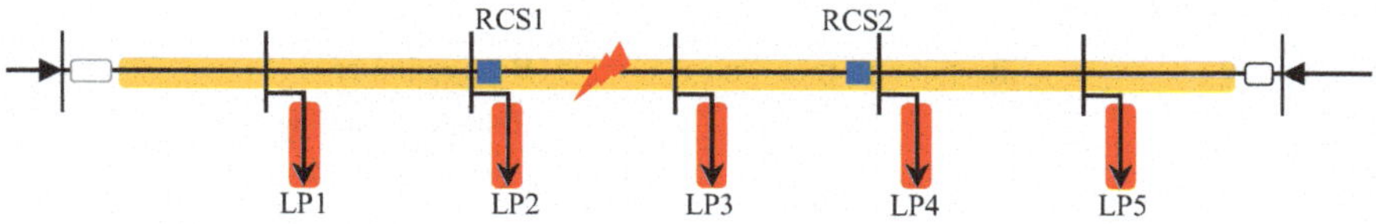


**Fig. 8.13** Representative feeder of Example 8.1 in CASE III-step II

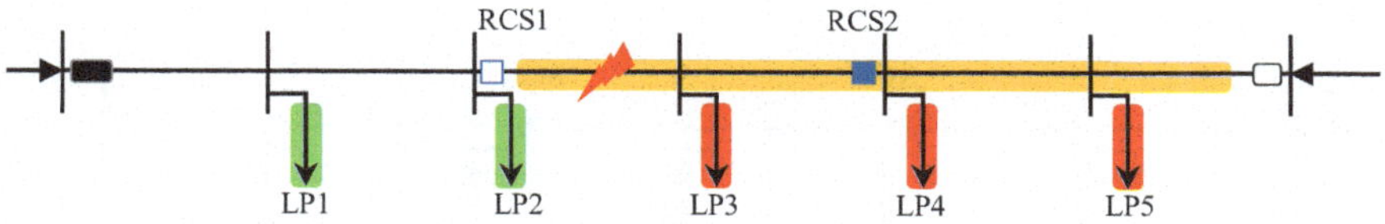


**Fig. 8.14** Representative feeder of Example 8.1 in CASE III-step III

Step I: The operators diagnose a fault occurrence in the network by the information obtained from the above-discussed approaches (e.g., customers contact).

Step II: The operators recognize that the fault has happened in the shown feeder. In this way, they are able to determine the fault zone by the above-mentioned approaches, e.g., fault indicator, (Fig. 8.13).

Step III: The operators may use trial and error approach to determine the suspicious area. To do so, they first open RCS1 remotely and close the CB. In this situation, since the fault has occurred downstream of RCS1, the CB does not operate. Hence, the operators recognize that the fault is somewhere after RCS1. This circumstance is shown in Fig. 8.14.
The next trial is to close RCS1, open RCS2, and close the CB. By doing so, the CB operates. Therefore, the operators recognize that the fault is somewhere between RCS1 and RCS2. This circumstance is shown in Fig. 8.15. The trial and error approach does not take considerable time which is neglected here.
Now, load points 1, 2, 4, and 5 can be restored remotely via the RCSs. This process takes 5 min. So, the customers whose connection points are

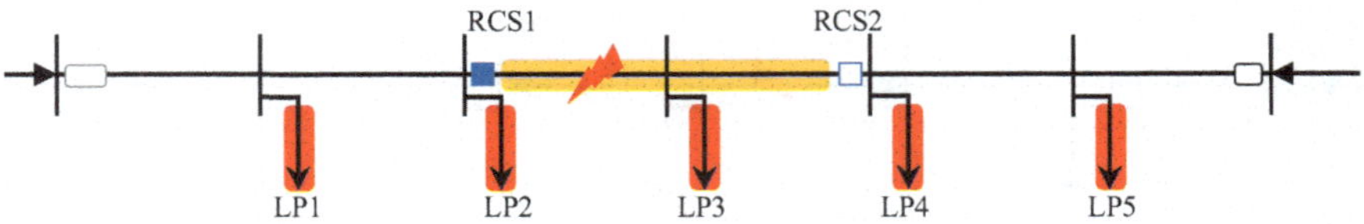


**Fig. 8.15** Representative feeder of Example 8.1 in CASE III-step III

**Table 8.3** Restoration time and mode of the load points in CASE III of Example 8.1

| Load point no. | Restoration time (min) | Restoration type |
|---|---|---|
| 1 | 5 | Remote switching |
| 2 | 5 | Remote switching |
| 3 | 200 | Fault location + repair action |
| 4 | 5 | Remote switching |
| 5 | 5 | Remote switching |

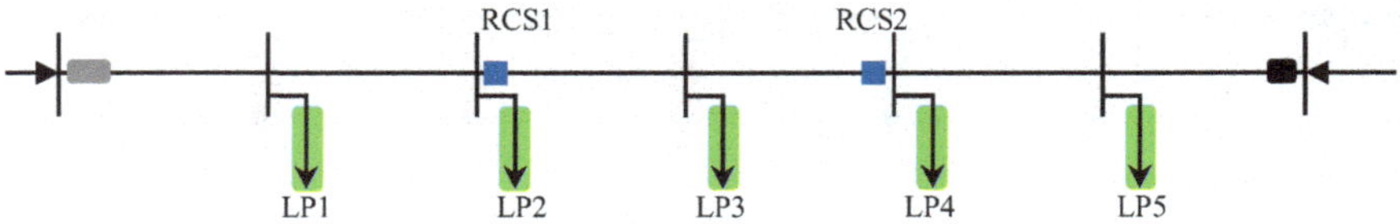


**Fig. 8.16** Representative feeder of Example 8.1 in CASE III-step VI

upstream of RCS1 and downstream of RCS2 experience 5 min interruption.

Step IV: The repair crews precisely patrol the suspicious area. By doing so, the crews determine the faulted section. This process takes 20 min.

Step V: The crews repair the faulted equipment which takes 180 min.

Step VI: Load point 3 should remain interrupted during the repair action. So, by taking into account the previous remedial actions, load point 3 retains interrupted for 200 (= 20 + 180) min. Finally, the network returns to the normal condition (Fig. 8.16).

The restoration times for the load points are represented in Table 8.3.

**CASE IV** In this case, both RCS and MS deployment is considered. The feeder is depicted in Fig. 8.17. The fault management steps are as follows:

Step I: The operators recognize that a fault has occurred somewhere in the network by the information obtained from the above-mentioned ways (e.g., customers contact).

Step II: The operators figure out that the fault has happened in the illustrated feeder. In this way, they are able to determine the fault zone by the above-mentioned approaches (e.g., fault indicator) (Fig. 8.18).

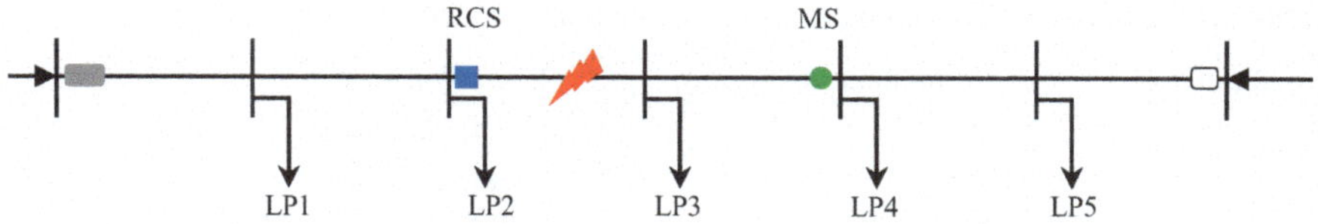


**Fig. 8.17** Representative feeder of Example 8.1 in CASE IV

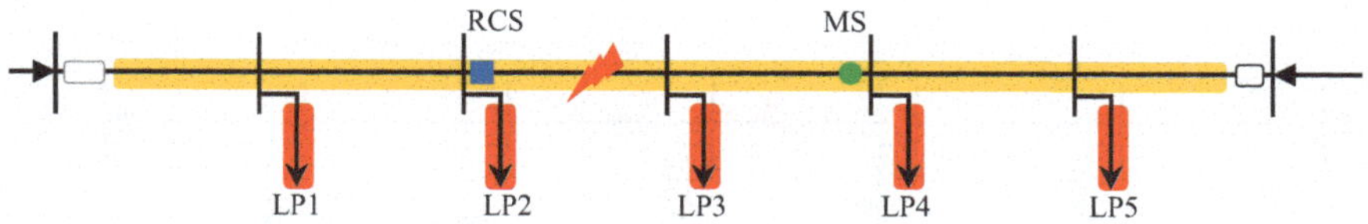


**Fig. 8.18** Representative feeder of Example 8.1 in CASE IV-step II

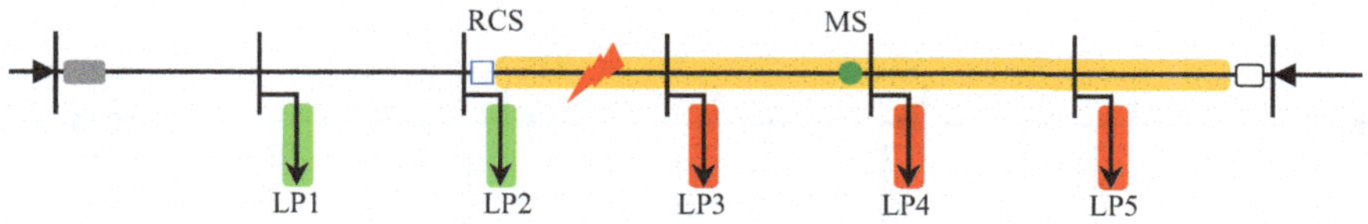


**Fig. 8.19** Representative feeder of Example 8.1 in CASE IV-step III

Step III: The operators apply trial and error approach to determine the suspicious area. To do this, the operators send a signal to RCS to be opened and close the CB. In this condition, the CB does not operate and load points 1 and 2 are re-energized, which means that the fault has originated from sections downstream the RCS. So, the RCS capability enables the operators to reduce the suspicious area and consequently to decrease the fault location duration. This circumstance is shown in Fig. 8.19.
Load points 1 and 2 are restored via remote switching action which takes 5 min. So, the customers whose connection points are upstream of the RCS experience 5 min interruption.

Step IV: The repair crews determine the faulted section by patrolling the suspicious area.

Step V: The crews restored load points 4 and 5 through opening the MS and closing the TS at the end of the feeder. In this situation, the load points experience 80 (= 20 + 60) min interruption (Fig. 8.20).

Step VI: The repair crews repair the faulted equipment, which takes 180 min.

Step VII: Finally, the network should return to its normal state. By doing so, load point 3 remains interrupted for 200 (= 20 + 180) min (Fig. 8.21).

**Table 8.4** Restoration time and mode of the load points in CASE IV of Example 8.1

| Load point no. | Restoration time (min) | Restoration type |
|---|---|---|
| 1 | 5 | Remote switching |
| 2 | 5 | Remote switching |
| 3 | 200 | Fault location + repair action |
| 4 | 80 | Fault location + manual switching |
| 5 | 80 | Fault location + manual switching |

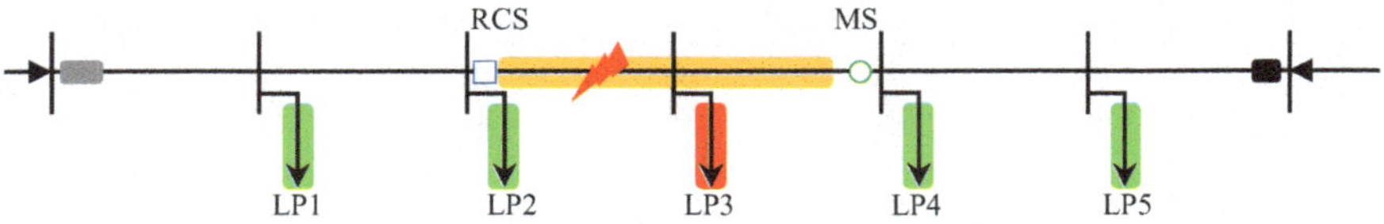


**Fig. 8.20** Representative feeder of Example 8.1 in CASE IV-step V

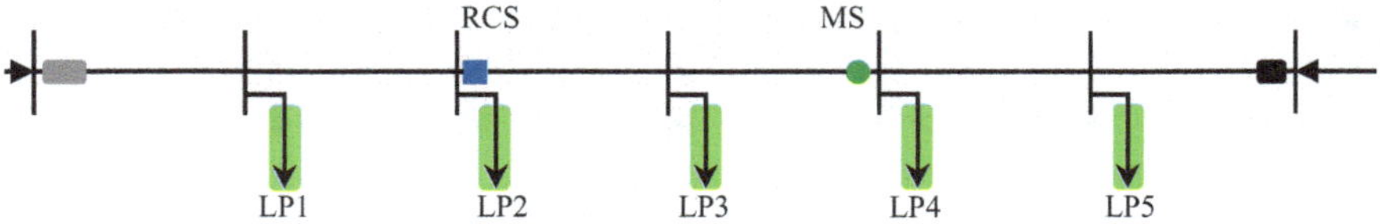


**Fig. 8.21** Representative feeder of Example 8.1 in CASE IV-step VII

The restoration times for the load points are represented in Table 8.4.

Table 8.5 summarizes the simulated cases. As can be seen, all customers should stay de-energized for a long duration in CASE I. In CASE II, the customers should be interrupted for shorter interval compared to the customers in CASE I. However, load point 3 experiences the same interruption duration in all of the cases. According to the results, the customer interruption duration of load points 1, 2, 3, and 4 is diminished by 120 min by installing MSs in CASE II. While, in CASE III where RCSs are employed in the feeder, the interruption duration of customers out

**Table 8.5** Customer interruption duration of CASE I–CASE IV in Example 8.1

| Load point no. | Customer interruption duration (min) | | | |
|---|---|---|---|---|
| | CASE I | CASE II | CASE III | CASE IV |
| 1 | 200 | 80 | 5 | 5 |
| 2 | 200 | 80 | 5 | 5 |
| 3 | 200 | 200 | 200 | 200 |
| 4 | 200 | 80 | 5 | 80 |
| 5 | 200 | 80 | 5 | 80 |

of the faulted section is reduced by 195 min. Also, in CASE IV, the same situation happens for load points 1 and 2 due to the capability of RCS by isolating these load points from the faulted section. Nevertheless, the customers who are fed through load points 4 and 5 should experience longer interruption time. More accurately, 75 min increment in interruption duration of these customers is the penalty of exploiting MS instead of RCS. The results of CASE I–CASE IV clearly demonstrate the significant impact of deploying RCS and MS on interruption duration decrement.

## 8.4 Switch Deployment Model

In the previous section, the impact of employing RCS and MS on customer interruption duration was explained. The purpose of this section is to present a mathematical model for SS deployment in distribution networks. SSs bring great benefits in diminishing system interruption costs through reducing customer interruption duration and decreasing the fault location time. However, deployment of SSs imposes considerable costs such as capital investment costs, installation costs, and maintenance costs. In this regard, employing of SSs in all possible locations in the system is neither essential nor cost-effective. In this regard, it is necessary to consider a trade-off between the benefits and costs of SS deployment. To do so, various heuristic and mathematical approaches have been developed in the literature. Most of the proposed methods try to minimize the system costs including interruption and SS deployment costs. In [10], the fuzzy decision approach was applied to solve the problem. In [11–13], heuristic optimization techniques such as genetic algorithm, simulated annealing algorithm, and ant colony algorithm were used to find the number and location of SSs. In [14], the authors extended the previous models by taking into account CBs in the problem. The relocation problem was proposed in [15] to locate SSs in a system. Reference [16] divided the candidate SS locations into several independent sets and the problem was solved for each set separately. The bellmen's principle was taken into account in [17] to determine the place of RCSs. Besides, the mathematical approaches were used in the format of MIP in [18–20]. In [18], authors determined the optimal number and location of RCSs by minimizing aggregated system costs. In addition, by extending the proposed model in [18], the impact of earth faults was taken into account in [19]. Furthermore, the budgetary limitation was considered in [20] by proposing a multi-stage planning model to determine the location of RCSs in each year. In addition, the joint RCS and MS placement is considered as reported in [20, 21]. The optimal number and location of SSs in switch placement problem are defined such that the overall system costs comprising of system interruption costs and related SS costs are reduced as [21]:

$$\text{Minimize}\, Cost^{int} + Cost^{SS} \tag{8.1}$$

where $Cost^{int}$ and $Cost^{SS}$ are respectively system interruption and SS costs, which are explained as follows.

### 8.4.1 Customer Interruption Costs

The customers' interruption costs depend on various parameters such as element failure rate, average customers' demands, and customers' damage function. Customers' damage function relies on the type of affected customers and the interruption duration. So, the system interruption costs are formulated as

$$Cost^{int} = \sum_{t \in T} \sum_{f \in F} \sum_{i \in I} \sum_{j \in J} \sum_{k \in K} \frac{(1 + q_{\lg})^{t-1}}{(1 + q_{dr})^{t}} \lambda_{f,i} L_{f,j,k} CDF_{f,i,j,k}(d^{int}_{f,i,j,k}) \tag{8.2}$$

where $\lambda_{f,i}$ is the failure rate of an element at location $i$ in feeder $f$. $L_{f,j,k}$ is the average demand of customers with type $k$ at location $j$ in feeder $f$. The CDF is represented with $CDF_{f,i,j,k}$ which indicates the damage costs of customers with type $k$ who are connected to load point at location $j$ in feeder $f$ when section $i$ is failed. $d^{int}_{f,i,j,k}$ is the interruption duration of customers with type $k$ at load point $j$ of feeder $f$ during failure in section $i$. The annual discount rate ($q_{dr}$) is deemed here to consider the present value of investment. In addition, without loss of generality, a constant load growth rate ($q_{\lg}$) is assumed here. In (8.2), the reliability data of network equipment and load data are assumed to be predetermined parameters. However, interruption duration and hence, CDF are function of the restoration mode. So, the relation between the location of the installed SSs and load points plays a fundamental role in determining CDF. Figure 8.22 shows a representative feeder for the developed mathematical model. As shown, CB is installed at the beginning and end of the feeder. Also, both sides of sections are candidate locations for SSs.

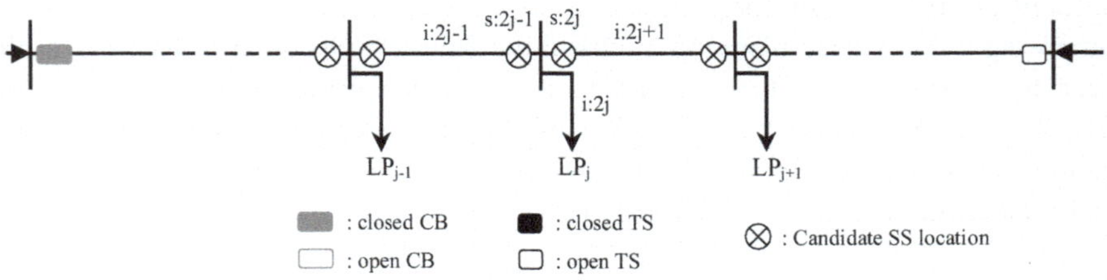


**Fig. 8.22** Sample feeder to illustrate the impact of switch location on interruption duration

According to the fault and customer locations in the system as well as the deployed SSs in the system, affected customers can be categorized into three groups. In case of existing any RCS between customers and the fault, they can be promptly isolated from the fault and re-energized through remote switching actions in a few minutes, while other customers would stay for fault location process. After locating the faulted section, other customers who can be isolated from the fault through MSs can be restored by proper switching actions. The rest of customers ought to tolerate interruption duration associated with repair time of the faulted section. Therefore, interruption duration of customers can be calculated based on the configuration of the SSs as

$$d_{f,i,j,k}^{int} \geq TTS_{f,s}^{RCS}; \quad \forall f \in F, \forall i \in I, \forall j \in J, \forall k \in K \tag{8.3}$$

$$d_{f,i,j,k}^{int} \geq \left(TTL_{f,i} + TTS_{f,s}^{MS}\right)\left[1 - \sum_{s=2j}^{i-1} X_{f,s}^{RCS}\right]; \quad \forall f \in F, \forall i \in I, \forall 2j<i, \forall k \in K \tag{8.4}$$

$$d_{f,i,j,k}^{int} \geq \left(TTL_{f,i} + TTS_{f,s}^{MS}\right)\left[1 - \sum_{s=i}^{2j-1} X_{f,s}^{RCS}\right]; \quad \forall f \in F, \forall i \in I, \forall 2j > i, \forall k \in K \tag{8.5}$$

$$d_{f,i,j,k}^{int} \geq \left(TTL_{f,i} + TTR_{f,i}\right)\left[1 - \sum_{s=2j}^{i-1} X_{f,s}^{RCS} - \sum_{s=2j}^{i-1} X_{f,s}^{MS}\right]; \quad \forall f \in F, \forall i \in I, \forall 2j<i, \forall k \in K \tag{8.6}$$

$$d_{f,i,j,k}^{int} \geq \left(TTL_{f,i} + TTR_{f,i}\right)\left[1 - \sum_{s=i}^{2j-1} X_{f,s}^{RCS} - \sum_{s=i}^{2j-1} X_{f,s}^{MS}\right]; \quad \forall f \in F, \forall i \in I, \forall 2j > i, \forall k \in K \tag{8.7}$$

$$d_{f,i,j,k}^{int} \geq TTL_{f,i} + TTR_{f,i}; \quad \forall f \in F, \forall i \in I, \forall 2j = i, \forall k \in K \tag{8.8}$$

where $X_{f,s}^{RCS}$ and $X_{f,s}^{MS}$ are the binary variables associated with existence of RCS and MS in location $s$ in feeder $f$. In case MS or RCS is installed in a location, the relevant binary variable will be set equal to 1, otherwise it takes the value of zero. In the above formulations, $TTS_{f,s}^{RCS}$, $TTS_{f,s}^{MS}$, $TTL_{f,i}$, and $TTR_{f,i}$ are respectively related to remote switching, manual switching, fault location, and repair time for interrupted customers restoration. Formulation (8.3) belongs to customers who can be isolated through remote switching action. The interruption duration of customers who can be isolated from fault point by manual switching is determined through expressions (8.4) and (8.5). These two formulations are respectively related to the customers whose location is upstream and downstream of the faulted section.

Constraints (8.6) and (8.7) are associated with customers who cannot be isolated from the faulted section through RCS and MS and should remain interrupted until the repair action is carried out. In addition, constraint (8.8) is related to faults occur in customers sections, thereby making sure that the customers are interrupted subsequent to repair time.

### 8.4.2 SS Costs

SS costs consist of capital investment, installation, and maintenance costs of deployed MSs and RCSs, which is expressed as

$$Cost^{SS} = CI + IC + MC \tag{8.9}$$

$$CI = \sum_{f \in F} \sum_{s \in S} \left( X_{f,s}^{RCS} CI_{f,s}^{RCS} + X_{f,s}^{MS} CI_{f,s}^{MS} \right) \tag{8.10}$$

$$IC = \sum_{f \in F} \sum_{s \in S} \left( X_{f,s}^{RCS} IC_{f,s}^{RCS} + X_{f,s}^{MS} IC_{f,s}^{MS} \right) \tag{8.11}$$

$$MC = \sum_{t \in T} \sum_{f \in F} \sum_{s \in S} \frac{1}{(1 + q_{dr})^t} \left( X_{f,s}^{RCS} MC_{f,s}^{RCS} + X_{f,s}^{MS} MC_{f,s}^{MS} \right) \tag{8.12}$$

where $CI_{f,s}^{RCS}$, $IC_{f,s}^{RCS}$, and $MC_{f,s}^{RCS}$ are the capital investment cost, installation cost, and annual maintenance cost of a RCS, respectively. $CI_{f,s}^{MS}$, $IC_{f,s}^{MS}$, and $MC_{f,s}^{MS}$ are the capital investment cost, installation cost, and annual maintenance cost of a MS. The mentioned SSs cost can vary for different locations. The capital costs can highly depend on the types of line (i.e., overhead or underground). Also, the installation costs are related to factors such as the geographical location (i.e., mountain region) and the types of the line. Moreover, RCS requires specific communication facilities to be controlled remotely from control center, thereby raising the RCS installation cost as the distance between the RCS location and control center increases.

### 8.4.3 Problem Constraints

DisCos own limited financial sources for equipping their system with SS. So, budget limitation plays a fundamental role in the optimal solution of the deployment problem. In this regard, to alleviate the DisCo's concerns regarding the financial restriction, some constraints are considered as

$$\sum_{f\in F}\sum_{s\in S} X_{f,s}^{RCS} \leq N_{RCS}^{\max} \tag{8.13}$$

$$\sum_{f\in F}\sum_{s\in S} X_{f,s}^{MS} \leq N_{MS}^{\max} \tag{8.14}$$

$$CI + IC \leq budget \tag{8.15}$$

$$\sum_{f\in F}\sum_{i\in I}\sum_{j\in J}\sum_{k\in K} \frac{N_{j,k}\lambda_{f,i}d_{f,i,j,k}^{int}}{N^{total}} \leq SAIDI \tag{8.16}$$

$$\sum_{f\in F}\sum_{i\in I}\sum_{j\in J}\sum_{k\in K} \frac{N_{j,k}\lambda_{f,i}L_{f,j,k}d_{f,i,j,k}^{int}}{N^{total}} \leq AENS \tag{8.17}$$

where the first and the second constraints define the maximum number of allocated RCSs ($N_{RCS}^{\max}$) and MSs ($N_{MS}^{\max}$) in the system, and the third one limits the investment cost, including capital investment and installation costs, to the company budget limitation (*budget*). In addition, the last two constraints consider the system reliability requirements. Broadly speaking, the system reliability indices serve to appraise the efficiency of a grid in emergency situations, e.g., fault occurrence, and DisCos are concerned to keep these indices below a defined level to avoid getting fined. In this regard, constraint (8.16) is associated with the SAIDI index. This index determines the expected value of customer interruption duration per year. $N_{j,k}$ denotes the number of customers in load point $j$ with type $k$, and $N^{total}$ represents the total number of customers in the network. Also, the expected value of energy not served by customers per year, defined as AENS, is regarded in constraint (8.17).

**Example 8.2** In this example, the proposed method is applied to Roy Billinton Test System (RBTS) bus 4, shown in Fig. 8.23, which has been broadly used for SS deployment problem [18–22]. This system consists of 7 feeders which feed 38 load points. The required data associated with load points, failure rates, and feeders' configuration are given in [23]. In addition, three types of customers including residential, small-user, and commercial are accommodated into this system. The CDF and other related information regarding the customers are provided in [24].

In this example, the capital investment and installation costs of RCS and MS are considered to be US k$4700 and US $500, respectively. The annual maintenance costs of RCS and MS are supposed to be 2% of capital and installation costs (i.e., US k$94 and US $10). The annual load growth rate and discount rate are set equal to 3 and 8% for a 15-year study horizon. In addition, remote and manual switching actions are assumed to take 5 and 60 min. The fault location time is deemed to be 20 min. Without loss of generality, it is presumed that TS, located at the end of the feeders, immediately operates whenever it is necessary. Also, the repair action for

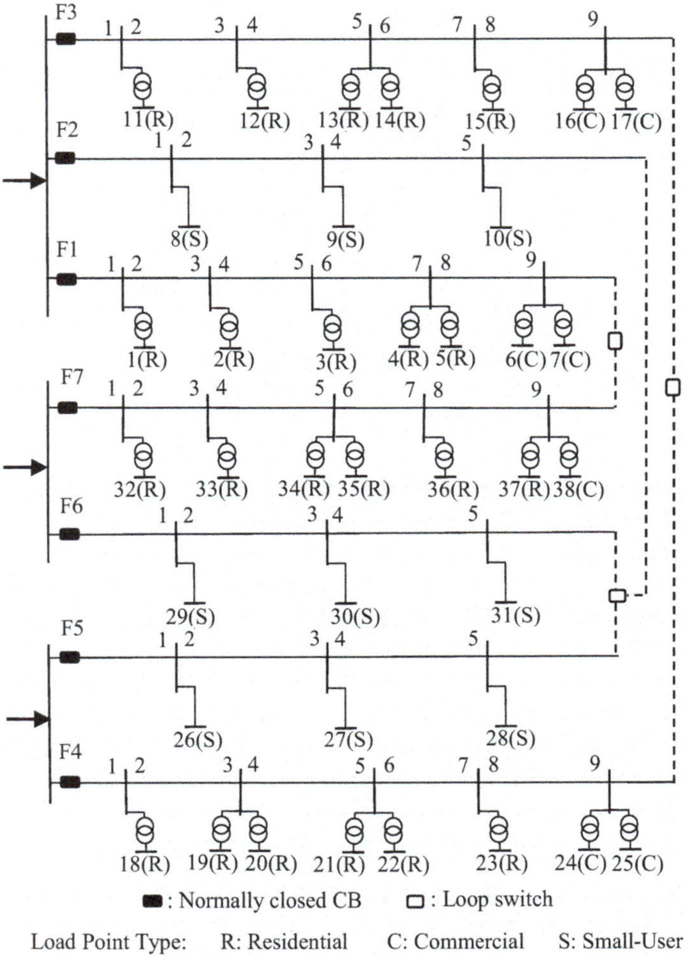


**Fig. 8.23** Single line diagram of RBTS-Bus4

each section takes 3 h. According to the test system shown in Fig. 8.23, the total number of candidate places for SSs is equal to 51.

**CASE I** This case provides information regarding original network not equipped with any SS. This case is a comparison benchmark to show the effectiveness of the developed model for SS deployment.

**CASE II** The optimal number and location of MSs in the network are determined. In this case, just MS deployment is considered and the allocation of RCSs is ignored. The results are represented in Table 8.6. As can be seen, 5 MSs are allocated to feeders 2, 5, and 6 where small-user customers are fed. In other words, MSs are installed in all possible locations. However, 4 MSs are deployed in other

**Table 8.6** Optimal location of MSs in CASE II

| Feeder | MSs location |
|---|---|
| 1 | 3,5,7,9 |
| 2 | 1,2,3,4,5 |
| 3 | 3,5,6,9 |
| 4 | 3,4,6,9 |
| 5 | 1,2,3,4,5 |
| 6 | 1,2,3,4,5 |
| 7 | 3,5,6,9 |
| Total | 31 |

**Table 8.7** Optimal location of RCSs in CASE III

| Feeder | RCSs location |
|---|---|
| 1 | 5,9 |
| 2 | 1,2,3,4 |
| 3 | 5,9 |
| 4 | 4,9 |
| 5 | 1,3,4,5 |
| 6 | 1,2,3,4,5 |
| 7 | 5,9 |
| Total | 21 |

feeders where both residential and commercial customers present. In these feeders, MSs are installed with approximately uniform distribution.

**CASE III** The optimal number and location of RCSs in the network are determined. In this case, the simulation is iterated to find the number and location of only RCSs without considering MSs. The results are provided in Table 8.7. As can be seen, 2 RCSs are installed in feeders 1, 3, 4, and 7 such that one of them is located in middle of the feeders and the other one is located at the end of the feeders where commercial customers are fed. However, at least 4 RCSs are utilized in feeders with small-user customers.

**CASE IV** The simultaneous placement of RCS and MS is simulated in this case. Table 8.8 gives the location of RCSs and MSs in each feeder. According to the table, 3 MSs are installed in each feeder. In feeders 1, 3, and 4, the first MS is installed in location 3 where the total load and CDF are higher than other feeders. However, the first MS is deployed in location 1 in other feeders. Regarding the RCS location, RCS is employed at the end of feeders 1, 3, 4, and 7 (location 9) where the commercial customers are fed.

The system costs including system interruption cost and switch costs are provided in Table 8.9. As can be seen, employing SSs in CASE II and CASE III leads to US k$459.58 and US k$721.62 reduction in system interruption costs (by 50.41 and 79.15%), respectively. Based on the results, using RCS provides higher achievements in system interruption costs instead of MS installation due to its

**Table 8.8** Optimal location of RCSs and MSs in CASE IV

| Feeder | MSs location | RCSs location |
|---|---|---|
| 1 | 3,5,7 | 9 |
| 2 | 1,2,5 | 3,4 |
| 3 | 2,5,6 | 9 |
| 4 | 3,4,6 | 9 |
| 5 | 1,3,4 | 2,5 |
| 6 | 1,2,4 | 3,5 |
| 7 | 1,5,6 | 9 |
| Total | 21 | 10 |

**Table 8.9** System costs (US k$) in CASE I–CASE IV of Example 8.2

| CASE | Equipment | | | | Interruption (US k$) | Total (US k$) |
|---|---|---|---|---|---|---|
| | Number of MSs | Number of RCSs | MSs (US k$) | RCSs (US k$) | | |
| I | – | – | – | – | 911.72 | 911.72 |
| II | 31 | – | 18.37 | – | 452.14 | 470.51 |
| III | – | 21 | – | 118.11 | 190.10 | 308.21 |
| IV | 21 | 10 | 12.44 | 56.25 | 208.10 | 276.79 |

significant capability in prompt isolating the healthy part from the faulted section. Applying simultaneously RCS and MS reduces the interruption cost from US k$911.72 to US k$208.10 by 77.18% saving in total interruption costs. Furthermore, although the system interruption costs in CASE III and CASE IV are roughly equal, the equipment costs in CASE IV are US k$49.42 smaller than that of CASE III. Hence, the DisCo can reduce system interruption costs by installing MS coupled with RCS without any increment in investment costs.

## 8.5 Affecting Parameters and Sensitivity Analyses

According to the SS deployment method, explained in the previous section, various parameters may affect the solution of SS deployment problem. This section intends to investigate the impact of key parameters on the SS deployment problem. To do so, the simulation is repeated to find the impact of different prominent parameters including CDF, failure rate and repair time of elements, financial constraints associated with number of allowable SS, and budget limitations.

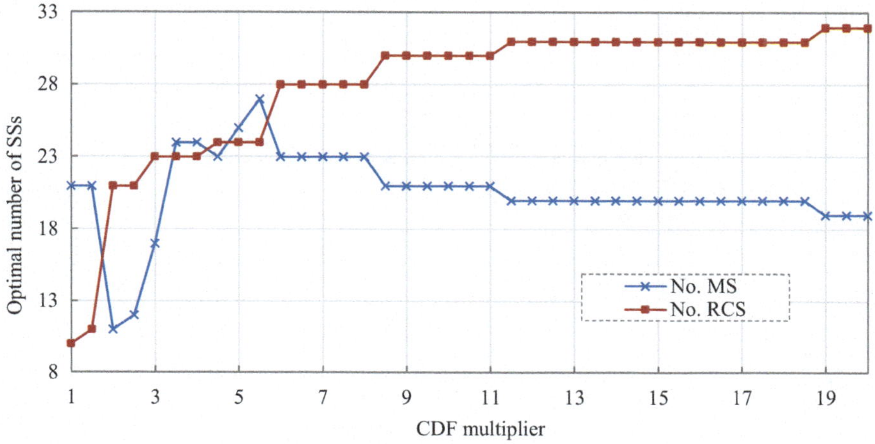


**Fig. 8.24** Optimal number of SSs versus different CDFs

## *8.5.1 Customer Damage Function (CDF)*

Here, the effect of CDF on the optimal solution of SS placement problem is scrutinized. In this regard, the simulations are conducted for different CDFs by considering a multiplier varying from 1 to 10. The optimal numbers of RCSs and MSs are depicted in Fig. 8.24. As can be seen, the number of RCSs increases as the CDF rises, thereby diminishing the customer interruption costs. However, the number of MSs does not follow the same pattern, because the number of MSs has a close relation with the number and location of RCSs. As the number of RCSs increases, the number of MSs decreases.

## *8.5.2 Failure Rate*

Elements in distribution networks are not exposed to failure with the same probability, such that sections with higher failure rates are more likely to undergo a fault. In this regard, it is valuable to investigate the impact of different failure rates on the solution of the problem. Also, according to the previous section, SSs are more likely to be installed in locations where the majority of customers can be isolated from sections with higher failure rate. With this in mind, the presented model of switch placement is simulated for different failure rates. The optimal numbers of RCSs and MSs are illustrated in Fig. 8.25. As can be seen, the number of RCSs is generally increased as the failure rate rises. Also, the more RCSs considerably restrict the suspicious fault area and thereby reducing the customer interruption time for customers whose connection point can be isolated prior to repair action. However, the number of installed MSs relies on the number and

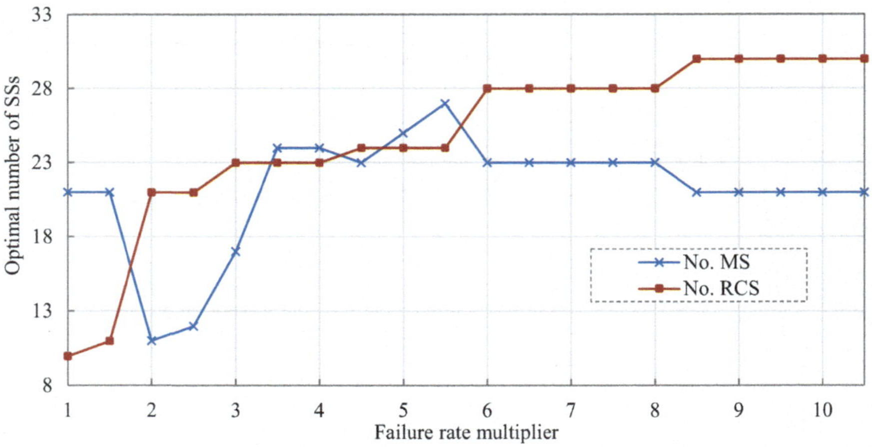


**Fig. 8.25** Optimal number of SSs versus different failure rate multiplier

location of RCSs such that as the failure rate is doubled, the number of MSs is extremely reduced while the number of RCSs is increased.

### *8.5.3 Repair Time*

According to several out of control factors such as geographical location (i.e., mountain regions and snow areas), the type of line (i.e., overhead and underground), and the number of repair crews, the repair time can varies in different systems. Therefore, as an affective parameter in SS deployment problem, the impact of repair time should be investigated on the SS allocation. In this regard, the simulation is repeated for different repair times. The numbers of allocated RCSs and MSs are depicted in Fig. 8.26. As can be seen, the optimal number of MSs is increased while the number of RCSs is constant as the repair time increases. In this regard, deploying more MSs remarkably reduces interruption duration of customers whose connection point cannot be restored via manual switching action. Hence, the customer interruption costs significantly waned when the repair time waxed.

### *8.5.4 Limited Number of SSs*

Although employing optimal number of SSs provides the cost effective solution, most DisCos may not be able to equip the system at the beginning of the planning due to the considerable expenses of switches. In this regard, they are interested to understand the extent to which their system efficiency increases as a limited number

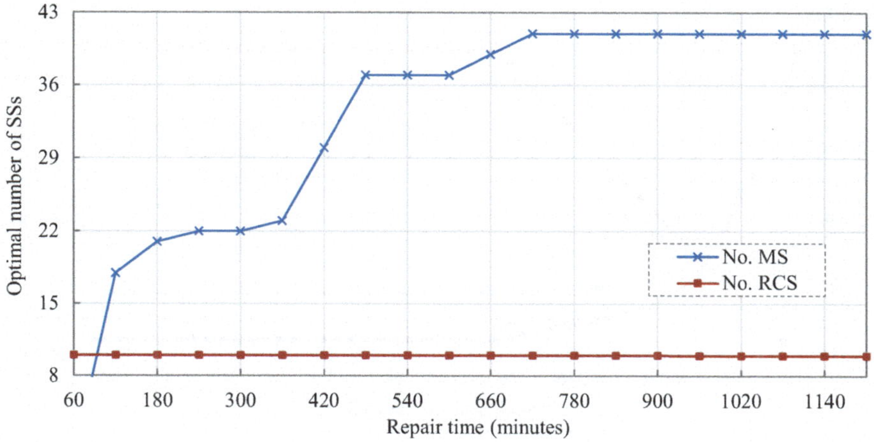


**Fig. 8.26** Optimal number of SSs versus different repair time

of SSs getting installed in the system. To this end, the SS deployment problem is run by restricting the number of available MSs and RCSs separately as shown in Figs. 8.27 and 8.28. The maximum number of SSs, i.e., RCSs or MSs, is gradually increased from 0 to 51. According to Fig. 8.27, installing the first MS leads to considerable reduction in customer interruption costs while the reduction is gradually declined when the solution converges to the optimal solution. As was mentioned in CASE II of Example 8.2, the optimal number of MSs is equal to 31, which is shown in the figure. Figure 8.28 represents the impact of the maximum number of RCSs on the system costs. As can be seen, the system interruption cost significantly decreases when the first RCS is installed. The problem is converged when

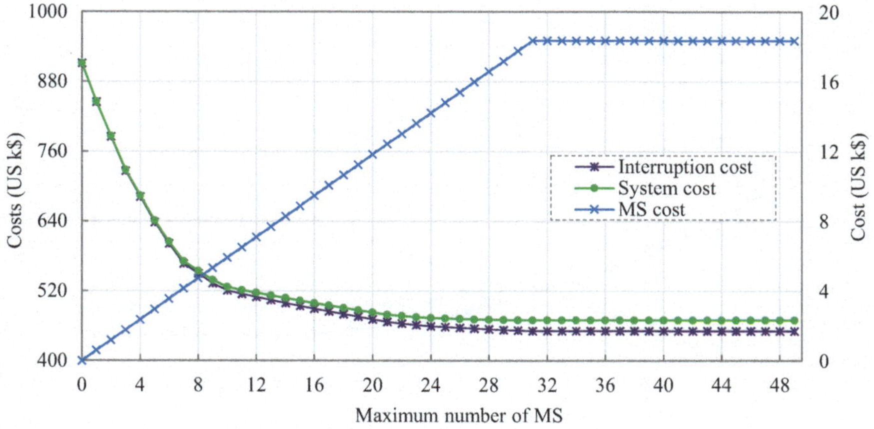


**Fig. 8.27** Impact of the maximum number of MS on the system costs

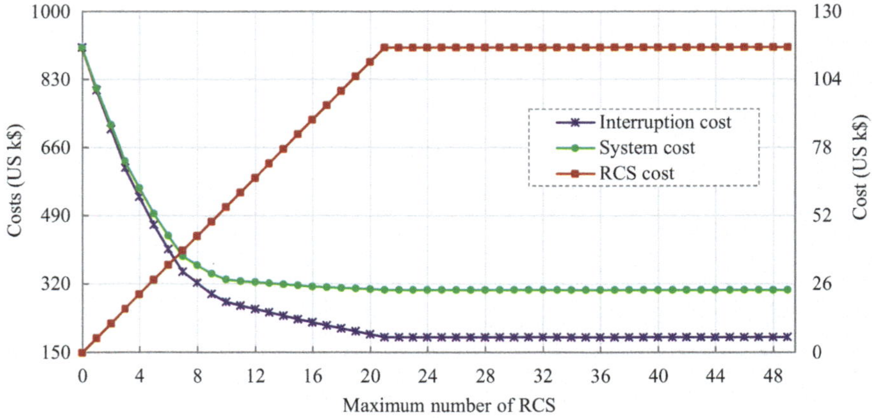


**Fig. 8.28** Impact of the maximum number of RCS on the system costs

the optimal solution reaches. According to the results, the optimal solution in this case is equal to 21 as was calculated in CASE III of Example 8.2.

## 8.5.5 *Restricted Budget*

Budget holds an utmost key in economic planning, such that most of companies are concerned about their financial resources for equipping their system. In this regard, in order to investigate the impact of budget limitation on the SS deployment problem, the economic constraint associated with a range of budget limitation is taken into account. The relevant results are shown in Fig. 8.29. As can be seen,

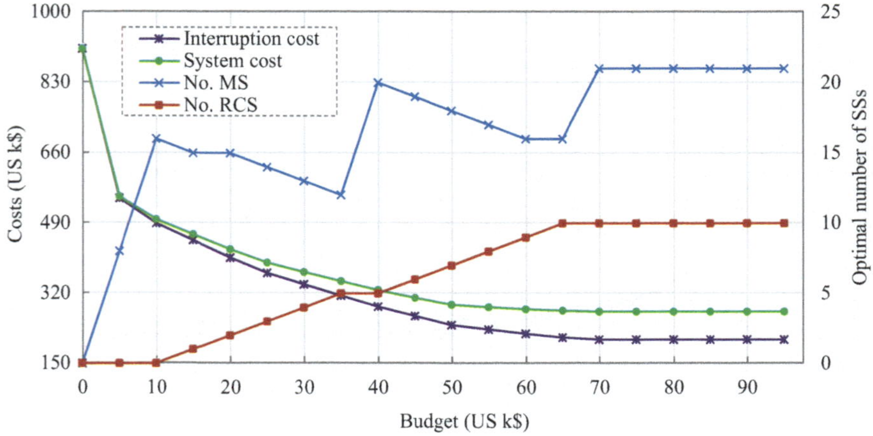


**Fig. 8.29** Impact of budget limitation on SS placement solution

only MSs are deployed in the system once allocated budget is less that US k$10. However, by allocating higher budget for equipping system with SS, the number of RCSs gradually increases, while the number of MSs does not follow a regular pattern. Broadly speaking, RCSs, due their quick operation for restoring customers, have greater ability than MS in decreasing customer interruption cost. However, they are more expensive than MSs, which might restrict their allocation in the system. In some cases, it might be more profitable to allocate several MSs than a few number of RCSs to improve the system reliability. As shown in Fig. 8.29, for some cases, by rising in the budget, the number of RCSs does not change, while more MSs are deployed in the system.

## 8.6 Switch Malfunction

Heretofore, it was assumed that SSs always operate properly. However, sometimes, SSs are not able to function as they are expected, which is referred to as SS malfunction. SS malfunction may degrade the SS worth for fault isolation procedure and consequently affects the optimal solution of the SS deployment problem. Various types of malfunction can be considered for SSs. Isolation capability malfunction of MS and RCS is referred to as their inability in isolating the faulted zone from the rest of system. In addition, the RCS may not respond to the signals sent by control center, referred to as remote controllability malfunction. In this type of malfunction, although RCS is not capable of operating remotely, the switching action can be done manually by the repair crews, treated as a MS. This section is aimed at considering the impact of the SS malfunctions on switch deployment problem. First, the impact of SS malfunction on the SS deployment problem is illustrated by an example. Then, the SS deployment problem, explained in Sect. 8.4, is reformulated by considering SS malfunctions. Finally, the relevant results are presented.

**Example 8.3** This example intends to describe the impact of various SS malfunctions, including MS and RCS isolation capability as well as RCS remote controllability malfunction, on the SS deployment problem explained in CASE IV of Example 8.1 (Fig. 8.30).

**CASE I:** MS isolation capability malfunction.

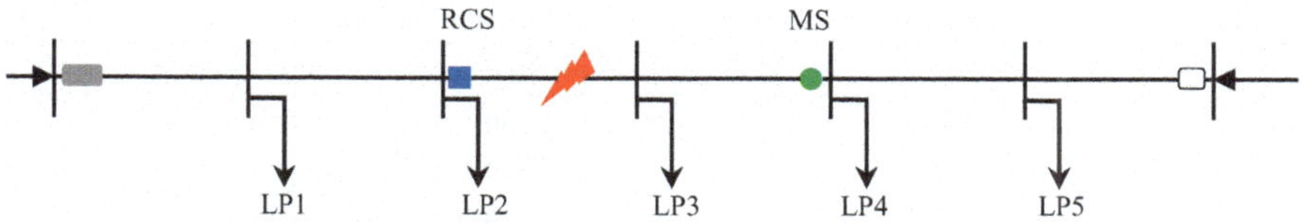


**Fig. 8.30** Representative feeder in Example 8.3

**Table 8.10** Restoration time and mode of the load points in CASE I of Example 8.3

| Load point no. | Restoration time (min) | Restoration type |
|---|---|---|
| 1 | 5 | Remote switching |
| 2 | 5 | Remote switching |
| 3 | 200 | Fault location + repair action |
| 4 | 200 | Fault location + repair action |
| 5 | 200 | Fault location + repair action |

In this case, it is assumed that the MS is not able to isolate the faulted section, while RCS operates properly. Since RCS can function successfully, steps from I to IV in this example are the same as those in Example 8.1. Therefore, the customers connected to load points 1 and 2 are assumed to be restored after remote switching action. However, after the field crews find the faulted section, load points 4 and 5 cannot be disconnected from the faulted section due to the failure in MS isolation capability. Hence, these load points should be kept interrupted during the repair action similar to load point 3. The restoration times for the load points are provided in Table 8.10.

As shown in Table 8.10, MS isolation capability malfunction leads to 120 min increment in customer interruption duration of load points 4 and 5. Hence, in this case, considering the MS malfunction can be interpreted as the absence of MS in the feeder. In addition, the interruption time of load points 1 and 2 does not change due to the successful operation of RCS between the load points and the faulted section. Also, the MS malfunction does not affect the interruption duration of load point 3 which already tolerates repair time without considering MS malfunction.

**CASE II** RCS isolation capability malfunction.

In this type of RCS malfunction, the RCS is not able to isolate faulted section. Therefore, it can be presumed that the RCS does not exist. In this regard, load points 1 and 2 cannot be isolated from the faulted section, and should remain interrupted subsequent to clearing the fault. However, the malfunction of RCS does not affect load points 4 and 5, and they can be disconnected from section 3 through the MS and restored from the adjacent feeder. With this in mind, Table 8.11 summarizes the interruption times of the load points.

As can be seen, the isolation capability malfunction of RCS increases the interruption duration of load points 1 and 2 by about 200 min. Accordingly, the

**Table 8.11** Restoration time and mode of the load points in CASE II of Example 8.3

| Load point no. | Restoration time (min) | Restoration type |
|---|---|---|
| 1 | 200 | Fault location + repair action |
| 2 | 200 | Fault location + repair action |
| 3 | 200 | Fault location + repair action |
| 4 | 80 | Fault location + manual switching |
| 5 | 80 | Fault location + manual switching |

**Table 8.12** Restoration time and mode of the load points in CASE III of Example 8.3

| Load point no. | Restoration time (min) | Restoration type |
|---|---|---|
| 1 | 80 | Fault location + manual switching |
| 2 | 80 | Fault location + manual switching |
| 3 | 200 | Fault location + repair action |
| 4 | 80 | Fault location + manual switching |
| 5 | 80 | Fault location + manual switching |

RCS isolation capability malfunction highly increases the interruption duration of customers whose connection points are supposed to be restored quickly, which results to a costly interruption.

**CASE III** RCS remote capability malfunction.

In this type of malfunction, it is assumed that the RCS cannot operate remotely, while it can isolate the fault manually, acting like a MS. Here, due to this type of malfunction, the RCS cannot receive the remote control signals, and the operator cannot limit the suspicious fault area. So, the field crews have to patrol all the feeder sections in order to find the faulted element. After determining the faulted section, the field crews open RCS and MS manually in order to restore load points 1, 2, 4, and 5 through manual switching actions. At the end, load point 3 is re-energized after that the faulted section is repaired. Table 8.12 gives the results in this case.

According to the table, due to the RCS remote controllability malfunction, the interruption duration of load points 1 and 2 increases by about 80 min. Although this growth in interruption duration is less than that of isolation capability case, i.e., 200 min, it is still much greater than remote restoration time, i.e., 5 min. Accordingly, RCS remote controllability malfunction can burden system with higher interruption costs.

By taking into account the results of the example, all types of SS malfunctions cause the interruption durations of some customers increase, thereby resulting in higher customer interruption costs. Therefore, malfunctions degrade SSs worth for reducing interruption cost and enhancing system reliability.

### *8.6.1 Switch Placement Model Considering Malfunctions*

As discussed in previous section, malfunctions detract from the SS worth in customer interruption cost reduction and improving service reliability. The effect of SS malfunctions on the system reliability and operation has been investigated in several literatures. The effect of devices failure in distribution networks was taken into account in [25]. In [26], the effect of malfunction of protective and automatic apparatuses in fault indicator deployment problem was deemed. The impact of SS malfunction on service reliability and SSs benefit was reported in [27].

As explained earlier, SS malfunction could rise the interruption duration of some customers which might result in higher interruption costs. Therefore, considering SS malfunctions can change the solution of SS deployment problem as it affects the interruption costs. To extend the SS deployment problem for considering SS malfunctions, the effect of different types of SS malfunctions on customers' interruption duration should be modeled. Here, three types of SS malfunctions including MS isolation capability malfunction, RCS isolation capability malfunction, and RCS remote controllability malfunction are considered. The impact of each type of malfunction on customers' interruption duration is expressed as follows.

#### 8.6.1.1 MS Isolation Capability Malfunction

As mentioned earlier, in MS isolation capability malfunction, the MS is not able to isolate the faulted section from the rest of the feeder. With this in mind, the following formulations calculate the customer interruption duration in case of MS isolation capability malfunction.

$$d_{f,i,j,k}^{int,m-MS} = \sum_{s' \in S} X_{f,s'}^{MS} d_{f,i,j,k,s'}^{int}; \quad \forall f \in F, \forall i \in I, \forall j \in J, \forall k \in K \tag{8.18}$$

$$d_{f,i,j,k,s'}^{int} \geq TTS_{f,s}^{RCS}; \quad \forall f \in F, \forall i \in I, \forall j \in J, \forall k \in K \tag{8.19}$$

$$d_{f,i,j,k,s'}^{int} \geq \left(TTL_{f,i} + TTS_{f,s}^{MS}\right)\left[1 - \sum_{s=2j}^{i-1} X_{f,s}^{RCS}\right]; \quad \forall f \in F, \forall i \in I, \forall 2j < i, \forall k \in K \tag{8.20}$$

$$d_{f,i,j,k,s'}^{int} \geq \left(TTL_{f,i} + TTS_{f,s}^{MS}\right)\left[1 - \sum_{s=i}^{2j-1} X_{f,s}^{RCS}\right]; \quad \forall f \in F, \forall i \in I, \forall 2j > i, \forall k \in K \tag{8.21}$$

$$d_{f,i,j,k,s'}^{int} \geq \left(TTL_{f,i} + TTR_{f,i}\right)\left[1 - \sum_{s=2j}^{i-1} X_{f,s}^{RCS} - \sum_{\substack{s=2j \\ s \neq s'}}^{i-1} X_{f,s}^{MS}\right]; \tag{8.22}$$

$$\forall f \in F, \forall i \in I, \forall 2j < i, \forall k \in K$$

$$d_{f,i,j,k,s'}^{int} \geq \left(TTL_{f,i} + TTR_{f,i}\right)\left[1 - \sum_{s=i}^{2j-1} X_{f,s}^{RCS} - \sum_{\substack{s=i \\ s \neq s'}}^{2j-1} X_{f,s}^{MS}\right]; \tag{8.23}$$

$$\forall f \in F, \forall i \in I, \forall 2j > i, \forall k \in K$$

$$d^{int}_{f,i,j,k,s'} \geq TTL_{f,i} + TTR_{f,i}; \quad \forall f \in F, \forall i \in I, \forall 2j = i, \forall k \in K \tag{8.24}$$

where $d^{int}_{f,i,j,k,s'}$ represents customer interruption duration when SS in location $s'$ encounters isolation capability malfunction. Also, $d^{int,m-MS}_{f,i,j,k,s'}$ is total customer interruption duration with considering all MSs malfunction in feeder $f$. The MS malfunction indicates that the MS cannot be able to disconnect customers from faulted section before repair action. Hence, for this type of malfunction, it can be assumed that the MS is not installed. So, to consider the malfunction of MS in location $s'$, $X^{MS}_{f,s'}$ is supposed to be excluded from (8.22) and (8.23).

### 8.6.1.2 RCS Isolation Capability Malfunction

Similar to the MS isolation capability malfunction, the impact of RCS isolation capability malfunction on the interruption duration can be formulated as

$$d^{int,ml-RCS}_{f,i,j,k} = \sum_{s' \in S} X^{RCS}_{f,s'} d^{int}_{f,i,j,k,s'}; \quad \forall f \in F, \forall i \in I, \forall j \in J, \forall k \in K \tag{8.25}$$

$$d^{int}_{f,i,j,k,s'} \geq TTS^{RCS}_{f,s}; \quad \forall f \in F, \forall i \in I, \forall j \in J, \forall k \in K \tag{8.26}$$

$$d^{int}_{f,i,j,k,s'} \geq \left(TTL_{f,i} + TTS^{MS}_{f,s}\right)\left[1 - \sum_{\substack{s=2j \\ s \neq s'}}^{i-1} X^{RCS}_{f,s}\right]; \quad \forall f \in F, \forall i \in I, \forall 2j < i, \forall k \in K \tag{8.27}$$

$$d^{int}_{f,i,j,k,s'} \geq \left(TTL_{f,i} + TTS^{MS}_{f,s}\right)\left[1 - \sum_{\substack{s=i \\ s \neq s'}}^{2j-1} X^{RCS}_{f,s}\right]; \quad \forall f \in F, \forall i \in I, \forall 2j > i, \forall k \in K \tag{8.28}$$

$$d^{int}_{f,i,j,k,s'} \geq \left(TTL_{f,i} + TTS^{MS}_{f,s}\right)\left[1 - \sum_{\substack{s=2j \\ s \neq s'}}^{i-1} X^{RCS}_{f,s} - \sum_{s=2j}^{i-1} X^{MS}_{f,s}\right]; \quad \forall f \in F, \forall i \in I, \forall 2j < i, \forall k \in K \tag{8.29}$$

$$d^{int}_{f,i,j,k,s'} \geq \left(TTL_{f,i} + TTR_{f,i}\right)\left[1 - \sum_{\substack{s=i \\ s \neq s'}}^{2j-1} X^{RCS}_{f,s} - \sum_{s=i}^{2j-1} X^{MS}_{f,s}\right]; \tag{8.30}$$
$$\forall f \in F, \forall i \in I, \forall 2j > i, \forall k \in K$$

$$d^{int}_{f,i,j,k,s'} \geq TTL_{f,i} + TTR_{f,i}; \quad \forall f \in F, \forall i \in I, \forall 2j = i, \forall k \in K \tag{8.31}$$

In (8.25), $d^{int,mI-RCS}_{f,i,j,k,s'}$ represents customer interruption duration with considering RCS isolation capability malfunction. In this type of malfunction, RCS is not able to disconnect fault, so it can be presumed that RCS does not exist. Therefore, to consider the malfunction of RCS in location $s'$, $X^{RCS}_{f,s'}$ should be removed from the RCS summation as shown in constraints (8.27)–(8.30).

#### 8.6.1.3 RCS Remote Controllability Malfunction

The impact of RCS isolation controllability malfunction on the interruption duration can be calculated as

$$d^{int,mR-RCS}_{f,i,j,k} = \sum_{s' \in S} X^{RCS}_{f,s'} d^{int}_{f,i,j,k,s'}; \quad \forall f \in F, \forall i \in I, \forall j \in J, \forall k \in K \tag{8.32}$$

$$d^{int}_{f,i,j,k,s'} \geq TTS^{RCS}_{f,s}; \quad \forall f \in F, \forall i \in I, \forall j \in J, \forall k \in K \tag{8.33}$$

$$d^{int}_{f,i,j,k,s'} \geq \left(TTL_{f,i} + TTS^{MS}_{f,s}\right)\left[1 - \sum_{\substack{s=2j \\ s \neq s'}}^{i-1} X^{RCS}_{f,s}\right]; \quad \forall f \in F, \forall i \in I, \forall 2j < i, \forall k \in K \tag{8.34}$$

$$d^{int}_{f,i,j,k,s'} \geq \left(TTL_{f,i} + TTS^{MS}_{f,s}\right)\left[1 - \sum_{\substack{s=i \\ s \neq s'}}^{2j-1} X^{RCS}_{f,s}\right]; \quad \forall f \in F, \forall i \in I, \forall 2j > i, \forall k \in K \tag{8.35}$$

$$d^{int}_{f,i,j,k,s'} \geq \left(TTL_{f,i} + TTR_{f,i}\right)\left[1 - \sum_{s=2j}^{i-1} X^{RCS}_{f,s} - \sum_{s=2j}^{i-1} X^{MS}_{f,s}\right]; \quad \forall f \in F, \forall i \in I, \forall 2j < i, \forall k \in K \tag{8.36}$$

$$d^{int}_{f,i,j,k,s'} \geq \left(TTL_{f,i} + TTR_{f,i}\right)\left[1 - \sum_{s=i}^{2j-1} X^{RCS}_{f,s} - \sum_{s=i}^{2j-1} X^{MS}_{f,s}\right]; \quad \forall f \in F, \forall i \in I, \forall 2j > i, \forall k \in K \tag{8.37}$$

$$d^{int}_{f,i,j,k,s'} \geq TTL_{f,i} + TTR_{f,i}; \quad \forall f \in F, \forall i \in I, \forall 2j = i, \forall k \in K \tag{8.38}$$

where $d^{int,mR-RCS}_{f,i,j,k,s'}$ is designed as customer interruption duration corresponding to consider RCS remote controllability malfunction in location $s'$. In this type of malfunction, the RCS fails to remotely isolate by the control signal from the control center. Hence, the customers who were supposed to be isolated through remote switching actions should remain interrupted until manual switching actions are done by the field crews. With this in mind, in order to consider the RCS with remote controllability malfunction in location $s'$, $X^{RCS}_{f,s'}$ should be excluded from the RCS summation as shown in (8.34) and (8.35). It is worth mentioning that the RCS which is not able to be isolated remotely should be included in RCS summation in constraints (8.36) and (8.37) since they can be opened manually and isolate the faulted section.

#### 8.6.1.4 Total Interruption Duration

So far, the customer interruption duration associated with all types of SS malfunctions was calculated. By considering the malfunction probability of the mentioned malfunctions, the following equation determines the total customer interruption duration.

$$\begin{aligned} d^{tot-int}_{f,i,j,k} &= (1 - P^{m-MS}_{f})(1 - P^{mI-RCS}_{f} - P^{mR-RCS}_{f})d^{int}_{f,i,j,k} \\ &+ P^{m-MS}_{f}(1 - P^{mI-RCS}_{f} - P^{mR-RCS}_{f})d^{int,m-MS}_{f,i,j,k} \\ &+ (1 - P^{m-MS}_{f})P^{mI-RCS}_{f}d^{int,mI-RCS}_{f,i,j,k} \\ &+ (1 - P^{m-MS}_{f})P^{mR-RCS}_{f}d^{int,mR-RCS}_{f,i,j,k} \end{aligned} \tag{8.39}$$

where $P^{m-MS}_{f}$, $P^{mI-RCS}_{f}$, and $P^{mR-RCS}_{f}$ represent the probability of MS isolation capability malfunction, RCS isolation capability malfunction, and RCS remote controllability malfunction, respectively. According to (8.39), the total customer interruption duration is the summation of four terms. The first term relates to the mode that no SS malfunction is considered, referred to Sect. 8.4. The second term is associated with MS isolation capability malfunction. Also, the third and fourth terms belong to RCS isolation capability and RCS remote controllability malfunction, respectively. Accordingly, the SS deployment problem can be extended to consider SS malfunction through considering the calculated customers interruption duration in (8.39).

To find the optimal number and location of SSs, it is necessary to minimize the system cost comprising of system interruption cost and SSs cost is minimized as

$$\text{Minimize } Cost^{int} + Cost^{SS} \tag{8.40}$$

where $Cost^{SS}$ is calculated according to (8.9)–(8.12). $Cost^{int}$ is determined as

$$Cost^{int} = \sum_{t\in T}\sum_{f\in F}\sum_{i\in I}\sum_{j\in J}\sum_{k\in K}\frac{(1+q_{\mathrm{lg}})^{t-1}}{(1+q_{dr})^{t}}\lambda_{f,i}L_{f,j,k}CDF_{f,i,j,k}(d_{f,i,j,k}^{tot-int}) \tag{8.41}$$

Above expression is similar to (8.2) except that the total interruption duration $(d_{f,i,j,k}^{tot-int})$ by considering the impact of SSs malfunction is taken into account.

**Example 8.4** In this example, the effect of SS malfunction on the SS deployment problem in RBTS-Bus4 is investigated using the extended model.

**CASE I** This case has to do with MS deployment problem by considering the MS isolation capability malfunction. The probability of MS malfunction is assumed to be 0.05. The optimal location of MSs is provided in Table 8.13. As can be seen, the number of installed MSs is reduced from 31, referred to CASE II of Example 8.2, to 21 in this case. Therefore, it can be concluded that considering MS malfunction causes the number of allocated MS in the system declines.

**CASE II** In this case, the optimal deployment of RCSs with considering RCS malfunctions is examined. The probabilities of RCS isolation capability malfunction and isolation controllability malfunction are considered to be equal to 0.015 and 0.02, respectively. The optimal location of RCSs is given in Table 8.14. According to the results, when RCSs fail to isolate the faulted section remotely or manually, the optimal number of RCSs does not change and is equal to the number of RCSs when they are able to operate properly.

**Table 8.13** Optimal location of MSs in CASE I of Example 8.4

| Feeder | MSs location |
|---|---|
| 1 | 7,9 |
| 2 | 1,2,3,4 |
| 3 | 5,9 |
| 4 | 5,9 |
| 5 | 1,2,4,5 |
| 6 | 1,2,3,4,5 |
| 7 | 5,9 |
| Total | 21 |

**Table 8.14** Optimal location of RCSs in CASE II of Example 8.4

| Feeder | RCSs location |
|---|---|
| 1 | 5,9 |
| 2 | 1,2,3,4 |
| 3 | 5,9 |
| 4 | 5,9 |
| 5 | 1,2,4,5 |
| 6 | 1,2,3,4,5 |
| 7 | 5,9 |
| Total | 21 |

**Table 8.15** Optimal location of RCSs and MSs in CASE III of Example 8.4

| Feeder | MSs location | RCSs location |
|---|---|---|
| 1 | 5 | 9 |
| 2 | 1,2,5 | 3,4 |
| 3 | 5 | 9 |
| 4 | 5 | 9 |
| 5 | 1,2,5 | 3,4 |
| 6 | 1,2,4 | 3,5 |
| 7 | 5 | 9 |
| Total | 13 | 10 |

**CASE III** In this case, the optimal number and location of MSs and RCSs are determined with considering all types of SS malfunction. The malfunction probabilities of MS and RCS are considered the same as in previous cases. Here, the optimal placement of RCS and MS is taken into account and the optimal solution is presented in Table 8.15. As can be seen, the optimal number of MSs is decreased from 21 to 13, while the optimal number of RCSs does not change and is equal to the number of RCSs when they operate properly.

Tables 8.16 and 8.17 summarize the number of SSs and system costs for the cases of considering and not considering SS malfunctions in SS deployment problem. By comparing the results, it can be noticed that considering the SSs malfunction reduces the number of deployed SSs and increases the customer interruption costs. Also, according to expression (8.39), the SSs benefit for enhancing system reliability relies on the malfunction probability of SSs and has

**Table 8.16** System costs (US k$) in CASE I–CASE III of Example 8.4

| CASE | Equipment | | | | Interruption (US k$) | Total (US k$) |
|---|---|---|---|---|---|---|
| | Number of MSs | Number of RCSs | MSs (US k$) | RCSs (US k$) | | |
| I | 21 | – | 12.44 | – | 472.04 | 484.48 |
| II | – | 21 | – | 118.11 | 197.59 | 315.70 |
| III | 13 | 10 | 7.70 | 56.25 | 225.55 | 289.50 |

**Table 8.17** System costs (US k$) in CASE I–CASE IV of Example 8.2

| CASE | Equipment | | | | Interruption (US k$) | Total (US k$) |
|---|---|---|---|---|---|---|
| | Number of MSs | Number of RCSs | MSs (US k$) | RCSs (US k$) | | |
| I | – | – | – | – | 911.72 | 911.72 |
| II | 31 | – | 18.37 | – | 452.14 | 470.51 |
| III | – | 21 | – | 118.11 | 190.10 | 308.21 |
| IV | 21 | 10 | 12.44 | 56.25 | 208.10 | 276.79 |

significant impact on the number of SSs. So, the SSs benefits wane as the malfunction probability is increased and consequently the allocated SSs are reduced.

## 8.7 Uncertainty and Financial Risk

In previous sections, the optimal solution of switch placement was determined by comparing the expected interruption costs and switch costs. The initial investment is known before the deployment of SSs while the system interruption costs rely on several parameters which are a function of uncertain parameters. The uncertainties alter the benefits of SSs in reducing the customer interruption costs which interprets that they impose undeniable financial risk on the investor as a private company. The purpose of this section is to assess the financial risk when the SSs are utilized in the network. To do so, the step by step algorithm is presented in the following sections. At the beginning, the main sources of uncertainties are defined which have direct correlation with the system interruption cost. Then, several scenarios are generated representing the status of selected uncertain parameters. Next, the optimal fault management, similar to optimal switch placement except that here, the location of SSs is known, is applied to each of the scenarios in order to calculate the system interruption cost. Finally, the final indices including DisCo profit and risk are reported. The financial risk is calculated through a pragmatic index which is commonly used in risky situations. The proposed approach is thoroughly discussed in the following subsections.

### *8.7.1 Uncertain Parameters*

The uncertainties in distribution systems are originated from various sources like the load forecast error, uncertain characteristic of renewable energy resources, stochastic nature of fault occurrence, and variable duration of repair actions. Among the proposed factors, stochastic nature of contingencies as well as repair time hold a substantial impact on the SSs achievements. In order to take into account the uncertainties, several approaches like mathematical and scenario based model, to name just a few, were presented. The mathematical approaches take the advantages of fuzzy concept or probability density function. The fuzzy approach is used when the values of uncertain parameters are not accessible. While, the other one considers the probability density function for each uncertain parameter in order to achieve the probability density function of the objective. The scenario based approach utilizes a set of scenarios that represent the status of uncertain parameters. Although the mathematical methods provide more precise solutions, their complexity leads to significant computation impediments. To avoid this issue, the scenario based approach is taken into account in this section. Also, the enough number of scenarios are generated to assure the accuracy of the method. It is worth mentioning that scenario reduction techniques can decline the number of generated scenarios with negligible impact on the accuracy of the final solution. In order to keep the reduced scenarios close to the original scenarios, they are determined based on the probability distance concept. The most common probability distance is the Kantorovich distance which is defined between two probability distributions. In [28], two scenario reduction techniques based on the Kantorovich distance were proposed. Among them, fast forward selection algorithm is taken into account in this chapter. The algorithm is a repetitive process where, at each iteration, one scenario is selected such that the Kantorovich distance between the original and the selected scenario sets is minimized. The algorithm is terminated when the number of selected scenarios reaches a predefined number. Finally, probability of the non-selected scenarios is transferred to the selected scenarios [29].

## *8.7.2 Financial Risk Indices*

As was mentioned heretofore, the uncertain parameters alter the RCS profit from the expected value and thus, induce financial risk. In this regard, various risk measures have been introduced to evaluate the financial risk [29]. The most common risk indices are volatility index (VolIn), shortfall probability (SP), expected shortage (ES), value at risk (VaR), and conditional value at risk (CVaR) to name just a few. The description and formulation of the mentioned indices are presented hereinafter.

### 8.7.2.1 Volatility Index (VolIn)

VolIn represents the variance of the profit from the expected value. In order to determine the VolIn, the expected value of the difference between the profit and the expected profit should be calculated. By assuming as the profit of each scenario, VolIn is formulated as

$$\mathrm{VolIn} = \varepsilon\left\{(\mathrm{Profit}_\omega - \varepsilon\{\mathrm{Profit}_\omega\})^2\right\} \tag{8.42}$$

where $\varepsilon$ is the expected value calculator and $\varepsilon\{\mathrm{Profit}_\omega\}$ is the expected profit. According to the definition of the index, the situation is more risky when VolIn value is higher.

#### 8.7.2.2 Shortfall Probability (SP)

SP represents the cumulative probability of profit smaller than a predefined profit. By considering the predefined value of profit as $\eta$, the mathematical formulation of SP can be expressed as

$$\mathrm{SP}_\eta = P(\omega|\mathrm{Profit}_\omega < \eta); \quad \forall \eta \in R \tag{8.43}$$

where $P$ is the cumulative probability of profit. As can be observed from the above formula, the situation is more risky when SP converges to 1. However, if SP value is near 0, the situation is less risky. Hence, the higher the index, the more risky is the situation.

#### 8.7.2.3 Expected Shortage (ES)

ES measures the expected profit for the scenarios with the profit less than a predefined value. In other words, ES is similar to SP, except that it calculates the expected profit while SP measures the probability. The formulation of ES is shown as

$$\mathrm{ES}_\eta = \varepsilon\{\mathrm{Profit}_\omega|\mathrm{Profit}_\omega < \eta\}; \quad \forall \eta \in R \tag{8.44}$$

According to the above formula, it is clear that the situation is less risky when ES value is higher.

#### 8.7.2.4 Value at Risk (VaR)

VaR indicates the maximum value of profit within a predefined percent of the worst scenarios. The worst scenarios have the least values of profit. VaR can be formulated as

$$\mathrm{VaR}_\alpha = \sup\{\eta|P(\omega|\mathrm{Profit}_\omega < \eta) \le 1 - \alpha\}; \quad \forall \alpha \in (0, 1) \tag{8.45}$$

where sup and $\alpha$ are the supreme value of a set and the predefined percent, respectively. VaR formulation reveals that the situation is less risky when its value is higher.

#### 8.7.2.5 Conditional Value at Risk (CVaR)

CVaR donates the expected value of profit in a predefined percent of the worst scenarios. CVaR can be expressed mathematically as

$$C\text{VaR}_{\alpha} = \varepsilon\{\text{Profit}_{\omega}|\text{Profit}_{\omega} < VaR_{\alpha}\}; \quad \forall\alpha \in (0, 1) \tag{8.46}$$

As can be expected, the financial risk decreases when CVaR value is increased.

By taking into account the above explanations, both SP and ES are calculated based on predefined values of profit while VaR and CVaR measure the risk based on predefined values of probability. Since determining the predefined value of profit is difficult, it is logical to consider a predefined percent of the most severe scenarios. Also, making decision based on the expected value is more comparative than the maximum value. So, the financial risk is represented with CVaR in this chapter.

### *8.7.3 Scenario Generation*

Here, the introduced uncertainties, i.e., fault occurrence and repair time, are considered to compose the situation of the system in each scenario. Then, several scenarios are generated by iterating the method presented in this subsection. A scenario represents the up/down status of elements in the system. Once a component fails to operate properly, its status changes from up to down. The time that the element fails has a direct correlation with the probability distribution function of the element failure. Various probability distribution functions have been employed to calculate the time to failure such as exponential, gamma, lognormal, and passion. Without loss of generality, the exponential distribution function is taken into account here. The Monte Carlo Simulation (MCS) is taken in use to generate scenarios. To do so, the following steps represent this approach [30].

Step 1: Time to failure of an element located in section $i$ and feeder $f$ is as

$$TTF_{f,i} = -MTTF_{f,i} \times \ln(u) \tag{8.47}$$

where $MTTF_{f,i}$ is the mean time to failure of the element and $u$ is a random variable with standard distribution. Time to failure indicates the time that the element fails to work properly. In other word, the status of the element changes from up state to down state.

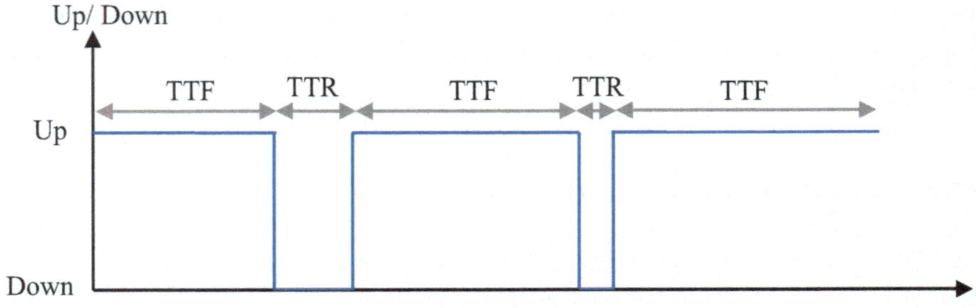


**Fig. 8.31** Up/down status of sample element

Step 2: After calculating time to failure of the faulted element, the time to repair of the element is calculated as

$$TTR_{f,i} = -MTTR_{f,i} \times \ln(u) \tag{8.48}$$

where $MTTR_{f,i}$ refers to mean time to repair of the element.

Step 3: Steps 1 and 2 are iterated for a period equal to or greater than the length of horizon study. Then, the up/down status of each element during the study horizon is determined. As an example, Fig. 8.31 shows the up/down status of a sample element.

Step 4: Steps 1–3 are repeated for all elements in the networks.

Step 5: In this step, the up/down status of elements is compared together to determine the up/down status of the network. By doing this, the location of faulted sections, time to failure as well as time to repair of them can be obtained from the up/down status of system.

Step 6: Steps 1–5 are repeated for specific number of scenarios.

The following example clarifies the scenario generation to product a scenario.

**Example 8.5** The up/down status of the feeder shown in Fig. 8.32 within a 15-year horizon is determined. Assume that time to failure and time to repair of the elements are 5 years and 3 h, respectively. To avoid complexity of the solution, it is

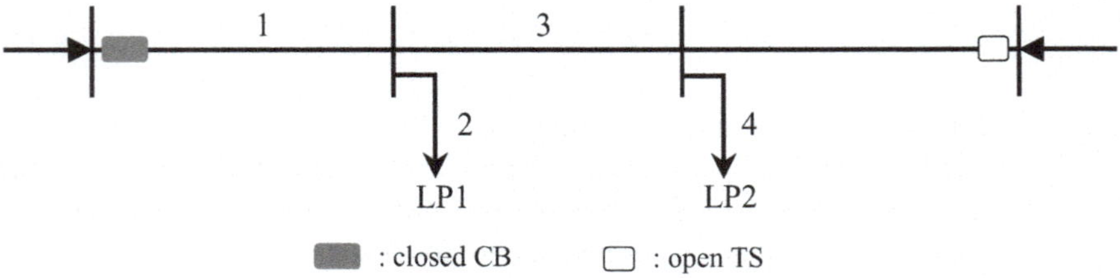


**Fig. 8.32** Representative feeder for scenario generation in Example 8.5

**Table 8.18** Time to failure and time to repair (h) of sections in the sample feeder in Example 8.5

| Section | Time to failure | Time to repair | Time to failure | Time to repair | Time to failure | Time to repair |
|---|---|---|---|---|---|---|
| 1 | 43,800 | 5 | 87,600 | 1 | – | – |
| 2 | 61,320 | 3 | 43,800 | 2 | 70,080 | 2 |
| 3 | 175,200 | 1 | – | – | – | – |
| 4 | 78,840 | 2 | 52,560 | 3 | – | – |

presumed that the random values are calculated and the corresponding time durations are determined.

To diminish the complexity of the MCS procedure, steps 1 to 3 of MCS is summarize in this example. So, random numbers are generated for each section and time to failure and time to repair for sections 1–4 are calculated. Since the length of study horizon is 15 years, the up/down status of elements should be calculated until the total time of each section reaches 15 years. Table 8.18 represents the up/down information of the sections.

According to the table, the first fault in section 1 occurs in time 43,800 h. The section requires 5 h to be repaired. Since the second fault in this section happens in time 131,405 h (= 43,800 + 5 + 87,600) which is larger than the length of the study horizon (131,400 h), just the first fault is considered in the up/down status of this section. In section 2, the first and the second fault occur in time 61,320 and 105,123 h (= 61,320 + 3 + 43,800), respectively. In section 3, since the first fault occurs after the length of study horizon, the status of the section remain up state during the study horizon. Finally, only one fault happens in section 4 which is in time 78,840 h and it takes 2 h to be repaired. By taking into account aforementioned points regarding the up/down status of sections, the up/down status of the sample system can be determined. In this regard, the first fault occurs in section 1 at time 43,800 h and the repair action takes 5 h long. The second one happens in section 2 at time 61,320 h and the repair action takes 3 h long. The third fault occurs in section 4 at time 78,840 h and 2 h is required to be repaired. The forth and the final fault happens in section 2 at time 105,123 h which 2 h is required to be repaired.

According to the example, generating time to failure and time to repair is based on the random values. So, it is possible that different scenarios have different numbers of fault. In order to achieve the proper accuracy in financial risk assessment, it is necessary to generate enough number of scenarios. When a fault occurs in the system, the fault management process should be conducted to reduce the consequences of fault occurrence. Hence, the next section presents the optimum fault management problem once a contingency occurs.

### 8.7.4 Optimum Fault Management Problem

By taking into account the points in Sect. 8.3, the aim of fault management process is to reduce the interruption duration of the customers once a contingency happens in the system. Also, the reduction in customer interruption duration is translated to reduction in customer interruption cost. In this regard, it is necessary to minimize the customer interruption cost when a contingency occurs in the system [31]. The optimum fault management problem for each contingency is as

$$\mathrm{Cost}_c^{int} = \sum_{f\in F}\sum_{i\in I}\sum_{j\in J}\sum_{k\in K}\frac{(1+q_{\lg})^{t_c-1}}{(1+q_{dr})^{t_c}}L_{f,j,k}CDF_{f,i,j,k}(d_{f,i,j,k}^{int}) \tag{8.49}$$

As can be seen in (8.49), the minimization of customers interruption cost is solved for each contingency. The costumer interruption cost depends on the average load of customers and the CDF. The average load is constant except that it grows with a predefined growth rate. However, the CDF relies on the customer interruption duration. The interruption duration depends on the correlation between the faulted section and the load point that feeds the customer. In case any RCS is present between the two, the customers can be isolated from the faulted equipment via remote switching actions. In this situation, the interruption duration is equal to remote switching action time. Other customers who can be isolated through manual switching actions will remain interrupted after locating the precise location of fault and prior repairing the faulted section. So, the interruption duration for these customers is longer than the customers whose service is restored remotely. Finally, the other customers who cannot be isolated from the faulted section should be kept interrupted until the faulted equipment is repaired. So, the customers can be restored in three ways as formulated in the following:

$$d_{f,i,j,k}^{int} \geq TTS_{f,s}^{RCS}; \quad \forall f\in F, \forall i\in I, \forall j\in J, \forall k\in K \tag{8.50}$$

$$d_{f,i,j,k}^{int} \geq \left(TTL_{f,i}+TTS_{f,s}^{MS}\right)\left[1-\sum_{s=2j}^{i-1}X_{f,s}^{RCS}\right]; \quad \forall f\in F, \forall i\in I, \forall 2j<i, \forall k\in K \tag{8.51}$$

$$d_{f,i,j,k}^{int} \geq \left(TTL_{f,i}+TTS_{f,s}^{MS}\right)\left[1-\sum_{s=i}^{2j-1}X_{f,s}^{RCS}\right]; \quad \forall f\in F, \forall i\in I, \forall 2j>i, \forall k\in K \tag{8.52}$$

$$d_{f,i,j,k}^{int} \geq (TTL_{f,i}+TTR_{f,i})\left[1-\sum_{s=2j}^{i-1}X_{f,s}^{RCS}-\sum_{s=2j}^{i-1}X_{f,s}^{MS}\right]; \quad \forall f\in F, \forall i\in I, \forall 2j<i, \forall k\in K \tag{8.53}$$

$$d_{f,i,j,k}^{int} \geq \left(TTL_{f,i} + TTR_{f,i}\right)\left[1 - \sum_{s=i}^{2j-1} X_{f,s}^{RCS} - \sum_{s=i}^{2j-1} X_{f,s}^{MS}\right]; \quad (8.54)$$
$$\forall f \in F, \forall i \in I, \forall 2j > i, \forall k \in K$$

$$d_{f,i,j,k}^{int} \geq TTL_{f,i} + TTR_{f,i}; \quad \forall f \in F, \forall i \in I, \forall 2j = i, \forall k \in K \quad (8.55)$$

The above expressions have the same meaning as (8.3)–(8.8) in Sect. 8.4. In this regard, to avoid repetition of what was discussed, the explanation of these equations is avoided. It is worth mentioning that $X_{f,s}^{RCS}$ and $X_{f,s}^{MS}$ are among known parameters whose values are obtained from the optimal SS placement problem in Sect. 8.4. By solving the fault management problem, the customer interruption cost is determined for each contingency. Then, the system interruption cost is calculated by summing up the customer interruption costs of contingencies within the scenario. Heretofore, scenario generation and interruption cost calculation methods were introduced. In the following, an approach is presented to achieve financial risk.

### 8.7.5 *Financial Risk Evaluation Approach*

In order to evaluate the imposed financial risk on a DisCo, it is necessary to obtain the probability density function of its profit. To do so, the profit of SS deployment is defined as the reduction in system interruption cost before and after SS deployment minus the SS costs. The step by step framework for deriving PDF of system interruption cost is shown in Fig. 8.33 [31]. The approach to calculate financial risk is as follows.

#### 8.7.5.1 Step 1: Preparing Input Data

The first step is to prepare input data including network data, study horizon, financial assumptions, and CDFs. The network data consists of failure rate and repair time of equipments, number and types of customers, the average load of customers, and the location of MSs and RCSs in the network. The location of SSs is known and it is determined from the optimal SS placement problem presented in Sect. 8.4. Study horizon is the time that the benefit of SS deployment is studied. Financial assumptions consist of switch costs, discount rate, and load growth rate. CDFs for different types of customers as function of interruption duration should be prepared in this step.

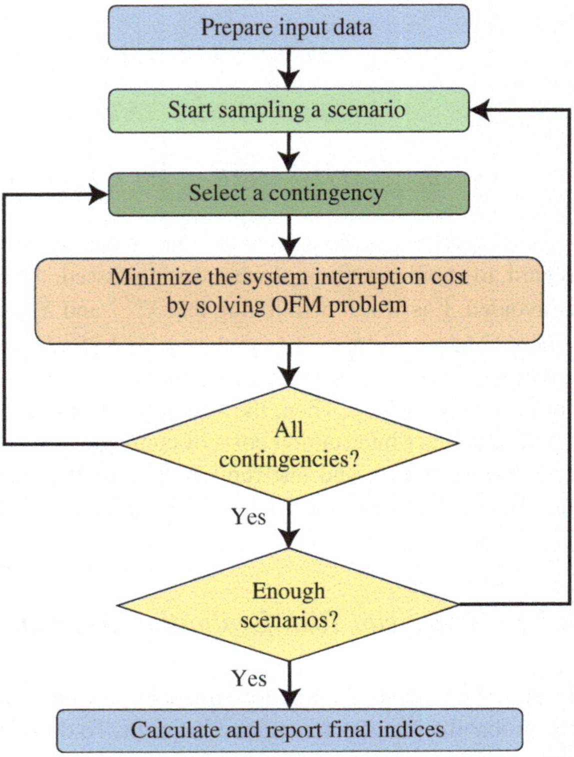


**Fig. 8.33** The proposed algorithm to calculate the financial risk under power system uncertainty

#### 8.7.5.2 Step 2: Scenario Selection

In this step, a scenario is selected from the generated scenarios. The scenario selection is iterated until the number of sampled scenarios reaches the maximum number of scenarios.

#### 8.7.5.3 Step 3: Contingency Selection

After a scenario is selected in previous step, a contingency within the sampled scenario is selected. According to the selected contingency, the location and the time to repair the faulted section are determined.

#### 8.7.5.4 Step 4: Optimum Fault Management Problem

Once the location and repair time of the faulted section are determined, the fault management problem is solved in order to minimize the system interruption cost. To do so, the proposed model of fault management is applied. It is worth mentioning that customers interruption duration and interruption costs are outputs of the problem.

#### 8.7.5.5 Step 5: Determining Probability Density Function

As mentioned heretofore, each scenario comprises of several contingencies within the study horizon. To calculate the system interruption costs, Steps 3–4 should be repeated for all contingencies within the sampled scenario. The system interruption cost for each scenario is the combination of interruption costs. Then, the process continues from Step 2 to iterate over all generated scenarios. Doing so, the probability density function of system interruption cost is obtained.

#### 8.7.5.6 Step 6: Calculation of Profit and Financial Risk Index

The previous steps should be conducted before and after installing SSs. The difference of the two PDFs is the PDF of gross profit of SS deployment. The PDF of net profit is calculated by subtracting SS costs from the gross profit. Then, the financial risk of SS placement can be calculated by taking into account the obtained PDF.

**Example 8.6** PDF of RCS profit for the network considered in Example 8.2 is determined here. To do so, 10,000 scenarios are generated to guarantee accuracy of the results. It is assumed that $\alpha$ in CVaR index calculation is equal to 0.99. Also, the location of RCS is determined based on the results of Example 8.2 as represented in Fig. 8.34.

Applying the presented approach for financial risk assessment, PDF of the profit is calculated and represented in Fig. 8.35. As can be seen, although the expected profit is US k$496.64, the profit varies from US k-21 to US k$1800. The wide range of the profit variation imposes a significant risk which should be taken into account in decision making.

The cumulative density function of the SS profit is depicted in Fig. 8.36. As can be seen, in 0.22% of the scenarios (i.e., 22 out of 10,000 generated scenarios), installing RCS leads to negative profits which means that the DisCo is confronted to loss.

The results associated with system costs with and without RCS deployment, RCS profit as well as risk index are provided in Table 8.19. As can be seen, the expected profit is 4.2 times the RCS cost. Also, the CVaR is 90% smaller than the expected profit which indicates the RCS deployment strategy as a risky investment plan.

As was mentioned, installing RCS reduces the interruption costs when a contingency occurs in the system. In this regard, it can be anticipated that the higher interruption costs prior to RCS installation, the more RCS profit is achieved. Besides, the interruption cost chiefly depends on several factors such as the number

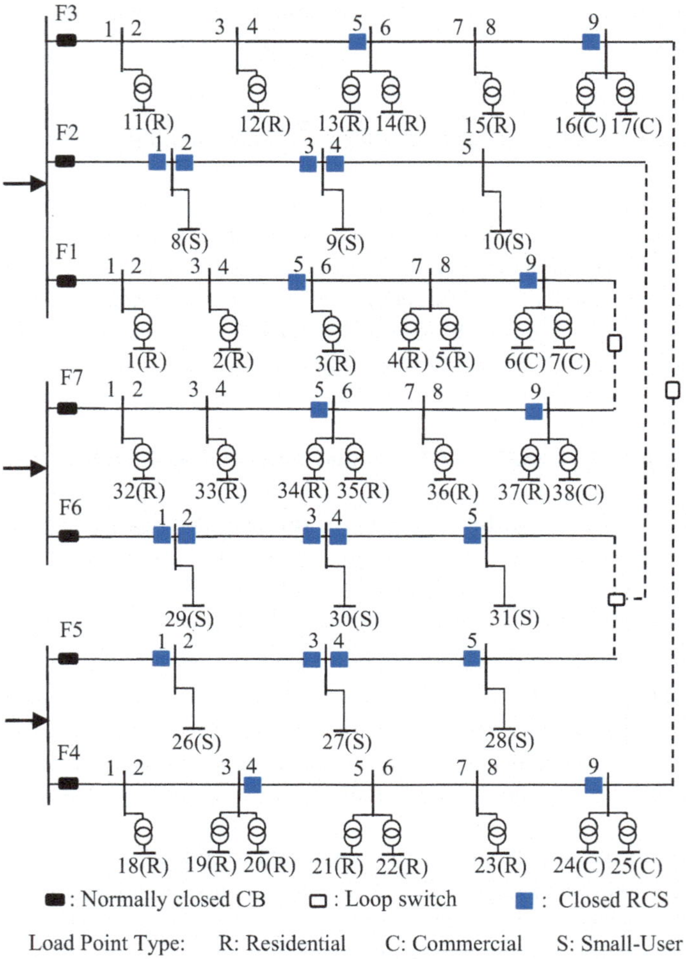


**Fig. 8.34** Location of installed RCSs in RBTS-Bus4

of fault occurrences, the repair action duration, and the energy curtailment. In this regard, the impact of the mentioned factors on the RCS profit is scrutinized and the results are shown in Fig. 8.37. As can be seen, the expected profit is near US k$340 when 16 faults occur or the interruption duration is about 48 h. Also, the profit is almost US k$650 when 26 faults occur or the interruption duration lasts 77 h. Furthermore, when the energy curtailment and interruption costs are 185 kWh and US k$520, respectively, the profit is near US k$270. Also, when the energy curtailment and interruption costs are 385 kWh and US k$1025, respectively, the net profit is near US k$700. Hence, the finding results indicate the positive relationship between the mentioned factors and RCS profit. It is worth pointing that although the RCS installation provides considerable profit, it is possible that the DisCo

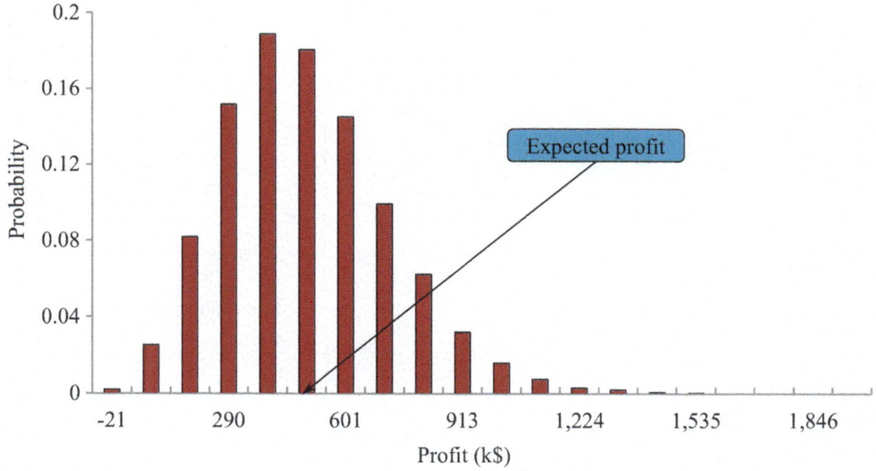


**Fig. 8.35** Individual probability of RCS deployment profit

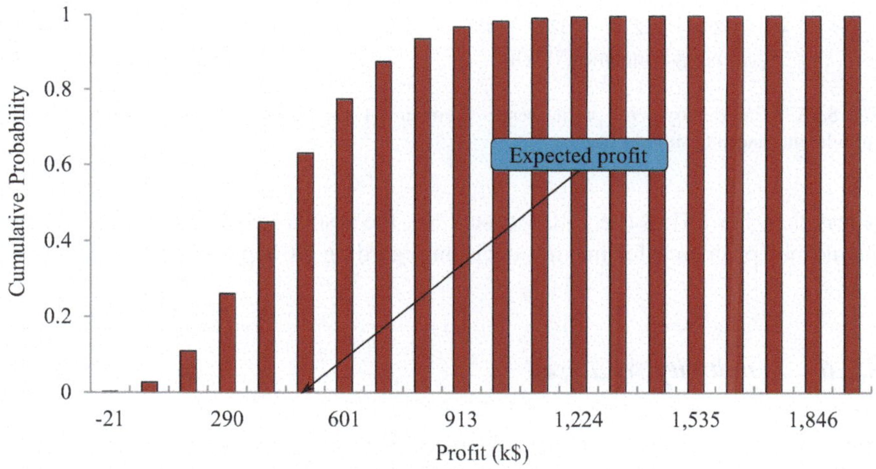


**Fig. 8.36** Cumulative density function of RCS deployment profit

**Table 8.19** Financial results (US k$) with and without RCS installation

| Costs without RCS | | | Costs with RCS | | | Expected profit | CVaR |
|---|---|---|---|---|---|---|---|
| RCS | Interruption | Total | RCS | Interruption | Total | | |
| – | 779.72 | 779.72 | 118.11 | 164.96 | 283.07 | 496.64 | 47.69 |

encounters with undeniable negative profit (loss). For example, the profit is negative when the number of faults is less than 7, system remains interrupted for less than 11 h, total energy curtailment is smaller than 40 kWh, or the interruption cost

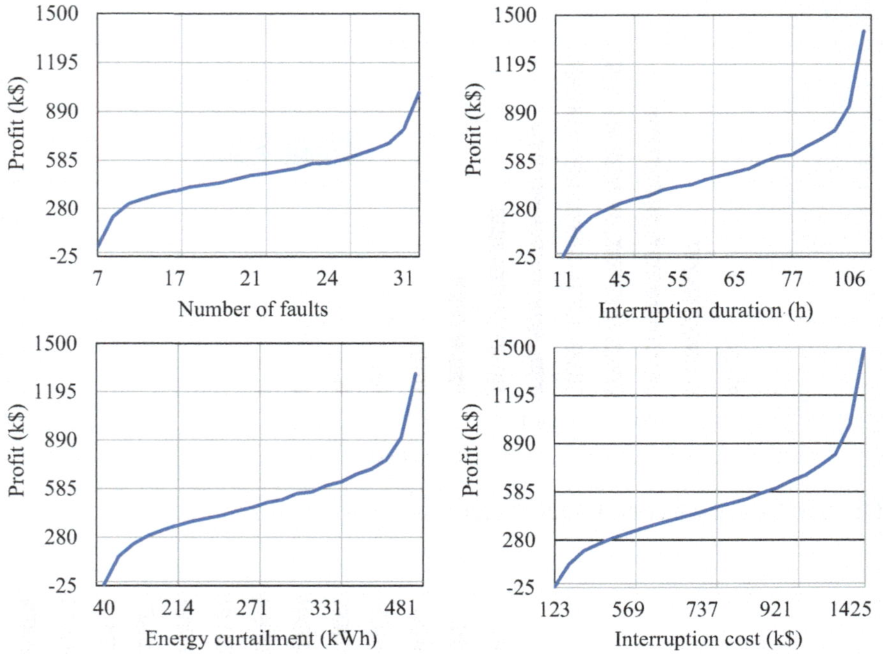


**Fig. 8.37** RCS deployment profit versus number of faults, interruption duration, energy curtailment, and interruption cost

is less than US k$118 (i.e., RCS costs). So, the results reveal that several factors should be considered for making decision regarding RCS placement.

## *8.7.6 Sensitivity Analyses*

Heretofore, it was revealed that installing RCS imposes significant risk on the DisCo. Here, the impact of key parameters such as the number of installed RCSs, length of study horizon, system size, RCS costs, CDF failure rate and repair time of equipments on the expected profit and financial risk is scrutinized.

### 8.7.6.1 Number of Installed RCSs

In previous simulations, it was assumed that 21 RCSs are allocated to the system. Here, in order to study the impact of the number of RCSs, the simulation is repeated for different number of RCSs. To do so, the number of RCSs is increased from 1 to 30 according to the priority order developed in Sect. 8.4. The expected profit and

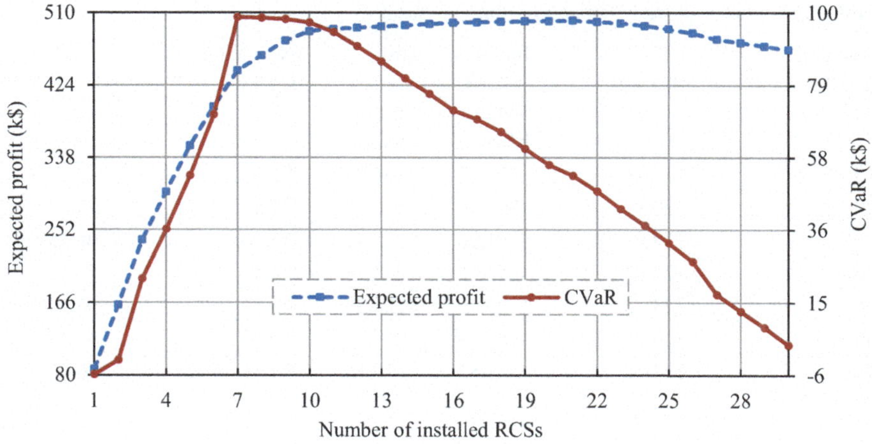


**Fig. 8.38** Expected profit and CVaR versus the number of installed RCSs

CVaR versus the number of embedded RCSs are depicted in Fig. 8.38. As can be seen, the maximum expected profit and CVaR are obtained when 21 and 7 RCSs are installed in the network, respectively. The maximum value of CVaR indicates the minimum value of financial risk. In this regard, the risk-taker DisCo prefers to install 21 RCSs in order to achieve the highest profit. However, the risk-averse DisCo allocates 7 RCSs in order to decline the imposed financial risk. So, the compromiser DisCo equips the system with any number of RCSs from 7 to 21 in order to achieve a tradeoff between the profit and risk. So, the DisCos with different risk awareness may make different decisions to equip their system.

### 8.7.6.2 Length of Study Horizon

In order to determine the impact of the length of study horizon, the simulation is repeated by increasing the study horizon duration. To do so, a set of scenarios is produced for 45 years and the impact of study horizon is observed from 15 to 45 years. For example, in case of study horizon equal to 30 years, the simulation is accomplished for contingencies occur before the 30th year and other contingencies are missed. The expected profit and CVaR for different lengths of study horizon are provided in Table 8.20. According to the results, the expected profit and CVaR are increased as the study horizon is extended since more contingencies are likely to occur for longer study horizons. Also, the value of CVaR is more sensitive to the study horizon than the expected profit. For example, when the study horizon is increased from 15 to 30 years by 2 times, the expected profit and CVaR are augmented by almost 1.6 and 5.4 times. So, considering financial risk in RCS deployment planning is essential in studies for short study horizons.

Table 8.20 Impact of study horizon on financial results

| Planning horizon (year) | RCS cost (US k$) | Interruption cost (US k$) | | Expected profit (US k$) | CVaR (US k$) |
|---|---|---|---|---|---|
| | | Without RCS | With RCS | | |
| 15 | 118.11 | 779.72 | 164.96 | 496.64 | 47.69 |
| 20 | 120.97 | 940.14 | 198.15 | 621.01 | 134.20 |
| 25 | 122.91 | 1061.96 | 223.99 | 715.06 | 205.38 |
| 30 | 124.23 | 1155.56 | 243.78 | 787.5 | 257.21 |
| 35 | 125.13 | 1229.05 | 259.34 | 844.58 | 305.69 |
| 40 | 126.75 | 1286.89 | 270.61 | 889.53 | 346.40 |
| 45 | 127.16 | 1331.82 | 280.16 | 924.51 | 376.63 |

#### 8.7.6.3 System Size

Since the size of typical networks is much larger than the size of the network used in this simulation, the effect of system size on the result of SS deployment is studied. To do so, a few RBTS-Bus4 networks are connected together to achieve larger networks. It is worth mentioning that the number and location of RCSs in each of them are based on the presented configuration of RCS in a single RBTS-Bus4. For example, 42 RCSs are installed in a system comprised of 2 RBTS-Bus4 networks. So, the investment costs of RCS is directly increased as the size of system is increased. In this regard, the simulation is repeated by increasing the system size and the results are provided in Table 8.21. As can be seen, the expected profit and CVaR wane as the system size wax. So, considering the financial risk is critical in small distribution systems while it is negligible in practical systems which have larger size (in several orders of magnitude). More accurately, when the system size is 2 times larger than the test system, the expected profit and CVaR are about 2 and 6.6 times, respectively. Also, when 4 RBTS-Bus4 networks are joined together, the expected profit and CVaR are near 4 and 20 times, respectively. Hence, the CVaR is more sensitive than the expected profit to the size of the system.

Table 8.21 Impact of system size on financial results

| Number of connected systems | RCS cost (US k$) | Interruption cost (US k$) | | Expected profit (US k$) | CVaR (US k$) |
|---|---|---|---|---|---|
| | | Without RCS | With RCS | | |
| 1 | 118.11 | 779.72 | 164.96 | 496.64 | 47.69 |
| 2 | 236.22 | 1560.01 | 325.94 | 997.85 | 315.63 |
| 3 | 354.33 | 2339.10 | 482.87 | 1501.90 | 625.41 |
| 4 | 472.45 | 3118.81 | 642.87 | 2004.31 | 954.86 |
| 5 | 590.5 | 3898.58 | 800.99 | 2507.02 | 1345.10 |

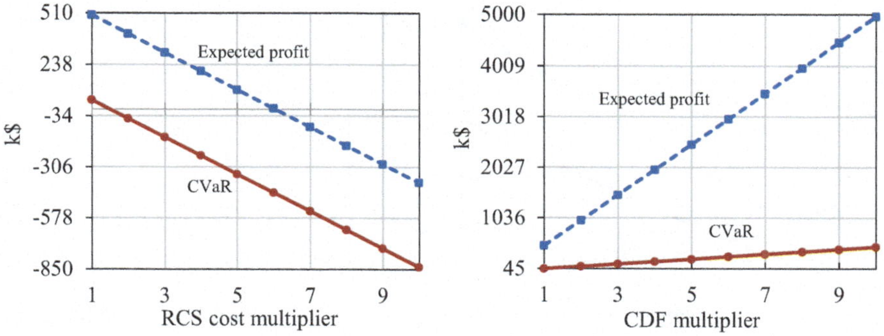


**Fig. 8.39** Profit and CVaR versus RCS costs and CDF multipliers

#### 8.7.6.4 RCS Costs

In order to scrutinize the impact of RCS costs, the simulation is iterated for different RCS costs and the results are shown in Fig. 8.39. As can be seen, the expected profit and CVaR are decreased when the RCS costs are increased. This means that the financial risk is less critical in future since the trend of RCS costs is decreasing.

#### 8.7.6.5 Customer Damage Function (CDF)

Here, the impact of CDF on the result of the simulation is studied. The results are depicted in Fig. 8.39 by increasing the CDF. As can be expected, higher CDF increases the expected profit and decreases the financial risk. According to Fig. 8.39, the expected profit is changed dramatically compared to CVaR. So, it is necessary to consider the financial risk in the system with less sensitive customers to interruption events.

#### 8.7.6.6 Failure Rate and Repair Time

There are several factors which have positive influence on the failure rate and repair time such as geographical conditions and type of lines used in the system, to name just a few. In this regard, the impacts of failure rate and repair time are individually studied in this subsection and the results are provided in Fig. 8.40. As can be seen, the expected profit is increased and financial risk is decreased when the failure rate and repair time are increased individually. So, considering financial risk is less important in system which is difficult to access the sections (e.g., mountain) or difficult to repair the equipments (e.g., underground system). In the other hand, considering risk in the network with smaller failure rate and repair time is critical.

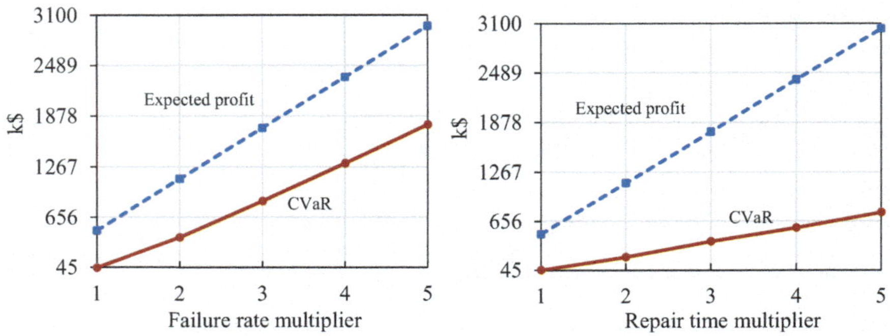


**Fig. 8.40** Profit and CVaR versus failure rate and repair time multipliers

# References


1. M.E. Raoufat, A. Taalimi, K. Tomsovic, R. Hay, Event analysis of pulse-reclosers in distribution systems through sparse representation, in *Proceedings of Intelligent System Application to Power Systems (ISAP)* (2017)
2. S. Akhlaghi, A.A. Ghadimi, A. Akhlaghi, A novel hybrid islanding detection method combination of SMS and Qf for islanding detection of inverter-based DG, in *Proceedings of Power and Energy Conference at Illinois (PECI)* (2014)
3. E. Foruzan, S. Asgarpoor, J.M. Bradley, Hybrid system modeling and supervisory control of a microgrid, in *Proceedings of North American Power Symposium (NAPS)* (2016)
4. A. Ghorbannia Delavar, M. Nejadkheirallah, M. Motalleb, A new scheduling algorithm for dynamic task and fault tolerant in heterogeneous grid systems using Genetic Algorithm, in *Proceedings of 3rd International Conference on Computer Science and Information Technology* (2010)
5. A. Eshraghi, R. Ghorbani, Islanding detection and transient over voltage mitigation using wireless sensor networks, in *Proceedings of 2015 IEEE Power & Energy Society General Meeting* (2015)
6. A. Shahsavari, M. Farajollahi, E. Stewart, C. Roberts, F. Megala, L. Alvarez, E. Cortez, H. Mohsenian-Rad, Autopsy on active distribution networks: a data-driven fault analysis using micro-PMU data, in *North American Power Symposium (NAPS)* (2017)
7. M. Farajollahi, A. Shahsavari, H. Mohsenian-Rad, Location identification of distribution network events using synchrophasor data, in *Proceediongs of North American Power Symposium (NAPS)* (2017)
8. M. Farajollahi, A. Shahsavari, H. Mohsenian-Rad, Location identification of high impedance faults using synchronized harmonic phasors, in *Proceedings of Innovative Smart Grid Technologies Conference (ISGT)* (2017)
9. M. Farajollahi, M. Fotuhi-Firuzabad, A. Safdarian, Deployment of fault indicator in distribution networks: a MIP-based approach. IEEE Trans. Smart Grid (2016)
10. V. Miranda, Using fuzzy reliability in a decision aid environment for establishing interconnection and switching location policies, in *Proceedings of CIRED* (1991)
11. G. Levitin, S. Mazal-Tov, D. Elmakis, Optimal sectionalizer allocation in electric distribution systems by genetic algorithm. Elect. Power Syst. Res. **31**(2), 97–102 (1994)
12. R. Billinton, S. Jonnavithula, Optimal switching device placement in radial distribution systems. IEEE Trans. Power Del. **11**(3), 1646–1651 (1996)
13. J.H. Teng, Y.H. Liu, A novel ACS-based optimum switch relocation method. IEEE Trans. Power Syst. **18**(1), 113–120 (2003)

14. A. Moradi, M. Fotuhi-Firuzabad, Optimal switch placement in distribution systems using trinary particle swarm optimization algorithm. IEEE Trans. Power Del. **23**(1), 271–279 (2008)
15. H. Teng, C.-N. Lu, Feeder-switch relocation for customer interruption cost minimization. IEEE Trans. Power Del. **17**(1), 254–259 (2002)
16. P.M.S. Carvalho, L.A.F.M. Ferreira, A.J.C. da Silva, A decomposition approach to optimal remote controlled switch allocation in distribution systems. IEEE Trans. Power Del. **20**(2), 1031–1036 (2005)
17. G. Celli, F. Pilo, Optimal sectionalizing switches allocation in distribution networks. IEEE Trans. Power Del. **14**(3), 1167–1172 (1999)
18. A. Abiri-Jahromi, M. Fotuhi-Firuzabad, M. Parvania, M. Mosleh, Optimized sectionalizing switch placement strategy in distribution systems. IEEE Trans. Power Del. **27**(1), 362–370 (2012)
19. O.K. Siirto, A. Safdarian, M. Lehtonen, M. Fotuhi-Firuzabad, Optimal distribution network automation considering earth fault events. IEEE Trans. Smart Grid **6**(2), 1010–1018 (2015)
20. M. Izadi, M. Farajollahi, A. Safdarian, M. Fotuhi-Firuzabad, A multistage MILP-based model for integration of remote control switch into distribution networks, in *Proceedings of 2016 International Conference on Probability Methods Applied to Power Systems (PMAPS),* Beijing (2016), pp. 1–6
21. M. Farajollahi, M. Fotuhi-Firuzabad, A. Safdarian, A joint manual and remote controlled switch placement in distribution system using MIP model, in *Proceedings of 2016 International Conference on New Research Achievements in Electrical and Computer Engineering* (ICNRAECE), Tehran, Iran
22. A. Shahsavari, A. Fereidunian, M. Mazhari, A joint automatic and manual switch placement within distribution systems considering operational probabilities of control sequences. Proc. Int. Trans. on Elect. Energy Syst. **25**(11), 2745–2768 (2015)
23. R.N. Billinton, I. Sjarief, A reliability test system for educational purposes-basic distribution system data and results. IEEE Trans. Power Syst. **6**(2), 813–820 (1991)
24. R.N. Billinton, E. Chan, G. Tollefson, G. Wacker, A Canadian customer survey to assess power system reliability worth. IEEE Trans. Power Syst. **9**(1), 443–450 (1994)
25. Y. He, *Modeling and evaluation effect of automation, protection, and control on reliability of power distribution systems, doctoral dissertation* (KTH, Royal Institute of Technology, Stockholm, 2002)
26. A. Shahsavari, S.M. Mazhari, A. Fereidunian, H. Lesani, Fault indicator deployment in distribution systems considering available control and protection devices: a multi-objective formulation approach. IEEE Trans. Power Syst. **29**(5), 2359–2369 (2014)
27. A. Safdarian, M. Farajollahi, M. Fotuhi-Firuzabad, Impacts of remote control switch malfunction on distribution system reliability, IEEE Trans. Power Syst. **99**, 1–1 (2016)
28. N. Growe-Kuska, H. Heitsch, W. Romisch, Scenario reduction and scenario tree construction for power management problems, in *Proceedings of 2003 IEEE Bologna Power Tech Conference* (2003)
29. A.J. Conejo, M. Carrión, J.M. Morales, *Decision Making Under Uncertainty in Electricity Markets* (Springer, NY, USA, 2010)
30. M. Izadi, A. Safdarian, Financial risk constrained remote controlled switch deployment in distribution networks. IET Gen. Trans. Dist. **12**(7), 1547–1553 (2018)
31. M. Izadi, A. Safdarian, Financial risk evaluation of RCS deployment in distribution systems. IEEE Sys. J. (2018)

# Chapter 9
# Cooperative Distributed Energy Scheduling in Microgrids

**Mehdi Rahmani-Andebili**

**Abstract** This chapter introduces a multi-time scale model predictive control (MPC) approach which is stochastically applied in the cooperative distributed energy scheduling problem of the microgrids (MG). The cooperative distributed approach is preferred, since a centralized one is not applicable in a competitive power market environment because it requires all the data of all the MGs, which is impractical. In this chapter, in order to deal with the variability and uncertainties associated with output power of the renewable energy resources (RES) and load demand, stochastic MPC is applied in distributed energy scheduling problem of MGs. Additionally, considering multi-time scale approach in the stochastic MPC is capable of simultaneously having vast vision for the optimization time horizon and precise resolution for the problem variables. Herein, each MG with a different set of sources is able to transact power with the electricity market and the neighboring MGs. The numerical study demonstrates that cooperation of the MGs in the distributed energy scheduling problem is beneficial, and also the multi-time scale MPC is advantageous compared to the single-time scale MPC in both non-cooperative and cooperative distributed energy scheduling problems.

**Keywords** Cooperative distributed energy scheduling · Microgrids
Multi-time scale approach · Stochastic model predictive control (MPC)
Renewables

## 9.1 Introduction

A microgrid (MG) is a group of interconnected loads and distributed energy resources with clearly defined electrical boundaries that acts as a single controllable entity with respect to the grid and can connect and disconnect from the grid to

---

M. Rahmani-Andebili (✉)
The Holcombe Department of Electrical and Computer Engineering,
Clemson University, Clemson, SC 29634, USA
e-mail: mehdir@g.clemson.edu



A. Arefi et al. (eds.), *Electric Distribution Network Management and Control*,
Power Systems, https://doi.org/10.1007/978-981-10-7001-3_9

enable it to operate in both grid-connected or island mode [1]. MGs have received considerable attention in the recent years because they can mitigate the growing problems in energy supplies [2]. In order to efficiently utilize an energy storage in a MG, the amount of energy generated or consumed in the MG in the current time step should not be independent of value of the energy generated or consumed in the previous and next time steps. In other words, the energy scheduling problem of a MG should be a multi-step decision making problem.

Due to presence of renewable energy resources in a system, the net generation of system has a lot of variability and uncertainty [3]. Therefore, to operate the MG in an efficient way, the generation sources and the energy storages must be optimally and dynamically scheduled considering value of the uncertain states of the MG over the optimization period as well as the technical constraints of all the sources.

In most of the previously published papers, the energy scheduling and economic dispatch problems have been investigated applying the centralized optimization approaches [4–12]. However, the centralized optimization techniques are not applicable in the restructured electricity industry due to lack of economic and technical information about all the market players [13, 14]. Moreover, the centralized optimization techniques are unable to solve huge and complex optimization problems due to a large number of variables and parameters of the problem [15]. Additionally, they are generally more vulnerable to single point failure that may stop the entire system from working [16].

Because of the above mentioned issues, distributed optimization techniques have been proposed to replace the centralized optimization approaches [17–21]. Table 9.1 presents a comprehensive summary of the literature review regarding the energy scheduling problems with both centralized and cooperative distributed optimizations. The literature review has been characterized based on several aspects and criteria. These aspects include type of control strategy (centralized or cooperative distributed), considering or ignoring stochastic nature of the uncertain states of the problem (solving the problem stochastically or deterministically), adaptively controlling the problem variables or avoiding it, multi-step or single-step decision making, multi-time scale or single-time scale approach, the duration of each time step, and modeling or disregarding the technical factors of the energy storages and generators.

In order to efficiently utilize an energy storage in a MG, the amount of energy generated/consumed in a MG in the current time step should not be independent of the value of energy generated/consumed in the previous and next time steps. This type of optimization problem is called multi-step decision making problem, since the value of system states and problem variables in different time steps of the optimization procedure are correlated.

In addition, in a multi-time scale optimization approach including more than one time scale, the optimization problem is solved for each time scale. In other words, the system state and problem variables are changed with different time scales.

As can be seen in Table 9.1, [4–12] have applied the conventional optimization technique; however, these techniques have several issues that hinder them to be applied in the power market and smart grid infrastructure. The studies done

**Table 9.1** Summary of the literature review regarding the energy scheduling problems

| Refs. | Optimization type | Stochastic approach | Adaptive control | Multi step decision making | Duration of time step | Modeling technical aspects of energy storage | Modeling technical aspects of generator |
|---|---|---|---|---|---|---|---|
| [4] | Centralized | | | ✓ | One hour | ✓ | |
| [5] | Centralized | ✓ | | | One hour | | ✓ |
| [6] | Centralized | | | | One hour | | ✓ |
| [7] | Centralized | ✓ | | | One hour | | ✓ |
| [8] | Centralized | | | | 5–15 min | | ✓ |
| [9] | Centralized | | | ✓ | One hour | | ✓ |
| [10] | Centralized | ✓ | | ✓ | One hour | ✓ | ✓ |
| [11] | Centralized | ✓ | ✓ | ✓ | One hour | ✓ | |
| [12] | Centralized | ✓ | | ✓ | One hour | ✓ | ✓ |
| [17] | Cooperative distributed | | | | One hour | | ✓ |
| [18] | Cooperative distributed | | | | One hour | | ✓ |
| [19] | Cooperative distributed | | ✓ | | One hour | | ✓ |
| [20] | Cooperative distributed | ✓ | ✓ | | One hour | | ✓ |
| [21] | Cooperative distributed | | | ✓ | One hour | ✓ | ✓ |

in [17–21] include cooperative distributed optimization techniques. Nevertheless, the above mentioned criteria have not been considered simultaneously in these studies. In addition, no study has applied multi-time scale approach. Furthermore, in most of the studies, the duration of the time step has been considered to be about one hour that is not realistic due to high variation of load demand and renewables power within a one hour period of time.

In this chapter, the cooperative distributed optimization technique is applied in the energy scheduling problem of MGs taking into account all of the above mentioned criteria and aspects. In order to deal with variability and uncertainties concerned with the power of the renewables and the load demand, stochastic model predictive control (MPC) technique is applied. In addition, in order to have vast vision for the optimization time horizon and precise resolution for the problem variables, MPC is considered to be a multi-time scale MPC (with both small and large time scales). Herein, the scale of time steps are considered about one-minute and one-hour and the duration of the optimization time horizon with any time scale is assumed to be ten time steps. Moreover, the investigated energy scheduling problem is a multi-step decision making problem, since value of the energy generated or consumed in different time steps of the optimization time horizon are correlated. Furthermore, several technical and economic factors for the generators and the batteries are considered. In this chapter, a combination of genetic algorithm (GA) and linear programming (GA-LP) is applied to solve the energy scheduling problem of each MG. Herein, the GA addresses the non-convexity and nonlinearity of the problem and the LP quickly finds the optimal solution.

The chapter is organized as follows. In Sect. 9.2, the proposed technique for solving the energy scheduling problem of the MGs is presented. In Sect. 9.3, the problem is formulated. The numerical study is accomplished in Sect. 9.4.

## 9.2 Proposed Technique

In this section, the proposed technique for solving the energy scheduling problem of the MGs is described.

### *9.2.1 Stochastic Approach*

#### 9.2.1.1 Forecasting the Value of Uncertain States

In this chapter, in every MG, values of the wind turbine power, the PV panel power, the load demand, and the market electricity price as the uncertain states of the problem are predicted for every time step of the optimization time horizon and for every time scale ($m+1,\ldots,m+N\tau$ and $h+1,\ldots,h+N\tau$) using neural network trained with Levenberg-Marquardt back-propagation algorithm available in MATLAB. Every applied neural network uses the historical data of the uncertain state as its entries. The set of predicted values for every uncertain state of the problem in every MG over the optimization time horizon are presented in (9.1).

$$\tilde{X}_{mg,t} = \left\{ \begin{array}{ccc} \tilde{P}^{W}_{mg,t+1} & \cdots & \tilde{P}^{W}_{mg,t+N\tau} \\ \tilde{P}^{PV}_{mg,t+1} & \cdots & \tilde{P}^{PV}_{mg,t+N\tau} \\ \tilde{D}^{L}_{mg,t+1} & \cdots & \tilde{D}^{L}_{mg,t+N\tau} \\ \tilde{\pi}^{M}_{t+1} & \cdots & \tilde{\pi}^{M}_{t+N\tau} \end{array} \right\}, \forall mg \in S^{MG}, \forall t \in \left\{S^{M}, S^{H}\right\} \tag{9.1}$$

$$S^{MG} = \{1,\ldots,NMG\},\ S^{M} = \{1,\ldots,NM\},\ S^{M} = \{1,\ldots,NH\}$$

#### 9.2.1.2 Modeling Uncertainties of the Forecasted Data

Figure 9.1a illustrates the predicted and measured data for the current time step and for the past time steps, and also the predicted data for every time step of the optimization time horizon. As can be seen in Fig. 9.1a, the forecasted data are compared with the real data (measured data) and value of the error of the

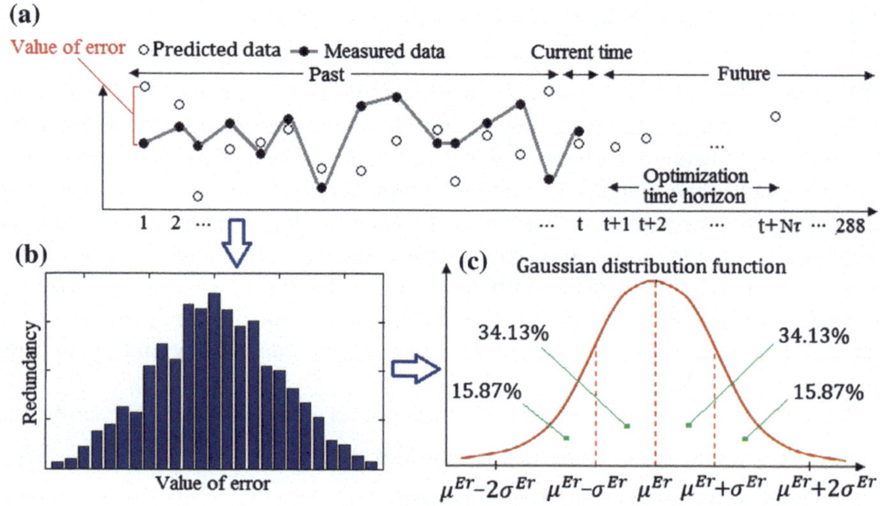


**Fig. 9.1** **a** Predicted data, measured data, and value of prediction error **b** Redundancy of the prediction errors respect to the value of the prediction errors. **c** Gaussian probability density function related to the prediction errors

predictions are calculated. Then, as can be seen in Fig. 9.1b, redundancy of the prediction errors respect to the value of the prediction errors are plotted on a chart. After that, an appropriate probability density function curve is fitted for the prediction errors, as can be seen in Fig. 9.1c. It is observed that the predication errors can be precisely fitted on a Gaussian probability density function with an appropriate mean ($\mu^{Er}$) and standard deviation ($\sigma^{Er}$) [22]. Finally, the curve is divided into four areas to define four distinct values for the prediction inaccuracy with occurrence probabilities about 15.87, 34.13, 34.13, and 15.87% related to $-\mu^{Er} - 2\sigma^{Er}$, $-\mu^{Er}$, $-\mu^{Er}$, and $-\mu^{Er} + 2\sigma^{Er}$, respectively. The values of the $-\mu^{Er}$ and $\sigma^{Er}$ are updated in the next predictions in the optimization procedure of the problem.

Although considering more scenarios in every problem results in more accurate outcomes for the problem, it may lead to an unmanageable optimization problem in some cases. With presence of renewables, this phenomenon happens because of short time step (one minute) considered in the optimization procedure of the problem, and also because of application of MPC (the problem must be optimized at every time step). Therefore, in order to avoid dealing with an unmanageable optimization problem, four scenarios corresponding to $\tilde{X}_t - \mu^{Er} - 2\sigma^{Er}, \tilde{X}_t - \mu^{Er}, \tilde{X}_t + \mu^{Er}, \tilde{X}_t + \mu^{Er} + 2\sigma^{Er}$ with occurrence probabilities 15.87, 34.13, 34.13, and 15.87%, respectively, are considered for every uncertain state of the problem.

### 9.2.2 Multi-time Scale Model Predictive Control

MPC as a well-established method in control engineering is capable of controlling a multi-variable constrained system by taking the control actions from the solution of an online optimization problem and predicting the system behavior repetitively [23]. The multi-time scale MPC with one-minute and one-hour scales and ten time steps as the optimization time horizon is illustrated in Fig. 9.2. As can be seen, at every time step, the optimization time horizon for both one-minute and one-hour scales is updated, and then the problem is solved for the updated time horizons; however, just the signals of the next time steps are accepted as the candidates for the decision signals. This procedure demonstrates the adaptability and dynamism features of the MPC.

Next, the decision signals for the problem variables are identified by comparing the value of one-minute scale stochastic forward-looking objective function $\left(\mathbb{F}^{FL}_{mg,m}\right)$ and the value of weighted one-hour scale stochastic forward-looking objective function $\left(\frac{1}{60} \times \mathbb{F}^{FL}_{mg,h}\right)$, as can be seen in (9.2). This process is repeated for every minute of the operation period $(1, \ldots, 1440)$. Considering small time scale (one-minute scale) and large time scale (one-hour scale) in the multi-time scale MPC contribute to have vast vision in the optimization time horizon and precise resolution in values of the variables, respectively. A forward-looking objective function with any time scale is presented in (9.3). As can be seen, the forward-looking objective function is sum of the value of the time step objective functions over the optimization time horizon.

$$\mathbb{Y}_{mg,m} = \begin{cases} \mathbb{Y}_{mg,m} & \mathbb{F}^{FL}_{mg,m} < \frac{1}{60} \times \mathbb{F}^{FL}_{mg,h} \\ \mathbb{Y}_{mg,h} & else \end{cases}, \forall mg \in S^{MG}, \forall m \in S^{M} \tag{9.2}$$

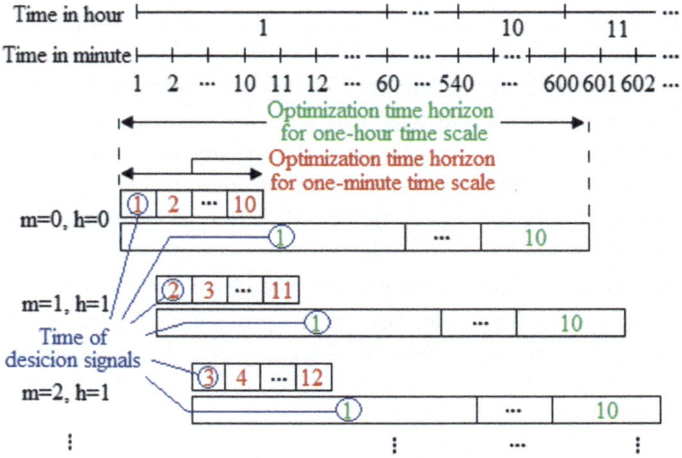


**Fig. 9.2** Concept of the multi-time scale MPC

where

$$F_{mg,t}^{FL} = \sum_{\tau=1}^{N\tau} F_{mg,t+\tau}, \forall mg \in S^{MG}, \forall t \in \{S^M, S^H\} \tag{9.3}$$

### 9.2.3 Cooperative Distributed Optimization

In this chapter, a cooperative distributed optimization technique is applied instead of a centralized optimization approach. Since in a competitive market environment the economic and technical data about all the market players is not complete, application of a centralized optimization is not practical [13, 14]. Moreover, applying the cooperative distributed technique can remove the curse of dimensionality that arises in the optimization of huge and complex problems due to large number of variables, parameters, and multiple inputs and outputs [15]. In fact, cooperative distributed optimization significantly reduces the complexity of problem and makes it scalable. Furthermore, it is robust for single point failure compared to a centralized optimization technique [16].

Figure 9.3 illustrates implementation and application of the cooperative distributed optimization for a system with five MGs. In this chapter, it is assumed that just the neighboring MGs can cooperate with one another. Thus, based on this assumption and the given sequence in Fig. 9.3, first, MG 1 solves its own energy scheduling problem for the given optimization time horizon with any time scale $(t+1,\ldots,t+N\tau, \forall t \in \{S^M, S^H\})$ considering the proposed electricity price and available power of MG 2 (as its neighbor) in every time step of the optimization time horizon. Then, MG 1 reports its proposed electricity price and available power in every time step of the optimization time horizon to MG 2. To complete one loop of the cooperative distributed optimization, this process is done for every MG based on the given sequence, that is, MG 1, MG 2, MG 3, MG 4, and MG 5. Next, the loop is repeated several times until no significant improvement is observed in value of the objective function of every MG.

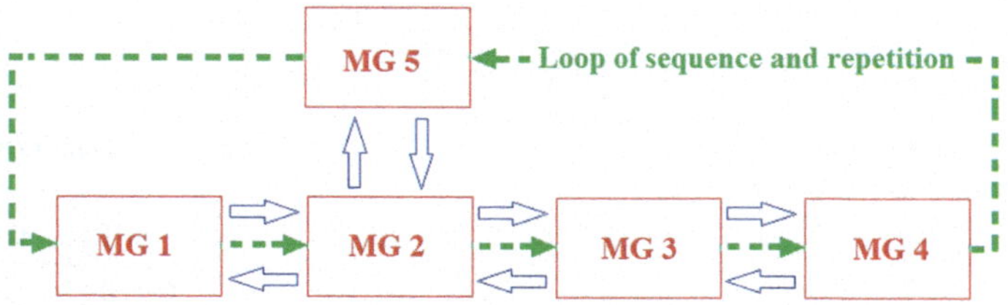


**Fig. 9.3** Applying cooperative distributed optimization on a system with five MGs

#### 9.2.3.1 Optimization Tool

The cooperative distributed energy scheduling problem of each MG is a non-convex problem. In this chapter, a combination of genetic algorithm and linear programming (GA-LP) is applied to solve the energy scheduling problem of each MG for one-minute and one-hour scales. In fact, the GA is applied to address the non-convexity and nonlinearity of the problem and the linear programming is applied to quickly find the optimal solution. Herein, the GA and the linear programming techniques deal with the discrete variables and the continuous variables of the problem, respectively.

The discrete variables of the problem handled by the GA include statuses of the generator and the battery in every time-step of the optimization time horizon for the given time scale, as can be seen in (9.4). Herein, $y^G$ indicates the status of generator that values "0" and "1" mean "off" and "on", respectively. Also, $y^B$ indicates the status of battery, where values "−1", "0", and "1" mean charging, idle, and discharging, respectively. Based on this, the dimensions of the defined chromosome in the GA are considered to be $N\tau \times 3$, as can be seen in Fig. 9.4. Herein, one bit for indicating status of the generator ("on" or "off") and two bits for indicating status of the battery (charging, idle, or discharging) are considered.

$$y_{mg,t} = \left\{ \begin{matrix} y^G_{mg,t} & \cdots & y^G_{mg,t+N\tau} \\ y^B_{mg,t} & \cdots & y^B_{mg,t+N\tau} \end{matrix} \right\}, \forall mg \in S^{MG}, \forall t \in \left\{S^M, S^H\right\} \tag{9.4}$$

Also, the continuous variables of the problem optimized by the LP include the value of power of the generator, the value of generated or consumed power of the battery, the value of transacted power with the neighboring MGs, and the value of transacted power with the electricity market in every time-step of the optimization time horizon for the given time scale.

$$Y_{mg,t} = \left\{ \begin{matrix} P^G_{mg,t} & \cdots & P^G_{mg,t+N\tau} \\ P^B_{mg,t} & \cdots & P^B_{mg,t+N\tau} \\ P^{NMG}_{mg,t} & \cdots & P^{NMG}_{mg,t+N\tau} \\ P^M_{mg,t} & \cdots & P^M_{mg,t+N\tau} \end{matrix} \right\}, \forall mg \in S^{MG}, \forall t \in \left\{S^M, S^H\right\}$$

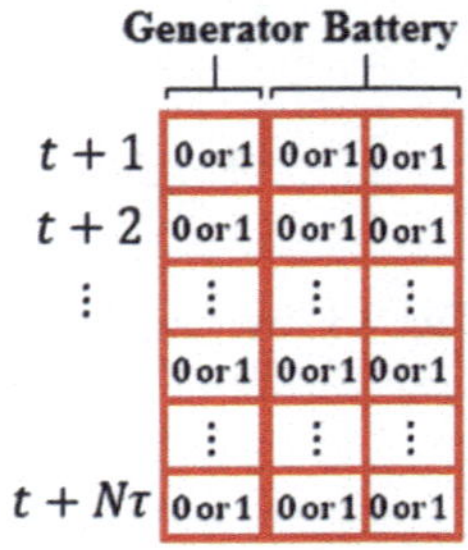


**Fig. 9.4** Structure of the defined chromosome in GA

In the following, different steps for applying the GA-LP in the cooperative distributed energy scheduling problem of a MG for the given time scale (one-minute scale or one-hour scale) are presented and described.

- *Step 1: Obtaining the primary data*

*Parameters for applying GA-LP:* These parameters include mutation probability of the genes $\left(\theta^{Mut}\right)$ and size of the population $\left(N^{GA}\right)$.

*Parameters of the system under study:* The values of all the system parameters and value of the defined scenario over the optimization time horizon for every uncertain state of the problem are obtained.

*Initial population:* The chromosomes of the population are initialized with random binary values.

- *Step 2: Updating the population*

*Applying crossover operator:* Three crossover points are randomly selected for every pair chromosomes, and then crossover operator is applied on the two chromosomes to reproduce two new chromosomes as the offspring.

*Applying mutation operator:* This operator is applied on every gene of every chromosome of the population with the definite probability $\theta^{Mut}$.

- *Step 3: Selecting new population*

*Evaluating fitness of every chromosome:* For every chromosome, the LP is executed and if all the constraints are satisfied, fitness of the chromosome (the inverse of value of the forward-looking objective function) is measured.

*Applying selection process:* As can be seen in (9.6), new chromosomes are selected via the probabilistic fitness-based selection (PFBS) technique, where fitter chromosomes are more likely to be chosen. The value of the selection probability of every chromosome is determined using (9.7), which is proportional to fitness of the chromosome.

$$a_{ch} = \begin{cases} 1 & \theta_{ch}^{PFBS} > r_{ch} \\ 0 & \theta_{ch}^{PFBS} < r_{ch} \end{cases} \tag{9.6}$$

$$\theta_{ch}^{PFBS} = \frac{fit_{ch}}{Max(S^{Fit})} \times 100, \; S^{Fit} = \{fit_1, \ldots, fit_{Nch}\} \tag{9.7}$$

- *Step 4: Checking termination criterion*

In this step, the convergence status of the optimization procedure is checked. Based on this, values of the improvements in fitness of the chromosomes of the old and new populations are measured and if there are no significant improvements in them, the optimization process is finished, otherwise, the algorithm is continued from *Step 2*.

- *Step 5: Introducing the outcome*

The consequences include optimal value of the discrete and continuous variables for every time step of the optimization time horizon.

## 9.3 Problem Formulation

The goal of each MG is minimizing value of the stochastic forward-looking objective function over the optimization time horizon with one-minute and one-hour scales. As can be seen in (9.8), value of the stochastic forward-looking objective function is determined by summing value of the forward-looking objective functions weighted by the corresponding occurrence probability. The forward-looking objective function with any time scale has been presented in (9.3). The different cost and benefit terms of the time step objective function is presented in (9.9). These terms include fuel cost of the generator, carbon emission cost of the generator, start up cost of the generator, shut down cost of the generator, switching cost of the battery, cost or benefit due to power transactions with the energy market, and cost or benefit because of power transactions with the neighboring MGs.

$$\min \mathbb{F}^{FL}_{mg,t} = min \sum_{Sc^W_t} \sum_{Sc^{PV}_t} \sum_{Sc^D_t} \sum_{Sc^\pi_t} F^{FL}_{mg,t} \times \Omega^W_t \times \Omega^{PV}_t \times \Omega^D_t \times \Omega^\pi_t, \forall mg \in S^{MG}, \forall t \in \left\{S^M, S^H\right\} \tag{9.8}$$

$$F^{mg,t} = \left\{ \begin{array}{c} \left[y^G_{mg,t} \times FC^G_{mg,t}\right] + \left[y^G_{mg,t} \times EC^G_{mg,t}\right] \\ + \left[\left(1 - y^G_{mg,t-1}\right) \times y^G_{mg,t} \times STUC^G_{mg}\right] \\ + \left[y^G_{m,g,t-1} \times \left(1 - y^G_{mg,t}\right) \times SHDC^G_{mg}\right] \\ + \left[\acute{y}^B_{mg,t} \times SWC^B_{mg}\right] \\ + \left[P^M_{mg,t} \times \acute{\pi}^M_t\right] + \left[\sum_{\substack{mg' \in S^{MG} \\ mg' \neq mg}} \left(P^{NMG}_{mg',t} \times \pi^{MG}_{mg',t}\right)\right] \end{array} \right\} \forall mg \in S^{MG}, \forall t \in \left\{S^M, S^H\right\} \tag{9.9}$$

where

$$\acute{y}^B_{mg,t} = \begin{cases} 0 & y^B_{mg,t-1} = y^B_{mg,t} \\ 1 & y^B_{mg,t-1} \neq y^B_{mg,t} \end{cases}, \forall mg \in S^{MG}, \forall t \in \left\{S^M, S^H\right\} \tag{9.10}$$

$$\acute{\pi}^M_t = \begin{cases} \pi^M_t & P^M_t > 0 \\ \varphi \times \pi^M_t & P^M_t < 0 \end{cases}, \forall t \in \left\{S^M, S^H\right\} \tag{9.11}$$

In (9.11), $\varphi$ is the coefficient used to determine the price of selling power to the grid based on the net energy metering (NEM) plan [24]. In the NEM plan, every

MG can deliver its extra power to the grid and sell it to the market at a lower price compared to the market price [24].

The value of electricity price proposed by every exporter MG at every time step is determined based on the marginal cost of the installed generator in the MG. However, if there is no generator in the exporter MG, the proposed price is determined based on the marginal cost of the installed generator in the importer MG.

$$\pi_{mg,t}^{MG} = \frac{\partial\left(FC_{mg,t}^{G} + EC_{mg,t}^{G}\right)}{\partial P_{mg,t}^{G}}, \forall mg \in S^{MG}, \forall t \in \left\{S^{M}, S^{H}\right\} \tag{9.12}$$

The fuel cost function and carbon emission function of each generator are considered quadratic polynomials presented in (9.13) and (9.14), respectively.

$$FC_{mg,t}^{G} = z_{1,mg}^{Fu} \times \left(P_{mg,t}^{G}\right)^{2} + z_{2,mg}^{Fu} \times \left(P_{mg,t}^{G}\right) + z_{3,mg}^{Fu}, \forall mg \in S^{MG}, \forall t \in \left\{S^{M}, S^{H}\right\} \tag{9.13}$$

$$EC_{mg,t}^{G} = \beta^{Em} \times \left(z_{1,mg}^{Em} \times \left(P_{mg,t}^{G}\right)^{2} + z_{2,mg}^{Em} \times \left(P_{mg,t}^{G}\right) + z_{3,mg}^{Em}\right), \forall mg \in S^{MG}, \forall t \in \left\{S^{M}, S^{H}\right\} \tag{9.14}$$

The value of switching cost of each battery is determined based on value of total cumulative ampere-hours throughput of the battery in its life cycle and value of the initial price of the battery. In fact, considering this cost term prevents the battery from unnecessary switching that is harmful to its life span.

$$SWC_{mg}^{B} = \frac{\pi_{mg}^{B}}{\xi_{mg}^{B}}, \forall mg \in S^{MG} \tag{9.15}$$

In the following, the technical constraints of the problem that must be held in every MG and at every time step of the operation period with any time scale (one-minute scale or one-hour scale) are presented and described.

***Supply-demand balance:*** The sum of power of generator, the power of wind turbine, the power of PV panel, the power of battery, the transacted power with the neighboring MGs, and the transacted power with the electricity market must be equal to the demand of load in every MG and at every time step of the operation period with any time scale. Herein, the transacted power with the neighboring MGs is considered positive if the MG is importing power and it is negative if the MG is exporting power. Moreover, the transacted power with the electricity market is considered positive if the MG is purchasing power from the market and it is supposed to be negative if the MG is selling power to the market.

$$y^G_{mg,t} \times P^G_{mg,t} + P^W_{mg,t} + P^{PV}_{mg,t} + y^B_{mg,t} \times P^B_{mg,t} + P^{NMG}_{mg,t} + P^M_{mg,t} = D^L_{mg,t}, \forall mg \in S^{MG}, \forall t \in \{S^M, S^H\} \tag{9.16}$$

***Power limits of the generator:*** The maximum power and minimum power limits of every generator is presented in (9.17).

$$\underline{P^G_{mg}} \leq y^G_{mg,t} \times P^G_{mg,t} \leq \overline{P^G_{mg}}, \forall mg \in S^{MG}, \forall t \in \{S^M, S^H\} \tag{9.17}$$

***Minimum up/down time limits of the generator:*** The duration that the generator is continuously "on" and "off" must be more than the defined minimum up time and minimum down time, respectively.

$$\Delta t^{G\_ON}_{mg} \geq MUT^G_{mg}, \forall mg \in S^{MG} \tag{9.18}$$

$$\Delta t^{G\_OFF}_{mg} \geq MDT^G_{mg}, \forall mg \in S^{MG} \tag{9.19}$$

***Power limits of the battery:*** The installed battery in the MG can act as a load or generator by being charged or discharged, respectively; however, value of power of the battery must be in the defined range.

$$-\overline{P^B_{mg}} \leq y^B_{mg,t} \times P^B_{mg,t} \leq \overline{P^B_{mg}}, \forall mg \in S^{MG}, \forall t \in \{S^M, S^H\} \tag{9.20}$$

***Minimum up/down time limits of the battery:*** The duration that the battery is continuously "on" and "off" must be more than the given minimum up time and minimum down time, respectively.

$$\Delta t^{B\_ON}_{mg} \geq MUT^B_{mg}, \forall mg \in S^{MG} \tag{9.21}$$

$$\Delta t^{B\_OFF}_{mg} \geq MDT^B_{mg}, \forall mg \in S^{MG} \tag{9.22}$$

***State of charge limits of the battery:*** In order to prolong the life time of the battery, the battery must not be discharged more than the allowable depth of discharge (DOD). Moreover, the battery has a definite capacity that cannot be charged more than that.

$$DOD^B_{mg} \leq SOC^B_{mg,t} \leq 100, \forall mg \in S^{MG}, \forall t \in \{S^M, S^H\} \tag{9.23}$$

***Zero-net-energy constraint for the battery in the optimization interval:*** This constraint is considered to use the battery only for an energy storage and not for generation or load source. In other words, the cumulative value of charged energy must be equal to the cumulative value of discharged energy in every optimization interval with any time scale.

$$\sum_{\tau=1}^{N\tau} y^{B}_{mg,t} \times P^{B}_{mg,t+\tau} = 0, \forall mg \in S^{MG}, \forall t \in \{S^{M}, S^{H}\} \tag{9.24}$$

***Maximum accessible power from a neighboring MG:*** The power that the MG can import from a neighboring MG must be less than the available power of the neighboring MG at every time step of the optimization period with any time scale.

$$P^{NMG}_{mg,t} \le P^{AV}_{mg',t}, \forall mg, mg' \in S^{MG}, \forall t \in \{S^{M}, S^{H}\} \tag{9.25}$$

where

$$P^{AV}_{mg',t} = \begin{cases} y^{G}_{mg',t} \times \left(\overline{P^{G}_{mg'}} - P^{G}_{mg',t}\right) + \overline{P^{B}_{mg'}} - P^{B}_{mg',t} & y^{B}_{mg',t} = 1 \\ y^{G}_{mg',t} \times \left(\overline{P^{G}_{mg'}} - P^{G}_{mg',t}\right) & y^{B}_{mg',t} \neq 1 \end{cases} \tag{9.26}$$

## 9.4 Numerical Study

### *9.4.1 Characteristic of the System Under Study*

Figure 9.5a illustrates the configuration of distribution system under study that include five MGs with different set of sources. Moreover, Fig. 9.5b shows the

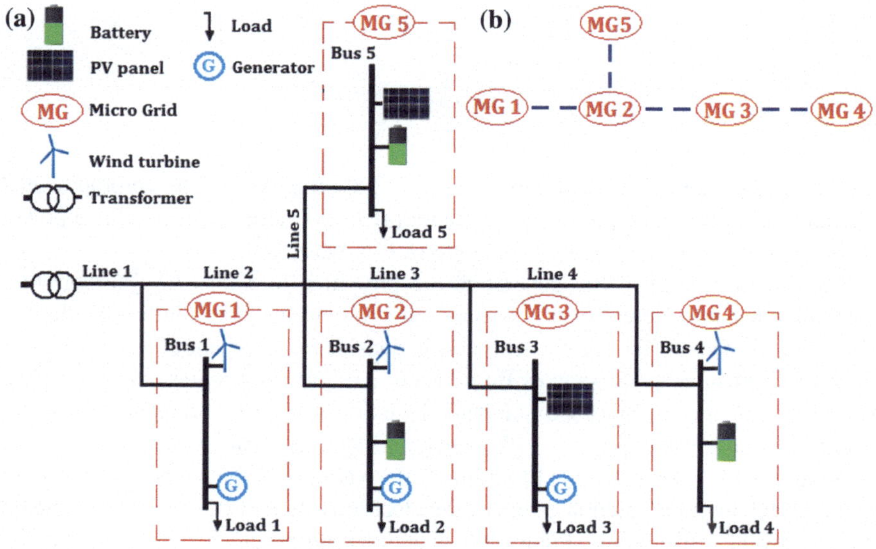


**Fig. 9.5** **a** Configuration of the distribution system under study, **b** Communication topology of the system

**Table 9.2** Technical data of the generators

| Parameter | MG 1 | MG 2 | MG 3 |
|---|---|---|---|
| $z_1^{Fu}$ (¢/kWh$^2$) | 0 | 0 | 0 |
| $z_2^{Fu}$ (¢/kWh) | 4.50 | 7.70 | 11.96 |
| $z_3^{Fu}$ (¢) | 0 | 0 | 0 |
| $z_1^{Em}$ (kg/kWh$^2$) | 0 | 0 | 0 |
| $z_2^{Em}$ (kg/kWh) | 0.610 | 0.890 | 1.356 |
| $z_3^{Em}$ (kg) | 0 | 0 | 0 |
| $\underline{P^G}$ (kW) | 20 | 15 | 10 |
| $\overline{P^G}$ (kW) | 120 | 100 | 70 |
| $MUT^G$ (min) | 5 | 5 | 5 |
| $MDT^G$ (min) | 5 | 5 | 5 |
| $STUC^G$ (¢) | 100 | 100 | 100 |
| $SHDC^G$ (¢) | 100 | 100 | 100 |

**Table 9.3** Value of the parameters of the system and problem

| | | | | | |
|---|---|---|---|---|---|
| $NM$ | 1440 | $\overline{P_4^W}$ (kW) | 40 | $MUT^B$ (min) | 5 |
| $NH$ | 24 | $\overline{P_3^{PV}}$ (kW) | 30 | $MDT^B$ (min) | 5 |
| $N\tau$ | 10 | $\overline{P_5^{PV}}$ (kW) | 30 | $\pi^B$ ($) | 0.000 |
| $\varphi$ | 0.5 | $\overline{P_2^B}$ (kW) | 10 | $\xi^B$ (Ah) | 100.000 |
| $\beta^{Em}$ (¢/kg) | 1 | $\overline{P_4^B}$ (kW) | 20 | $\rho^{Mut}$ (%) | 5 |
| $\overline{P_1^W}$ (kW) | 40 | $\overline{P_5^B}$ (kW) | 30 | $N^{GA}$ | 50 |
| $\overline{P_2^W}$ (kW) | 40 | $DOD^B$ (%) | 20 | | |

communication topology exist between the neighboring MGs. The communicated data include value of the proposed electricity price and value of the available power of the MG at every time step of the optimization period.

The technical data of the generators are presented in Table 9.2. Furthermore, value of other parameters of the system and problem are presented in Table 9.3. The value of penalty for carbon emission is considered to be about the introduced value by California Air Resources Board auction of greenhouse gas emissions [25]. The values of the average and standard deviation related to prediction errors of power of wind turbine, power of PV panel, and load demand are considered about 5 kW and 3 kW, respectively. In addition, the value of the average and standard deviation related to prediction errors of the electricity market price are considered 5 and 3 ¢/kWh, respectively. Figure 9.6 illustrates the predicted electricity market price in every minute of the operation period (one day).

### 9.4.2 Problem Simulation

The results implementing the proposed method on the test system is illustrated in Figs. 9.7, 9.8, 9.9, 9.10 and 9.11. As can be seen in Fig. 9.7a and b, MG 1 prefers to keep its generator off and purchase electricity from the market and import power from the neighboring MGs between the 1st-405th minute of the operation period. For the rest of the operation period, MG 1 starts up the generator and exports power to the neighboring MGs and sells power to the market.

As can be seen in Fig. 9.8a, MG 2 starts up its generator in the 545th minute and keeps it on for the rest of the operation period; however, in some periods, sets the power of the generator at minimum power limit and avoids shutting it down. Also, as can be seen in Fig. 9.8b, before the 945th minute, MG 2 purchases the needed power from the market and the neighboring MGs and after that, MG 2 mainly sells power to the market and exports power to the neighboring MGs. Moreover, the optimal charging and discharging pattern of the installed battery in MG 2 can be seen in Fig. 9.8a.

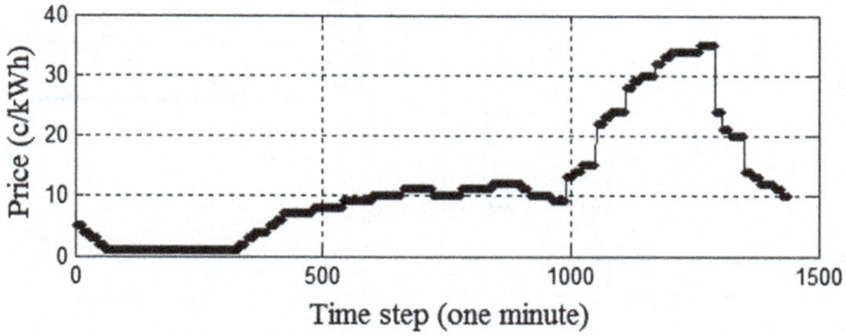


**Fig. 9.6** Predicted electricity market price in every minute of the operation period (one day)

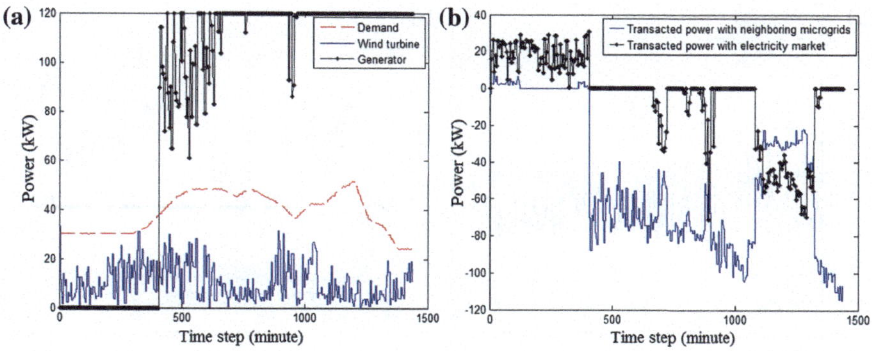


**Fig. 9.7** **a** The demand and power of wind turbine and generator installed in MG 1 at every minute of the operation period. **b** The transacted powers with the neighboring MGs and market by MG 1

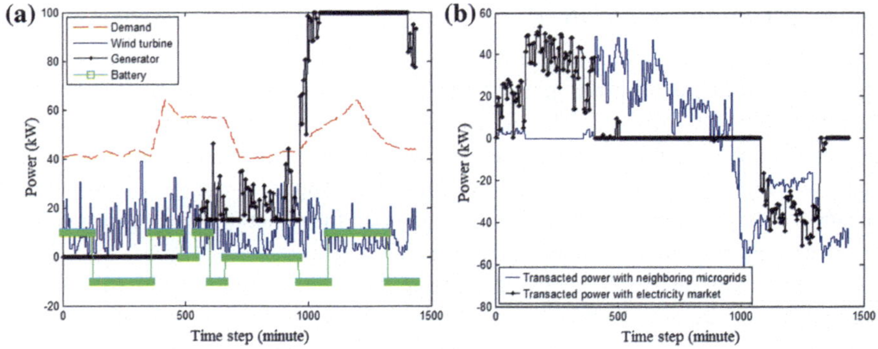


**Fig. 9.8** **a** The demand and power of wind turbine, generator, and battery installed in MG 2 at every minute of the operation period. **b** The transacted powers with the neighboring MGs and market by MG 2

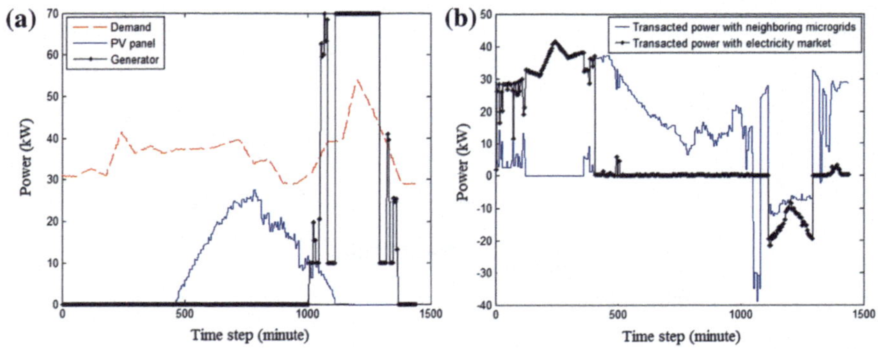


**Fig. 9.9** **a** The demand and power of PV panel and generator installed in MG 3 at every minute of the operation period. **b** The transacted powers with the neighboring MGs and market by MG 3

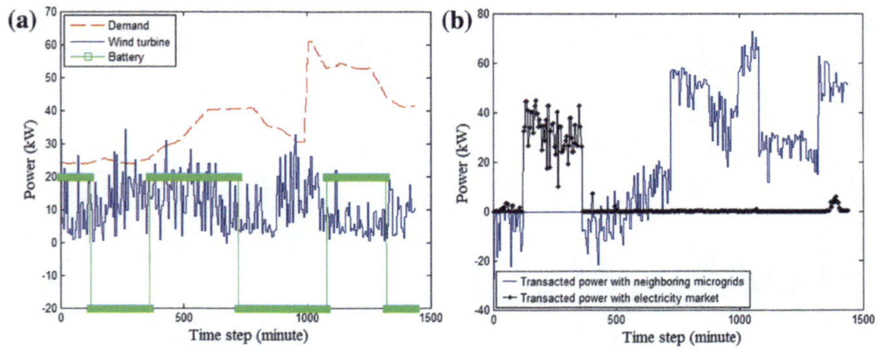


**Fig. 9.10** **a** The demand and power of wind turbine and battery installed in MG 4 at every minute of the operation period. **b** The transacted powers with the neighboring MGs and market by MG 4

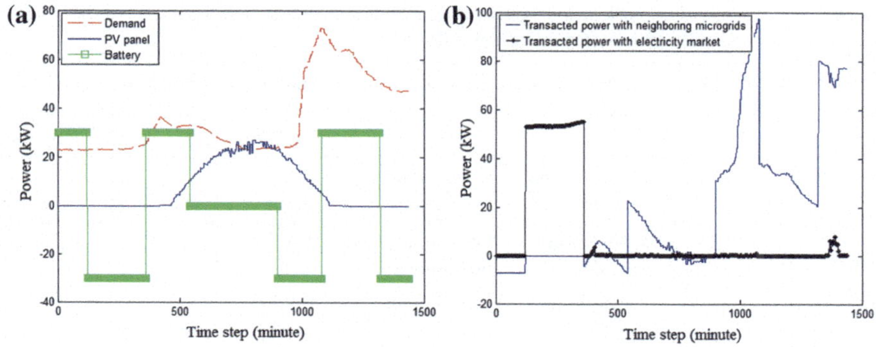


**Fig. 9.11** **a** The demand and power of PV panel and battery installed in MG 5 at every minute of the operation period. **b** The transacted powers with the neighboring MGs and market by MG 5

As can be seen in Fig. 9.9a and b, MG 3 defers using its generator until 1002th minute and, in return, purchases power from the market and imports power from the neighboring MGs. After that time, MG 3 mostly sells power to the market and exports power to the neighboring MGs.

The demand level and power of different sources at every minute of the operation period related to MG 4 and MG 5 are illustrated in Fig. 9.10a, b and in Fig. 9.11a, b, respectively. As can be seen in Figs. 9.10b and 9.11b, these MGs import power from the neighboring MGs and the market during most of the operation period. In addition, the optimal charging and discharging pattern of the installed battery in MG 4 can be seen in Fig. 9.10a.

The value of total operation cost of the system for the non-cooperative and cooperative distributed energy scheduling problems by single-time scale MPC (one-minute and one-hour scales) and multi-time scale MPC are presented in Table 9.4. As can be seen, applying multi-time scale MPC technique in the energy scheduling problem of the set of MGs has better result in both non-cooperative and cooperative distributed optimizations. In addition, cooperation of the MGs in the distributed energy scheduling problem results in lower operation cost for the set of MGs.

**Table 9.4** The operation cost of the system ($) for non-cooperative and cooperative distributed optimizations by single and multi-time scales MPC

| Type of distributed energy scheduling | Type of MPC | Time scale | Operation cost ($/day) | Batteries usage (%) | | |
|---|---|---|---|---|---|---|
| | | | | MG 2 | MG 4 | MG 5 |
| Non-cooperative | Single-time scale | One-minute | 318.31 | 0 | 0 | 0 |
| | | One-hour | 285.50 | 66 | 83 | 66 |
| | Multi-time scale | One-minute and one-hour | 281.76 | 41 | 66 | 58 |
| Cooperative | Single-time scale | One-minute | 199.81 | 0 | 0 | 0 |
| | | One-hour | 189.12 | 66 | 66 | 66 |
| | Multi-time scale | One-minute and one-hour | 184.62 | 75 | 100 | 75 |

# Appendix

## System parameters and variables

| | |
|---|---|
| $D^L$ | Load demand (kW) |
| $DOD^B$ | Depth of discharge limit for battery (%) |
| $EC^G$ | Emission cost of generator (¢) |
| $F$ | Time step objective function |
| $F^{FL}$ | Forward-looking objective function |
| $\mathbb{F}^{FL}$ | Stochastic forward-looking objective function |
| $FC^G$ | Fuel cost of generator (¢) |
| $MUT$ | Minimum up time of generator or battery (minute) |
| $MDT$ | Minimum down time of generator or battery(minute) |
| $N\tau$ | Number of time steps in optimization time horizon |
| $P^{AV}$ | Available power of a neighboring microgrid (kW) |
| $P^B, \overline{P^B}$ | Power and rated power of battery (kW) |
| $\underline{P^G}, \overline{P^G}$ | Minimum and maximu power limits of generator (kW) |
| $P^W, \overline{P^W}$ | Power and rated power of wind turbine (kW) |
| $P^{PV}, \overline{P^{PV}}$ | Power and rated power of PV panel (kW) |
| $P^{NMG}$ | Transacted power with neighboring microgrids (kW) |
| $P^M$ | Transacted power with electricity market (kW) |
| $SOC^B$ | State of charge of battery (%) |
| $STUC^G$ | Start up cost of generator (¢) |
| $SHDC^G$ | Shut down cost of generator (¢) |

(continued)

(continued)

| | |
|---|---|
| $SWC^B$ | Switching cost of battery (¢) |
| $z^{Fu}$ | Fuel cost coefficients of generator |
| $z^{Em}$ | Emission coefficients of generator |
| $\pi^M$ | Predicted value for electricity market price (¢/kWh) |
| $\pi^{MG}$ | Proposed electricity price of microgrid (¢/kWh) |
| $\pi^B$ | Initial price of a battery (¢) |
| $\mu^{Er}, \sigma^{Er}$ | Mean and standard deviation of prediction errors |
| $\Omega$ | Occurrence probability of a scenario (%) |
| $\varphi$ | Market price coefficient based on the NEM plan |
| $\xi^B$ | Total cumulative ampere-hours throughput of battery in its life cycle (Ah) |

## GA-LP parameters

| | |
|---|---|
| $a$ | Acceptance indicator of a chromosome |
| $fit$ | Fitness of a chromosome |
| $N^{GA}$ | Number of chromosomes in population |
| $r$ | A random number between [0,100] |
| $S^{Fit}$ | Set of fitness of chromosomes of population |
| $\theta^{Mut}$ | Value of mutation probability of genes (%) |
| $\theta^{PFBS}$ | Value of selection probability of a chromosome (%) |

# References


1. Department of Energy [Online], Available: http://energy.gov/oe/services/technology-development/smart-grid/role-microgrids-helping-advance-nation-s-energy-system
2. F. Valencia, J. Collado, D. Sáez, L. G. Marín, Robust energy management system for a microgrid based on a fuzzy prediction interval model. IEEE Trans. Smart Grid. https://doi.org/10.1109/tsg.2015.2463079
3. C. Opathella, B. Venkatesh, Managing uncertainty of wind energy with wind generators cooperative. IEEE Trans. Power Syst. **28**(3), 2918–2928 (2013)
4. C. Chen, S. Duan, T. Cai, B. Liu, G. Hu, Optimal allocation and economic analysis of energy storage system in microgrids. IEEE Trans. Power Elect. **26**, 2762–2773 (2011)
5. E. Arriagada, E. Lopeza, M. Lopezc, R. Blasco-Gimenezd, C. Roaa, M. Poloujadoff, A probabilistic economic dispatch model and methodology considering renewable energy, demand and generator uncertainties. Elect. Power Syst. Res. **121**, 325–332 (2015)
6. G.J. Osorio, J.M. Lujano-Rojas, J.C.O. Matias, J.P.S. Catalao, A probabilistic approach to solve the economic dispatch problem with intermittent renewable energy sources. Energy **82**, 949–959 (2015)
7. J. Hetzer, D.C. Yu, K. Bhattarai, An economic dispatch model incorporating wind power. IEEE Trans. Energy Convers. **23**, 603–611 (2008)

8. S.S. Reddy, P.R. Bijwe, A.R. Abhyankar, Real-time economic dispatch considering renewable power generation variability and uncertainty over scheduling period. IEEE Syst. J. https://doi.org/10.1109/jsyst.2014.2325967
9. S.M. Hakimi, S.M. Moghaddas-Tafreshi, Optimal planning of a smart microgrid including demand response and intermittent renewable energy resources. IEEE Trans. Smart Grid **5**, 2889–2900 (2014)
10. T.A. Nguyen, M.L. Crow, Stochastic optimization of renewable-based microgrid operation incorporating battery operating cost. IEEE Trans. Power Syst. (2015). https://doi.org/10.1109/TPWRS.2015.2455491
11. N. Holjevac, T. Capuder, I. Kuzle, Adaptive control for evaluation of flexibility benefits in microgrid systems. Energy **92**, 487–504 (2015)
12. W. Su, J. Wang, J. Roh, Stochastic energy scheduling in microgrids with intermittent renewable energy resources. IEEE Trans. Smart Grid **5**, 1876–1883 (2014)
13. F. Nishimura, R.D. Tabors, M.G. Llic, J.R. Lacalle-Melero, Benefit optimization of centralized and decentralized power systems in a multi area environment. IEEE Trans. Power Syst. **8**, 1180–1186 (1993)
14. B. Singh, R. Mahanty, S.P. Singh, Centralized and decentralized optimal decision support for congestion management. Int. J. Electr. Power Energy Syst. **64**, 250–259 (2015)
15. L. Chua, L. Yang, Cellular neural networks. IEEE Int. Symp. Circuits Syst. **2**, 985–988 (1998)
16. C. Marshal, *Enterprise Modeling with UML-Designing Successful Software Through Business Analysis*, 2nd edn. (Addison-Wesley, 1999)
17. S. Yang, S. Tan, J.X. Xu, Consensus based approach for economic dispatch problem in a smart grid. IEEE Trans. Power Syst. **28**, 4416–4426 (2013)
18. G. Binetti, A. Davoudi, D. Naso, B. Turchiano, F.L. Lewis, A distributed auction-based algorithm for the nonconvex economic dispatch problem. IEEE Trans. Ind. Elect. **10**, 1124–1132 (2014)
19. A.J. del Real, A. Arce, C. Bordons, Combined environmental and economic dispatch of smart grids using distributed model predictive control. Int. J. of Electr. Power Energy Syst. **54**, 65–76 (2014)
20. C. Gong, X. Wang, W. Xu, A. Tajer, Distributed real-time energy scheduling in smart grid: stochastic model and fast optimization. IEEE Trans. Smart Grid **4**, 1476–1489 (2013)
21. Q.H. Wu, J. Zheng, Z. Jing, Coordinated scheduling of energy resources for distributed DHCs in an integrated energy grid. CSEE J. Power Energy Syst. **1**, 95–103 (2015)
22. M. Rahmani-andebili, Spinning reserve supply with presence of plug-in electric vehicles aggregator considering compromise between cost and reliability. IET Gener. Transm. Distrib. **7**, 1442–1452 (2013)
23. J.B. Rawlings, D.Q. Mayne, *Model Predictive Control: Theory and Design* (Nob Hill Publishing, LLC, Madison, WI, 2009). [Online] Available: http://jbrwww.che.wisc.edu/home/jbraw/mpc/electronic-book.pdf
24. Federal Energy Regulatory Commission [Online], Available: http://www.ferc.gov/industries/electric/indus-act/section-1241.pdf
25. U.S. Energy Information Administration (EIA) [Online], Available: http://www.eia.gov/todayinenergy/detail.cfm?id=9310. Accessed Oct 2015

# Chapter 10
# Protection System Considerations in Networks with Distributed Generation

**Gabriel Albieri Quiroga, Carlos Frederico Meschini Almeida, Henrique Kagan and Nelson Kagan**

**Abstract** This chapter presents methodologies to assess the impact of distribution generation on electric distribution network protection systems for an integrated network planning. The distributed generation (DG) alternative is seen as a shift in the paradigm of energy generation in the world. The adoption of renewable sources of energy for residential or commercial production brings not only environmental benefits, but also an opportunity to ease the supplying difficulties found in many countries. Despite solving some of the energy suppling problems, this paradigm change caused by the insertion of DG in electric distribution networks can bring some undesired technical impacts. The typical impacts assessed in the electric distribution networks planning involve the expansion of the network, such as losses, power factor, line loading, and voltage profiles, among others. However, a massive insertion of DG may also cause significant problems to the network protection, which involves the protection planning. In this way, it is necessary to identify potential protection issues and design means to model, diagnose and mitigate such issues. This chapter aims at describing the importance of predicting the potential impacts of high penetration of renewable sources on the protection system from electric distribution networks, in order to achieve an integrated network planning. Thus, in the beginning of this chapter it is discussed the main protection system issues that may arise from the high penetration of distributed generation, such as loss of protection coordination, overvoltage, loss of protection

G. A. Quiroga (✉) · H. Kagan · N. Kagan
Electrical Energy and Automation Engineering Department,
University of Sao Paulo, Sao Paulo, Brazil
e-mail: gabriel.quiroga@usp.br; quirogaa.gabriell@gmail.com

H. Kagan
e-mail: henrique.kagan@sinapsisenergia.com

N. Kagan
e-mail: nelsonk@pea.usp.br

C. F. M. Almeida
Mining and Petroleum Engineering Department,
University of Sao Paulo, Sao Paulo, Brazil
e-mail: cfmalmeida@usp.br



A. Arefi et al. (eds.), *Electric Distribution Network Management and Control*,
Power Systems, https://doi.org/10.1007/978-981-10-7001-3_10

sensitivity, directional false tripping, unwanted fuse blowing, beside others. The chapter than demonstrates possible models utilized to assess those impacts, such as network modeling and renewable sources modeling, possible approaches such as the probabilistic and deterministic perspectives, regarding DG allocation algorithms and the possible methodologies for assessment such as scenarios or sensitivity analysis. The methodologies evaluate the protection impacts and are based on several short circuit calculations and for the probabilistic approach, the Monte Carlo Method. The results shown in the chapter may represent the calculation of such impacts for electric distribution networks. Those may contain loss of sensitivity, directional false tripping and unwanted fuse blowing impact calculations for some networks to illustrate the methodologies. To conclude the chapter, there will be more discussions regarding the results presented and the potential benefits of including this analysis on the integrated distribution network planning. In the end, the chapter illustrates the application of the presented methodologies with a case study using a real distribution network.

**Keywords** Distributed generation · Protection planning · Integrated planning

## 10.1 Introduction

The DG may introduce new possibilities for the electric distribution networks, while making planning and operation activities more challenging. Among these possibilities, the DG allows commercial consumers to enjoy an extensive set of Reliability versus Price combinations. Therefore, the DG could appear as an autonomous electric distribution network, which meets consumer requirements such as reactive power and harmonic compensation, peak shaving and power factor correction, as well as providing reserve generation means and improvement in reliability [1]. However, the existing electric distribution networks were not developed to accommodate these new technologies. So, the connection of DG to the electric system may violate existing planning and operating practices. In this sense, it should be noted that the integration of DG, as well as other storage devices, in the distribution networks can alter the existing practice of having a unidirectional power flow, and this considerably affects the coordination of the utilities' protection systems [2–4].

In the context of the DG impact on the utilities' protection system, several protection problems, such as unintentional islanding, overvoltage, loss of sensitivity, directional false tripping, and unwanted fuse blowing were identified and documented [2–4]. Yet, critical issues, such as DG insertion limits, considering the influence of location, capacity and technology remains unanswered. The criticality of this issue stems from the enormous infrastructure of the related utilities, which contribute to making any large-scale change impractical in order to promote the DG accommodation. Thus, the development of a generic procedure for determining the impact on energy distribution practices, as well as for the evaluation of the insertion limits in terms of location, capacity, and technology of DG becomes interesting. This will help to facilitate the secure integration of DG units into existing electric distribution networks by better understanding the requirements and enhancements needed to achieve this goal. The direct result of this development is the financial savings expectation to the utilities by capturing the main aspects in the implementation of DG in existing networks.

The innovation presented in the chapter is based on the use of a probabilistic methodology to create scenarios for insertion of small scale DG units into the LV networks and to analyze the possible impacts on the MV networks protection systems. The typical methodologies used to analysis the DG impact are presented initially and serve to guide the evaluation of possible impacts on the protection system. Commonly, these methodologies are based on a deterministic analysis, evaluating the impact of a few large generating units, usually with a nominal power of more than 1 MW as seen in [5–9].

In the chapter the approach is different, showing that the same analysis techniques can be used to assess the impacts of the massive penetration of small distributed generation units scattered over the network. Thus, the innovation relies on the probabilistic approach given to the problem, since in the planning of distribution networks; it becomes interesting to have a methodology capable of simulating different scenarios of penetration of small DG units, usually with a nominal power between 1 kW and 1 MW.

## 10.2 DG Protection Impact Assessment for an Integrated Planning

In the context of the electric distribution networks old paradigm, network planning consists of solving the impacts caused by the demand growth. In this way, planning horizons are used to assess the impact on electric distribution networks that may arise in electric distribution networks, such as voltage problems and line overloading issues. They usually range from 5 to 10 years, and consider horizontal and vertical demand projections.

These projections consist of statistical studies to predict the demand growth. Thus, in the planning problem, the planner evaluates the impact of the demand growth, within a planning horizon, and projects necessary reinforcements, expansion of existing networks or the development of new networks in order to solve those issues.

The typical impacts assessed in the electric distribution networks planning are line loading, voltage profiles, and losses. The usual solutions to these impacts consist of cable replacement, transformer replacement, installation of voltage regulators, installation of capacitor banks and even the construction of new feeders and new substations, among others, in order to meet the new demand for electric power.

However, in the new smart grid paradigm, there are other factors that must be taken into account in electric distribution network planning, such as distributed energy resources (DER), more specifically the appearance of the DG. So now, in addition to the projection of demand growth, there is a necessity to project the supply growth, which derives from the DG insertion.

It is known that DG causes not only impacts to the network expansion, but also to the network protection. Therefore, during the network planning, it is important to quantify the actions to solve protection issues. The main impacts to the protection of electrical distribution networks, which are discussed in this chapter, are loss of protection coordination, overvoltage, loss of protection sensitivity, directional false tripping, and unwanted fuse blowing. Thus, the main electric distribution network protection planner actions to solve such impacts are the elaboration of plans that aim at the possible sensitivity adjustment of protection devices, the installation of protection devices with directional function, the replacement of fuses, the modification of coordination and selectivity studies, and even the development of restriction of DG size in certain network areas. This chapter presents the main impacts caused by DG to the protection of electrical distribution networks, methodologies to assess such impacts, electrical models that aid in the assessment and possible approaches that can be used in the process of protection planning.

## 10.3 Potential DG Impacts on the Protection System

The main protection system issues that arise from the high penetration of distributed generation must be defined and addressed in order to determine the limits of DG insertion that can trigger it. Thus, the main issues that come with the DG insertion are discussed in this section. Those are loss of protection coordination, loss of protection sensitivity, directional false tripping, overvoltage, unwanted fuse blowing and unintentional islanding. This section also presents a generic method to evaluate the impact of DG on the protection of electric distribution networks. In order to guide the impact assessment, each protection problem is discussed individually, making possible the determination of the insertion limits that trigger the issues addressed.

### *10.3.1 Loss of Protection Coordination*

In normal operation, protective devices are coordinated so that the primary protection scheme operates before the rear scheme. The interconnection of DGs increases the level of short circuit. Depending on the initial coordination parameters, along with the location, size and type of DG, it is possible to reach a situation where the coordination cannot be achieved anymore. In such cases, the back-up protection device operates before the primary protection, resulting in inadequate operation for part of the network loads. The main protective devices used in the protection coordination are fuses, reclosers and relays. To illustrate the loss of coordination, it will be shown the effects of the DG in the fuse-fuse coordination.

The Following Two characteristics determine the behavior of a fuse:

- The fuse minimum melting (MM);
- The total clearing (TC).

The MM determines the time from the initiation of an overcurrent to the instant arcing begins inside a fuse. The TC is the total opening time of a fuse from the occurrence of an overcurrent until the fuse stops current flow. This is the sum of link melting and arcing time. To illustrate this behaviour, it is shown in Fig. 10.1 how fuse-fuse coordination is traditionally done in a radial distribution system without DG.

In order to be coordinated for any fault downstream of fuse 2, fuse 2 should operate before fuse 1. This would be achieved if TC characteristic of fusel 2 is below the MM characteristics of fuse 1 by a safe margin for any fault. The coordination graph shows that the fuses are coordinated for all fault currents within the coordination range. The coordination range is the range between the minimum and maximum fault current for any fault downstream of fuse 2.

The DG impact into the protection system point of view derives from the change in the values of fault currents flowing in the system for any given fault. Even a "back-flow" is possible to happen, resulting in other problems such as directional false tripping and unwanted fuse blowing, which will be addressed in the following sections. Due to the insertion of the DG, the minimum and maximum fault current values may increase from source side due to all upstream DG [10]. In this case,

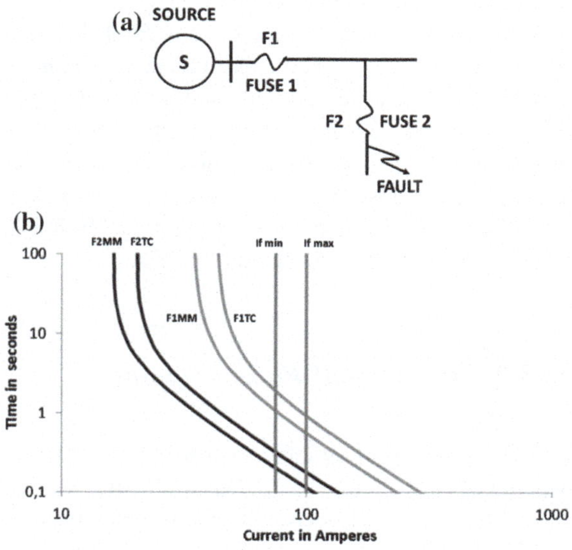


**Fig. 10.1** **a** Schematic diagram of a traditional fuse-fuse coordination without DG. **b** Coordination graphs of a traditional fuse-fuse coordination without DG

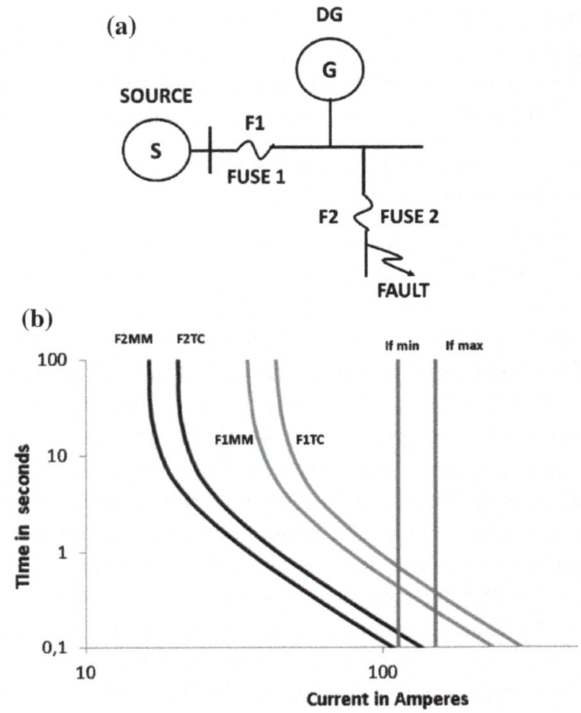


**Fig. 10.2** **a** Schematic diagram of a fuse-fuse coordination with DG. **b** Coordination graphs of the impact of DG on fuse-fuse coordination

the back-up protection ends up opening, instead of the main protection. To illustrate this behaviour, it is shown in Fig. 10.2 how fuse-fuse coordination can be affected in a radial distribution system with DG.

In order to be coordinated for any fault downstream of fuse 2, fuse 2 should operate before fuse 1. The coordination graph shows that the fuses are no longer coordinated for all fault currents within the coordination range, since the maximum fault current may sensitize fuse 1 instead of fuse 2. Depending on the configuration of the system, loss of coordination may occur between any pair of protection devices, i.e., recloser-fuse, fuse-fuse, and so on. In each case, the minimum short circuit current causing the loss of coordination helps in identifying the penetration level allowed for the installation of DGs.

### *10.3.2 Directional False Tripping*

The DG installation on electric distribution networks can lead to directional false tripping of protection devices. This problem is the malfunction of protective devices caused by unwanted operation of a protection device and the disconnection of a "healthy" part of the system which is unnecessarily interrupted. The directional false tripping is easily observed in mesh systems supplied by a same source.

Moreover, this problem is relevant in radial systems, since it allows the fault to be supplied by DG installed in adjacent side electric distribution networks.

To illustrate such a situation, consider two parallel radial electric distribution networks fed from the same source (same HV/MV substation transformer). Without the DG, if a short circuit occurs in one of the two electric distribution networks, the short circuit will be completely fed by the utility substation transformer and the other electric distribution network will not contribute to the short circuit current as shown in Fig. 10.3. The current direction in the circuit breaker CB 2 will not be reversed and its relay will not operate in response to the short circuit in the electric distribution network EDN 1.

Alternatively, if a DG is installed in the "healthy" electric distribution network (END 2), the contribution of END 2 to the short circuit current in END 1 will not be null, as shown in Fig. 10.4. In this case, if the circuit breaker CB 2 has a protection tripping curve faster than circuit breaker CB 1, the circuit breaker CB 2 can respond

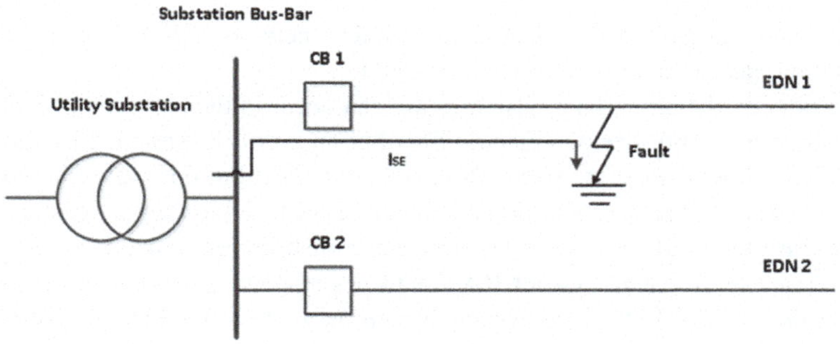


**Fig. 10.3** Short circuit current through an adjacent electric distribution network without DG, EDN: electric distribution network

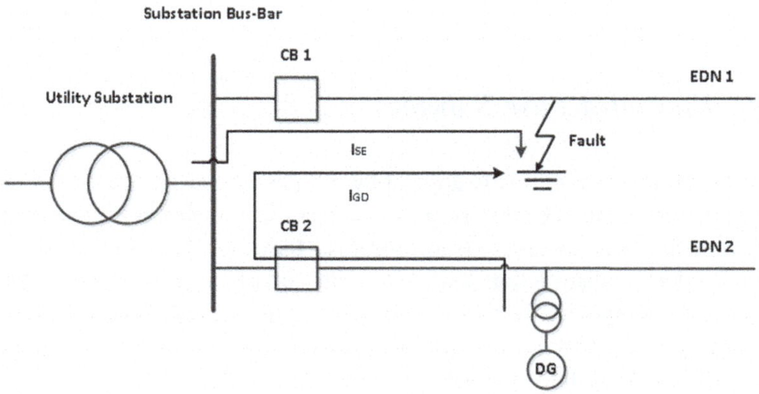


**Fig. 10.4** Short circuit current through an adjacent electric distribution network with DG, EDN: electric distribution network

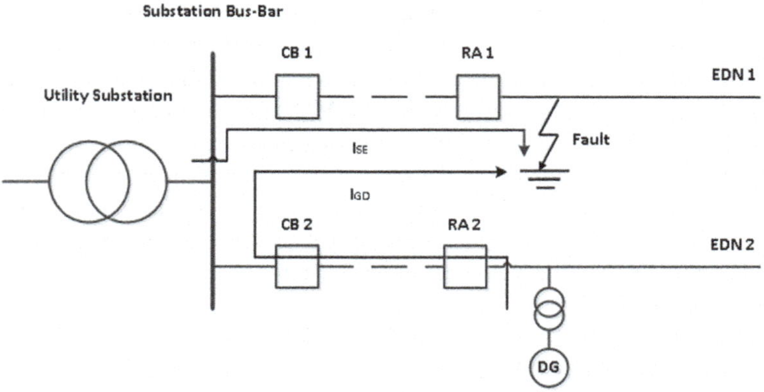


**Fig. 10.5** Short circuit current through an adjacent electric distribution network with DG, considering recloser effects, EDN: electric distribution network

faster to short circuit in the electric distribution network END 1, unnecessarily interrupting electric distribution network END 2.

Additionally the directional false tripping can occur in the presence of reclosers. To illustrate that, consider two electric distribution networks supplied by the same source, as shown in Fig. 10.5. Both electric distribution networks contain circuit-breakers and reclosers to prevent unnecessary fuse blowing downstream due to temporary faults. If there is no DG in electric distribution network EDN 2, both circuit breaker CB 2 and recloser RA 2 will not respond to a fault in the electric distribution network EDN 1. However, if there is a DG in the electric distribution network EDN 2, the DG will contribute to the short circuit current. This contribution to the short circuit current can be perceived by the recloser RA 2, which responds rapidly to current levels above its tripping level. Thus, the recloser RA 2 may operate before the recloser RA 1, causing an unnecessary interruption of a "healthy" portion of the system.

### *10.3.3 Unwanted Fuse Blowing*

Fuse saving strategies are usually applied by utilities to prevent fuses from burning due to temporary faults. For this purpose, the automatic reclosers de-energized and re-energized the line, in order to disable fuse blowing. This strategy is widely adopted in electric distribution networks with suburban or rural characteristics. Many faults are temporary by nature, so the role of the recloser is to try to eliminate the temporary faults without the need of a permanent interruption, in addition to saving fuses, avoiding unnecessary fuse blowing.

However, the DG installation in electric distribution networks could affect the coordination between the recloser and the fuses due to the additional DG

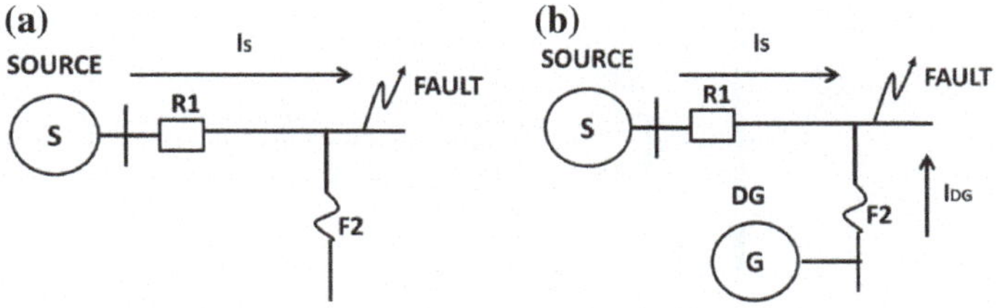


**Fig. 10.6** **a** Fault contribution upstream the fuse F2 without DG. **b** Fault contribution upstream the fuse F2 with DG

contribution to the short circuit current. In such cases, the coordination between recloser and fuses may be lost and fuses may burn before the recloser operation. Or both the fuse may burn and the recloser may operate at the same time, adding more difficulty to fault location.

To illustrate such a situation, consider two circuits. The first circuit is a radial network supplied by a single source. The second circuit is the same network, but this time supplied by both the source and a DG as shown in Fig. 10.6.

Without the DG, a fault upstream the fuse 2 is supplied only by the network source and no coordination between the recloser R1 and fuse F2 is needed, since there is no current passing through fuse F2. In the second situation the short circuit contribution comes both from the network source and the DG. This may cause a loss of coordination between the recloser R1 and fuse F2 and to fuse F2 to blow without necessity.

## *10.3.4 Loss of Sensitivity*

The addition of a DG in a distribution system may reduce short circuit current contribution from the network main source. This affects the operation of network protection devices, such as circuit breakers, reclosers and fuses, disturbing their ability to "sense" the fault current. This aspect is quite dependent on the type, size and location of DG.

To illustrate the DG installation impact on the protection devices sensitivity, consider a generic radial system, as shown in Fig. 10.7. A generic radial system without the installation of any DG is shown in Fig. 10.7a. In this network, under short circuit conditions, the short circuit current is completely supplied by the utility's substation. In this case, the network protection relay must be set to respond to the lowest short circuit current value.

It is shown in Fig. 10.7b the same circuit with a DG installed. For a fault after the DG location, the short circuit current is supplied by both the substation and the DG. The contribution of each source will depend on the impedances between the source and the point of failure.

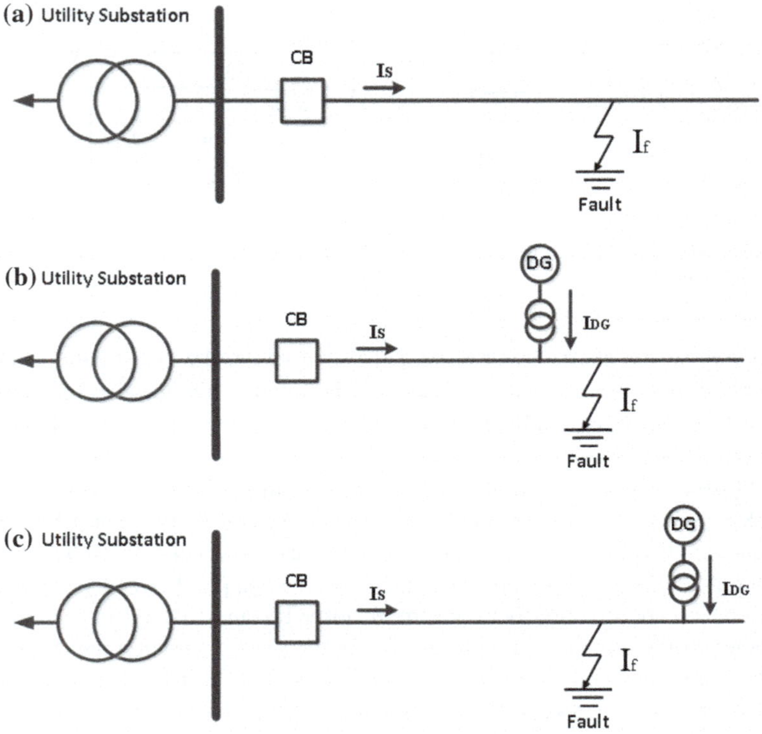


**Fig. 10.7** **a** Fault supplied only by the main source. **b** Fault supplied by both the main source and the DG located between the substation and the fault. **c** Fault supplied by both the main source and the DG located afterward the fault

For a fault located between the substation and the DG, as shown in Fig. 10.7c, the contribution of the substation to the short circuit current is independent of DG. The extreme effect of DG is to decrease the network's sensitivity to a limit where it does not "sense" the fault.

### *10.3.5 Overvoltage*

The main problems related to overvoltage and associated with DG include [4, 11]:

- Temporary overvoltage due to phase-to-ground short circuit conditions caused by improper application of DG grounding or coupling transformer connection.
- Overvoltage originated on the utility and adjacent environment, which affects the distributed generation scheme.
- Resonant overvoltages that may arise during islanding conditions.
- Surges that may arise due to high power injection by distributed resources.

Temporary overvoltages occur due to ground faults, caused by improper DG grounding conditions or coupling transformers. The correct dimensioning of the lightening arresters can prevent damage to the DG installation caused by the high voltage surges originated in the adjacent environment.

Resonant overvoltages can arise due to the interaction between the combination of the coupling transformer and the DG inductance and the system capacitance. This is a complex operational problem, which must be approached in a different context using detailed dynamic simulation programs such as EMTDC or EMTP. Similarly, overvoltages caused by a high power injection by the DG are also a complex operational problem, which must be approached in a different context.

There are five typical transformer connections widely used for the coupling of DG to utility systems, as seen in Table 10.1.

For the first three connections, the main advantages are:

- There are no impacts on the coordination of the utility neutral relay;
- Any ground fault on the DG side will not be perceived by the utility's protection system, i.e., coordination involving ground faults will not be affected.

However, the major concern of coupling transformer connections with the ungrounded side of the utility is that after opening the utility network's circuit breaker due to a permanent fault, the system will be powered by a non-grounded supply. This submits the undamaged phases to surges that will approach the line voltage level of the system.

Regarding the fourth type of connection, grounded wye (MT)/Delta (DG), for a ground fault on the DG side, the utility does not contribute to the zero sequence current. This prevents the utility protection to act without necessity. However, the disadvantage of this type of connection is that it acts as a source of zero sequence current, thus establishing a zero sequence current for faults involving the ground in the distribution system. This can significantly impact the coordination of the relay responsible for the utility neutral protection.

In addition, the zero sequence current flow in the coupling transformer primary (utility side) will cause a high current to flow in the secondary (DG side) and this may lead to heating problems in the transformer. One solution commonly used to overcome this problem is the installation of ground impedances on the neutral side of the primary side, in order to limit the excessive flow of current. In such cases, the ground impedance must be high enough to limit current flow but low enough to ensure the effectiveness of DG grounding.

**Table 10.1** DG coupling transformer connections

| MV—Utility side | LV–DG side |
|---|---|
| Delta | Delta |
| Delta | Grounded Wye |
| Wye | Delta |
| Grounded Wye | Delta |
| Grounded Wye | Grounded Wye |

Another impact refers to the presence of unbalanced loads in the distribution circuit. Normally, the currents of these loads return to earth through the utility transformer neutral grounding, located in the substation. With the insertion of a coupling transformer with this type of connection, the current due to this imbalance will be divided between the utility transformer and the DG coupling transformer. During drastic unbalance conditions, such as fuse blowing at electric distribution network branches, the charging capacity of the coupling transformer can be reduced. The use of grounded wye connection on the utility side and a delta connection on the DG side has the advantage of limiting overvoltages that may be caused when the utility's circuit breaker opens, thereby preventing lightning arrestors and loads from being damaged.

The fifth type of connection, the grounded wye (MT)/grounded wye (DG side), has the advantage that surges will not be originate when the utility circuit breaker opens (Considering the neutral solidly grounded). However, the major disadvantage is that this type of connection functions as an undesirable source of zero sequence current, similar to that described for the prior connection type.

## 10.4 Electric Models

The objective of this chapter is, in addition to demonstrate possible impacts of the DG in the protection of distribution networks, offer alternatives for the analysis of such impacts. With this purpose, power flow and short circuit tools are used. Thus, in order to evaluate the impact of the insertion of distributed generators into distribution networks, it is necessary to model the distribution networks as well as the prosumers. The models used will be presented below.

### *10.4.1 DG Electric Modelling*

The methodology considers the use of traditional tools for electric distribution networks analysis such as the power flow and short circuits simulation. It is known that the contribution of inverters during faults is not zero and varies by design. For most fault conditions, several inverters continue supplying current to the network, subsequent to a fault for a period ranging from 4 to 180 ms. For most vendor-specific inverter models, the inverter current during the fault may range from 100 to 200% of the inverter-rated current [12, 13]. Thus, the DG model for a source coupled by inverters, considers a short circuit contribution of 2 times its nominal current ratio, while the ratio considered for rotating machines is 10 times its nominal current as defined in [13].

The short circuit sequence model used to represent the inverter based DG considers no contribution from negative or zero sequence as seen in [14], while model used to represent the machines considers its connection, such as delta or wye.

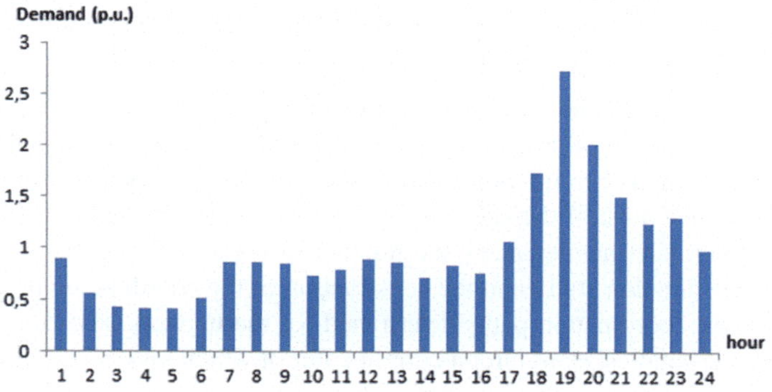


**Fig. 10.8** Client characteristic load curve

### *10.4.2 Network Electric Model*

The methodology require the electric model of network equipment such as MV and LV networks, MV and LV distribution transformers, protection devices and consumers as represented in [15]. The MV and LV networks are modeled based on the symmetrical components theory of a PI circuit model. The MV and LV transformer are modelled based on the symmetrical components theory.

The clients in the distribution network model are based on load curves. The load curves are calculated based on the clients typical load behavior and their monthly energy consumption as seen in [15]. A residential characteristic curve is detailed in Fig. 10.8.

## 10.5 DG Impact Methodologies

The methodologies to assess the impacts on the network protection are based on the risk of not considering the DG contribution to the fault currents. To asses this risk, it is necessary to evaluate each of the impacts previously introduced, with and without the presence of the DG. Thus, the next sections will present generic procedures to assess the penetration limits of DG for each one of the protection issues raised.

### *10.5.1 Generic Procedure for Assessing the DG Limits for Loss of Coordination*

The objective of this section is to develop a generic procedure to guide the determination of the penetration limits of DGs in terms of size, location and technology from the point of view of the loss of coordination. The procedure consists in five steps as it follows.

1. In a given system, define different coordination paths. A coordination path may be defined as the set of protection devices located along the circuit path from the main circuit breaker to the protection device located further downstream of the network. Since most of the side fuses (or sub-side) are selected to be similar (in order to facilitate maintenance), the number of coordination paths are limited.
2. Do the coordination study and the construction of the coordination diagrams for the different coordination paths determined for the network studied.
3. Observe the smallest short circuit current value at which loss of coordination can occur between the coordination paths. This current value may be the intersection of the coordination curves of two successive protective devices. It should be noted that such a current value may not exist if there is no intersection between the coordination curves. In these cases, there will be no limit to the installation of any DG capable of promoting the violation of the system coordination.
4. Define candidate points where the DG could be installed. Insertion limits should be calculated for these specific points. The candidate points could be defined by means of planning studies, whose objective would be the optimal allocation DG, or determined by the consumers, or even be located randomly to calculate the risk of the DG installation.
5. Simulate the installation of the DG at the candidate points. Increasing the DG penetration level until it achieves the smallest short circuit current that promotes loss of coordination. It is necessary to repeat this step for every candidate point. To understand the influence of the DG technologies it is necessary to repeat this step to other technologies such as: synchronous generators, inverter generators and induction generators.

### *10.5.2 Generic Procedure for Assessing the DG Limits for Directional False Tripping*

The objective of this section is to develop a generic procedure to guide the determination of the limits of DG insertion in terms of size, location and technology, from the point of view of protection devices directional false tripping. The procedure is very similar to the one developed for coordination and can be summarized as follows.

1. Define the different coordination paths in the system under study.
2. Conduct the coordination study and the construction of coordination diagrams for the different coordination paths determined for the study system, without considering the presence of DG.
3. Define candidate points where DG can be installed.

4. Simulate the DG installation in the candidate points by increasing the DG insertion level until the lowest short circuit current that prompts the occurrence of the directional false tripping problem occurs. It is necessary to repeat this step for every candidate point.

### *10.5.3 Generic Procedure for Assessing the DG Limits for Unwanted Fuse Blowing*

The purpose of this section is to develop a generic procedure to guide the determination of the DG insertion limits, from the point of view of the unexpected burning of fuses. The procedure can be summarized as follows.

1. Define the different coordination paths in the system under study.
2. Run the protection study and construct the coordination diagrams for the coordination paths without considering the DG installation.
3. Observe the short circuit level for which unexpected fuse blow-off may occur. This current value is obtained by observing the intersection between the fast recloser curve and the arc extinction curve of the fuse.
4. Define the candidate points for the installation of DG.
5. Simulate the DG installation at the candidate points, and then increase the DG insertion level until the current reaches the value for unexpected fuse firing. It is necessary to repeat this step for every candidate point.

### *10.5.4 Generic Procedure for Assessing the DG Limits for Loss of Sensitivity*

The objective of this section is to develop a generic procedure to guide the determination of the DG insertion limits from the point of view of loss of sensitivity. The procedure can be summarized as follows.

1. Determine the pick-up currents (three-phase and phase-to-ground) for the each circuit breaker and relay.
2. Determine the protection zone of each protection device. This can be approximated by the point with the lowest short circuit current that is protected by the protection device.
3. Define the candidate points where a DG could be installed. Insertion limits should be calculated for these specific points. The candidate points could be defined by means of planning studies, whose objective would be the optimal allocation DG, or determined by the consumers, or even be located randomly to calculate the risk of the DG installation.

4. Simulate the DG installation in the candidate points and establish the fault conditions for the points defined in Step 1. Then increase the DG insertion level until it reaches the level of sensitization (pick-up value). It is necessary to repeat this step for every candidate point. To understand the influence of the DG technologies it is necessary to repeat this step to other technologies such as: synchronous generators, inverter generators and induction generators.

### *10.5.5 Generic Procedure for Assessing the DG Limits for Overvoltage*

As can be seen from the discussion presented in Sect. 10.3.5, not grounding the DG coupling transformer exposes the network and consumers connected to it to dangerous temporary overvoltage levels. On the other hand, solid groundings may limit the sensitivity of the protection to unacceptable levels during ground faults. The increase in ground impedance must meet these two aspects simultaneously. That is, increasing the ground impedance can be done in such a way that the dimensioned value allows the overvoltage levels to be acceptable (about 25% higher than the nominal value) with a small reduction in sensitivity (around 5%) for the smallest fault current. Reference [3] suggests that a reactor installed at the interconnection transformer ground with a typical value between 1.0 and 1.5 of the zero-sequence impedance value of the transformer itself may be the point of equilibrium between the overvoltage levels and loss of protection sensitivity. A generic procedure to guide the determination of the DG insertion limits from the point of view of temporary overvoltages can be summarized as it follows.

1. Define the candidate points where the DG can be installed, including the details for the interconnection transformer and for grounding.
2. For not grounded coupling transformers the second step is to define the DG protection scheme, which must ensure that the DG is disconnected from the system before the circuit breaker opening, for each candidate point.
3. For grounded connections, the second step is to simulate the phase-to-ground short circuit and calculate the percentage change in ground fault sensitivity. If the sensitivity change is greater than 5%, insert a grounding reactor into the coupling transformer grounding.
4. For grounded connections, the third step is to calculate the overvoltage due to the insertion of the grounding reactor and check if it is within the acceptable range (e.g., 25% higher than the nominal voltage). Then repeat the second step until the sensitivity and overvoltage level are acceptable, for each candidate point.

## 10.6 Insertion Level Approaches

The methodologies presented take into account the concept of raising the penetration levels as means to increase the generator power or the allocation of more generators in the distribution networks. This occurs in order to achieve the penetration limits that are harmful to the protection system. The following sections will demonstrate mechanisms for raising these levels of penetration with both deterministic and probabilistic bias.

### *10.6.1 Deterministic DG Allocation Approach*

In the deterministic DG allocation approach the raising of the penetration levels means the increase of a DG power installed at the network. But this only is not enough to evaluate the DG impact, since the location is an important variable to the analysis.

Thus, in the deterministic DG allocation approach, the location of the DG in determined by arbitrary means. One possibility of the DG allocation proposition is to define the candidate points by means of planning studies, whose objective would be the optimal allocation of DG to promote the minimization of losses and to improve the voltage profile.

For the generic procedure for assessing the DG limits for loss of sensitivity, one can allocate the DG in the place of greater impact. In that way one can get the most conservative analysis possible. The conservative point can be defined analytically. Thus, an example of how to calculate the optimal position for the generator allocation is shown in Fig. 10.9.

It is shown in Fig. 10.9 a circuit breaker and its protection area, the defined point of shortest short circuit current, as point B and a generator unit, with delta wilding, to be allocated without a coupling transformer, in point A. The equivalent impedance of the supply is defined as Zsupply (Ohm), the DG impedance is defined as

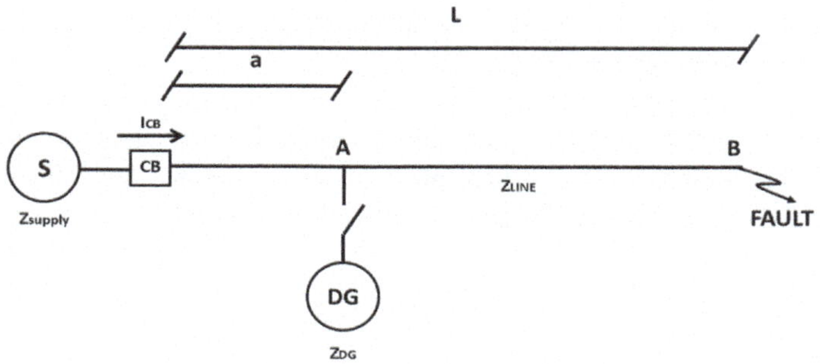


**Fig. 10.9** Example of conservative allocation for loss of sensitivity impact analysis

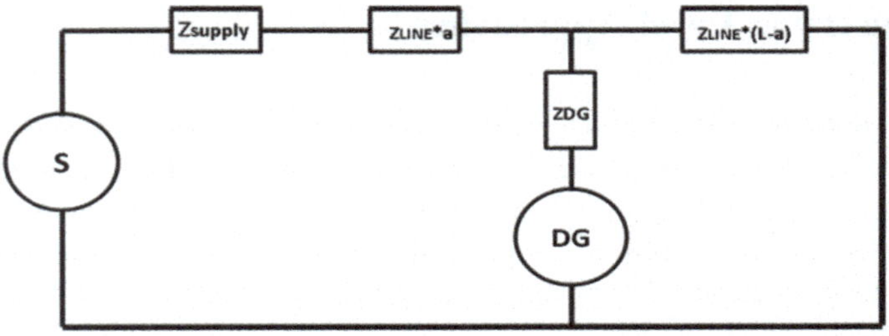


**Fig. 10.10** Short circuit model

Zdg (Ohm) and the line impedance is defined as Zline (Ohm/km). For a three-phase short circuit at point B, the short circuit model is shown in Fig. 10.10, the equation that defines the short circuit current seen by the circuit-breaker is given by (10.1) and the maximum impact point is defined in (10.2).

$$I_{CB} = \frac{V}{a^2 * \left(\frac{-Z_{line}^2}{Z_{DG}}\right) + a * \left(\frac{Z_{line}^2 * L - Z_{line} * Z_{supply}}{Z_{DG}}\right) + \left(\frac{Z_{LINE} * L * Z_{supply}}{Z_{DG}} + Z_{LINE} * L + Z_{supply}\right)} \tag{10.1}$$

$$Minimum(I_{CB}) = {I_{CB}}' = 0 \rightarrow a = \frac{L - \frac{Z_{supply}}{Z_{LINE}}}{2} \tag{10.2}$$

For the generic procedure for assessing the DG limits for directional false tripping and unwanted fuse blowing, one can allocate the DG in the place of greater impact. In that way one can get the most conservative analysis possible. The conservative point of connection in this case is always the closest point to the fault between the protection device and the point of highest short circuits current possible. Thus, an example of optimal position for the generator allocation is shown in Fig. 10.11.

In the deterministic approach the increase in the DG insertion level can be interpreted as the growth in the DG power. It should be noted that increasing the DG size, the coupling transformer inherently increases the short circuit capacity. The impedance of the DG as well as the transformer's connection can be estimated according to procedures seen at [11].

Despite its practicality, this approach is not always enough to assess the influence of the DG location on the impacts addressed. In this way, the following section deals with this topic with a statistical point of view to define the influence of the location of DG on the impact of DG on network protection.

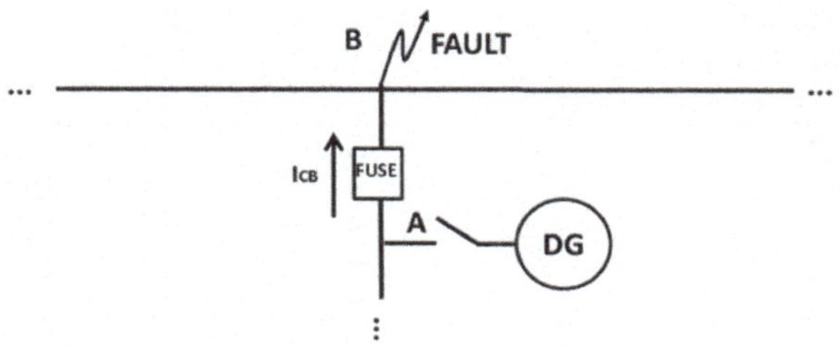


**Fig. 10.11** Example of conservative allocation for directional false tripping impact analysis

## *10.6.2 Probabilistic DG Allocation Approach*

In the probabilistic DG allocation approach the raising of the penetration levels means the increase of DG installed at the network. The DG insertion level can be determined as a factor of the prosumer energy consumption. In that way, any generic procedure to define the network DG insertion limit is a sensitivity analysis.

In the deterministic DG allocation approach, the location of the DG in determined randomly among the network consumers. The first possibility of DG allocation proposition is to define the same probability to each DG consumer as a candidate points.

Other possibility is, in order to model the assignment of the DG to distribution networks clients, create a probabilistic assumption based on the attractiveness of the DG alternative for each client. Such assumption is capable of creating a distinction on the prosumers probability of adoption. In [16, 17], this assumption is made observing the distribution network consumer disparities, such as type of load (rural, residential, commercial, industrial, etc.) and energy consumption. Therefore the attractiveness function developed was sensitive not only to the client's economic profile, represented by its consumption, but also to financial benefits from the installation of distributed generation, which may vary for different types of clients. It is shown in Fig. 10.12 the probability of DG adoption for each consumer of a real network, represented by a probabilistic attractiveness function.

The purely probabilistic simulation, as presented before, is not necessarily accurate, since the location of the DG influences in a relevant way the impacts assessed. In that way, two simulations with the same insertion level can result in

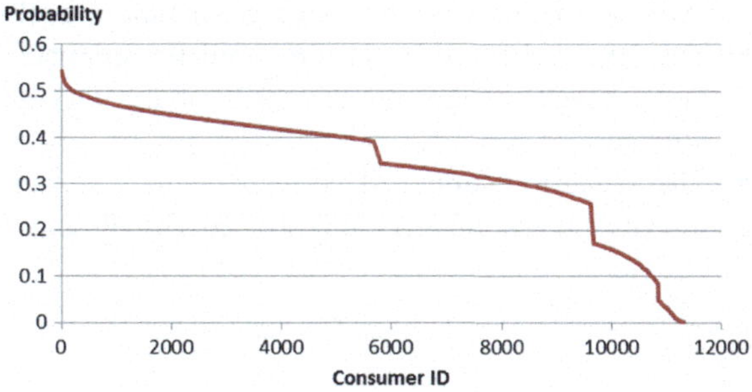


**Fig. 10.12** Example of a probabilistic attractiveness function used to asses DG insertion limits as seen in [16]

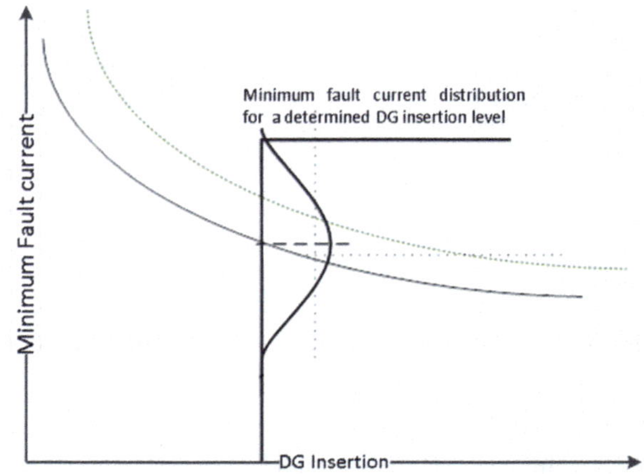


**Fig. 10.13** Example of the Monte Carlo method utilization in a sensitivity analysis

different results. At this point the Monte Carlo method can help significantly. The Monte Carlo method is utilized by the massive random sampling of the simulated fault currents, in order to obtain the distribution of these currents to a certain level of penetration. An example of the Monte Carlo method utilization in a sensitivity analysis for the loss of sensitivity assessment can be seen in Fig. 10.13.

## 10.7 Case Study

The case study presents an application of the methodologies discussed in the previous sections. Thus, this section illustrates the application of the presented methodologies, considering a probabilistic approach to the analysis of the impact caused by the massive penetration of small DG units scattered across a real network. The main characteristics of the case study are:

- It uses a real electric distribution network, composed by MV and LV networks and all its possible prosumers.
- It addresses three main issues presented in this chapter that are the directional false tripping, the unwanted fuse blowing and the loss of sensitivity.
- It considers the DG electric models addressed in the sections before, both the inverter coupled and the rotating machine models.
- The approach used is the probabilistic approach where probabilistic assumptions, based on the attractiveness of the DG alternative for each client, are utilized.
- The scenarios simulated correspond to a sensitivity analysis considering the insertion levels of 0, 1, 3, 5, 10, 20, 30, 40 and 50%.

### 10.7.1 Electric Distribution Network Data

The network data utilized to build the MV and LV networks was obtained from a Brazilian utility's GIS system. The case study's simulation environment was a real electric distribution network, which has approximately 26.589 bus bars, 9.481 clients, the peak demand of 10.5 MW and 154 km of extension. The electric distribution network is shown at Fig. 10.14. The blue lines show the MV networks while the red lines detail the LV networks.

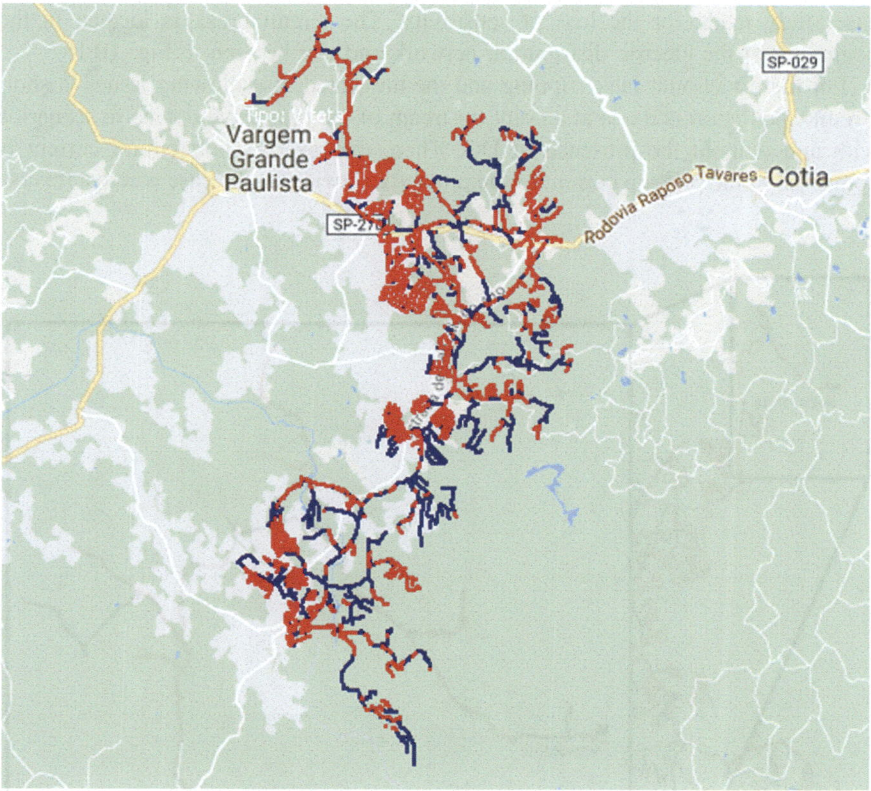


**Fig. 10.14** Real electric distribution network used in the case study

## *10.7.2 Assessment Methodology*

In the study case, three main issues are assessed. They are the directional false tripping, the unwanted fuse blowing and the loss of sensitivity issues.

For the loss of sensitivity issue, several short circuits are calculated at points downstream of the device protection zone, in scenarios with and without the presence of DG. Then, the minimum short circuit current is compared to the pick-up current of the protective device. In the case where a minimum short circuit current is inferior to the pick-up current, there may be a loss of sensitivity problem.

The circuit breaker of the electric distribution network is used to exemplify the case study made for the loss of sensitivity. The circuit break is located in the beginning of the electric distribution network and can be seen in Fig. 10.15.

For the directional false tripping and the unwanted fuse blowing issues, a short circuit calculation is made at a point upstream of the protective device in scenarios with and without the presence of DG. Then, the reverse short circuit current is compared to the pick-up current of the protective device. In the case where a reverse

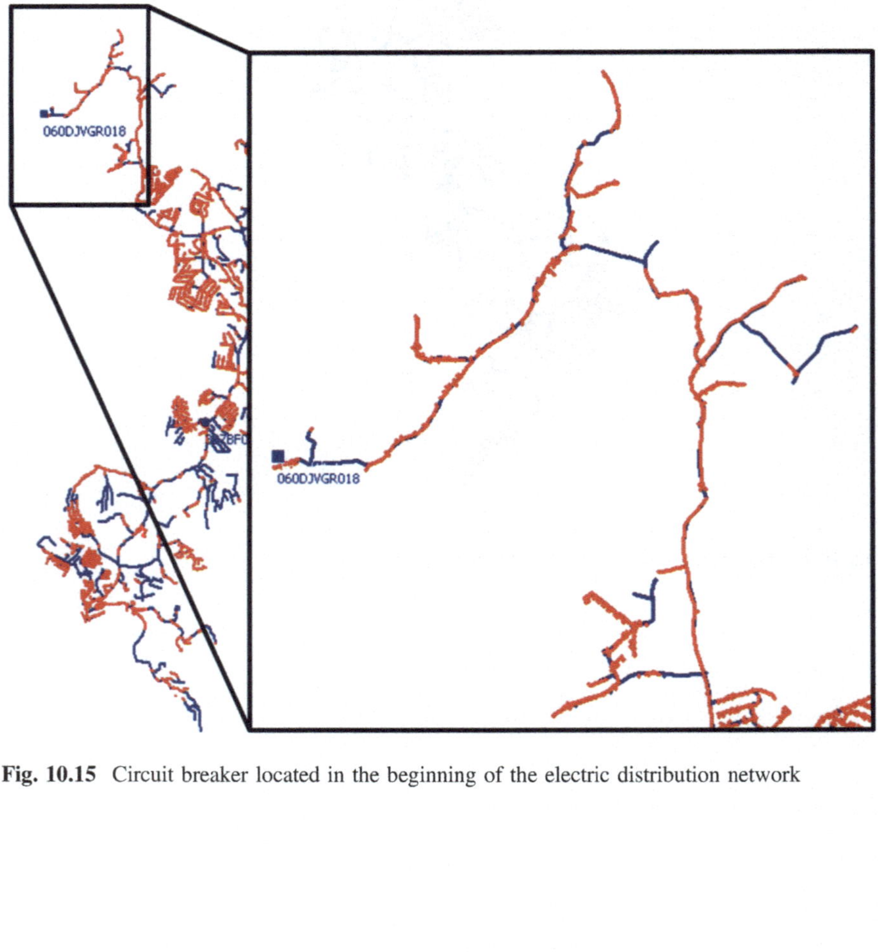


**Fig. 10.15** Circuit breaker located in the beginning of the electric distribution network

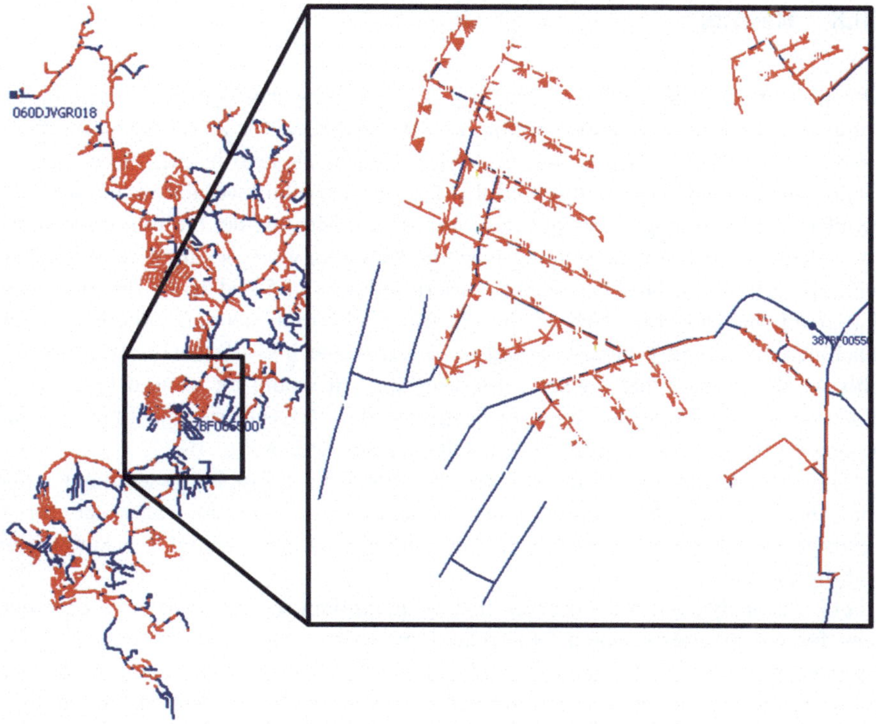


**Fig. 10.16** Fuse located in a branch in the middle of the electric distribution network

short circuit current is higher than the pick-up current, there may be a directional false tripping or an unwanted fuse blowing problem, depending on the protective device observed.

The circuit breaker of the electric distribution network is also used to exemplify the case study made for the directional false tripping issue. A fuse used to protect a network branch is used to exemplify the case study made for the unwanted fuse blowing issue. The fuse is located in the middle of the electric distribution network and can be seen in Fig. 10.16.

### *10.7.3 DG Electric Model*

In the study case, only small scale DG where considered. The DG installed capacity where defined by the prosumer chosen to install the technology. The DG installed capacity should be such that the energy produced would meet the prosumer monthly energy consumption.

Furthermore, both technologies presented in this chapter, inverter coupled and rotating machine, were assessed in this case study.

## 10.8 Results

The first analysis studied was the loss of sensitivity assessment. This analysis compared the minimum short circuit current, calculated for a contingency downstream of the electric distribution network circuit breaker, to its pick-up current.

For this analysis, it was observed that for any insertion level tested, the inverter coupled DG technology does not raise any issues. Meanwhile, the minimum short circuit current seen for the rotating machine technology crosses the circuit breaker pick-up current threshold between the insertion levels of 5 and 10%. This indicates that if all the DG connected to the electric distribution network where rotating machines, the circuity breaker could have a possible loss of sensitivity problem. This would indicate that for such insertion level, it would be advised that a new protection study would be done, to determine the necessity of adjustment of the protection device sensitivity. These results are detailed at Fig. 10.17.

The second analysis studied was the directional false tripping assessment. This analysis compared the maximum reverse short circuit current, calculated for a contingency upstream of the electric distribution network circuit breaker, to its pick-up current.

For this analysis, it was observed that the maximum reverse short circuit current seen for the inverter coupled DG technology crosses the circuit breaker pick-up current threshold close to the insertion level of 30%. This indicates that if all the DG connected to the electric distribution network where inverter coupled, the circuity breaker could have a possible directional false tripping problem. Meanwhile the rotating machine technology crosses the threshold for a significant small insertion level, between 0 and 3%. This would indicate that for such insertion level, it would be advised that a new protection study would be done, to determine the necessity the installation of protection devices with directional protection. These results are detailed at Fig. 10.18.

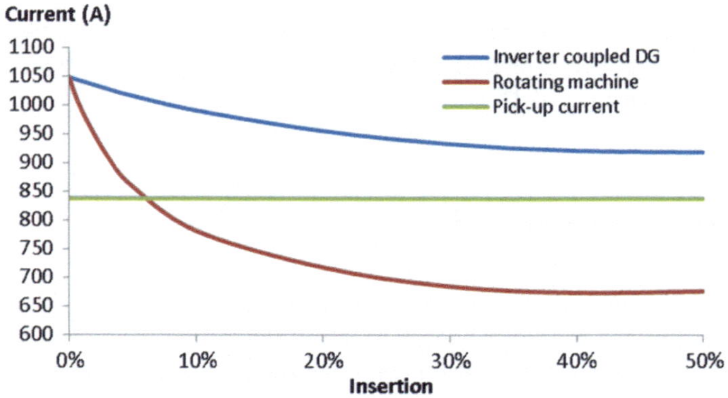


**Fig. 10.17** Circuit breaker loss of sensitivity assessment

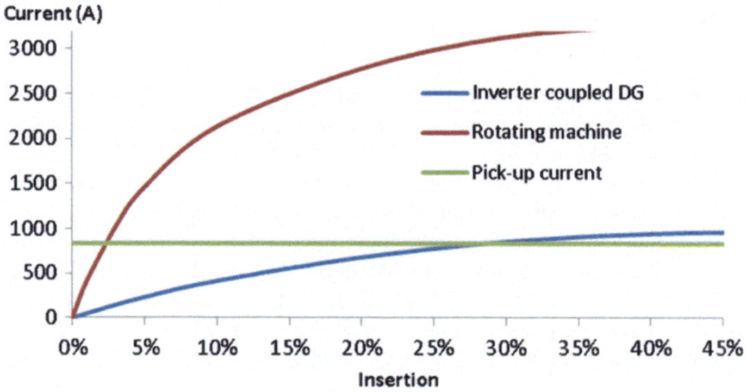


**Fig. 10.18** Circuit breaker directional false tripping assessment

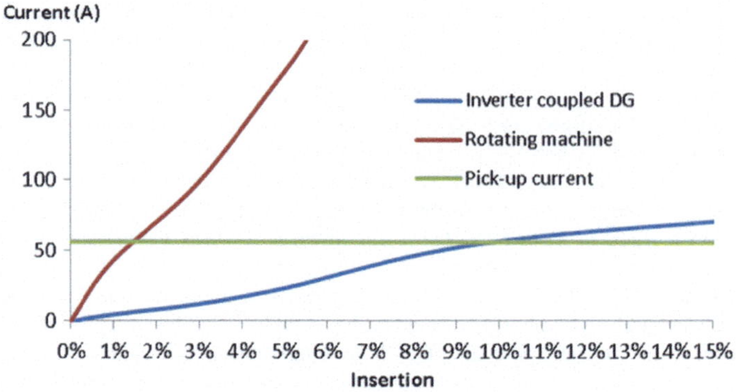


**Fig. 10.19** Fuse directional false tripping assessment

The third analysis studied was the unwanted fuse blowing assessment. This analysis compared the maximum reverse short circuit current, calculated for a contingency upstream of the electric distribution network fuse, to its pick-up current.

For this analysis, it was observed that the maximum reverse short circuit current seen for the inverter coupled DG technology crosses the circuit breaker pick-up current threshold close to the insertion level of 10%, while the maximum reverse short circuit current seen for the rotating machine technology crosses the threshold between the insertion levels of 1 and 3%. This would indicate that for such insertion level, it would be advised that a new protection study would be done, to determine the necessity the fuse replacement. These results are detailed at Fig. 10.19.

## References


1. J.L. Del Monaco, in *The role of distributed generation in the critical electric power infrastructure*, Proceedings of the Power Engineering Society Winter Meeting IEEE, vol. 1 (2001), pp. 144–145
2. C.J. Mozina, Update on the Current Status of DG Interconnection Protection-What IEEE 1547 Doesn't Tell You About DG Interconnection Protection, Beckwith Electric Co., Inc., Ninth Annual Protection Seminar Sept. 24–28, 2006, Florida, USA
3. M. Nagpal, F. Plumptre, R. Fulton, T.G. Martinich, Dispersed generation interconnection—utility perspective. IEEE Trans. Ind. Appl. **42**(3), 864–872 (2006)
4. P. Barker, in *Overvoltage considerations in applying distributed resources on power systems*, Proceedings of Power Engineering Society Summer Meeting, 2002 IEEE, vol. 1 (2002), pp. 109–114
5. L.K. Kumpulainen, K.T. Kauhaniemi, Analysis of the impact of distributed generation on automatic reclosing, in *Power Systems Conference and Exposition* (2004), IEEE PES
6. S. Chaitusaney, A. Yokoyama, Impact of protection coordination on sizes of several distributed generation sources, in *Power Engineering Conference* (2005), IPEC 2005
7. S.M. Brahma, A.A. Girgis, Development of Adaptive Protection Scheme for Distribution Systems with High Penetration of Distributed Generation. IEEE Trans. Power Deliv. **19**(I.1) (2004)
8. M.T. Doyle, in *Reviewing the impacts of distributed generation on distribution system protection*, Power Engineering Society Summer Meeting (2002), IEEE
9. L.F. Ochoa, A. Padilha-Feltrin, G.P. Harrison, Evaluating distributed generation impacts with a multiobjective index. IEEE Trans. Power Deliv. **21**(I.3) (2006)
10. A. Girgis, S. Brahma, Effect of distributed generation on protective device coordination in distribution system, in *LESCOPE Large Engineering Systems Conference on*, 11–13 July 2001
11. IEEE Recommended Practice for Calculating Short Circuit Currents in Industrial and Commercial Power Systems, IEEE Standard 551-2006
12. F. Katiraei, et.al., Investigation of solar PV inverters current contributions during faults on distribution and transmission systems interruption capacity, in *Western Protective Relay Conference*, Oct 2012
13. D. Turcotte, F. Katiraei, Fault contribution of grid-connected inverters, in *IEEE Electrical Power Conference*, October 22–23, 2009, Montreal, Quebec, Canada
14. E. Muljadi, M. Singh, R. Bravo, V. Gevorgian, Dynamic model validation of pv inverters under short circuit conditions, in *Green Technologies Conference* (2013), IEEE
15. N. Kagan, C.C.B. Oliveira, E.J. Robba, Introdução aos Sistemas de Distribuição de Energia Elétrica, ed. by Edgar Blucher, São Paulo (2005)
16. G.A. Quiroga, H. Kagan, J.C.C. Amasifen, C.F.M. Almeida, E. Vicentini, Evaluation of distributed generation impacts on distribution networks under different penetration scenarios, in *Innovative Smart Grid Technologies Latin America (ISGT LATAM)* (2015)
17. G.A. Quiroga, H. Kagan, J.C.C. Amasifen, C.F.M. Almeida, E. Vicentini, Study of the distributed generation impact on distributed networks, focused on quality of power, in *International Conference on Harmonics and Quality of Power (ICHQP)* (2016)

# Chapter 11
# Stability of Distribution Networks with Wind Turbines

**Ahmed Rashad, Salah Kamel, Francisco Jurado and Shady H. E. Abdel Aleem**

**Abstract** This chapter presents modeling of generators that are used in wind farms such as squirrel cage induction generators (SCIG), doubly fed induction generators (DFIG) and, permanent magnet synchronous generator (PMSG). Installing wind farms must fulfill some rules or requirements. These requirements are developed by transmission system operator in order to guarantee the continuity and stability of the interconnected grid. This chapter presents the stability of two different types of combined wind farms. The first type is based on a combination of SCIG and DFIG wind turbines and known as combined wind farm (CWF). CWF collects the benefits of SCIG and DFIG where SCIG is cheaper compared with DFIG and PMSG. Despite DFIG is expensive, DFIG is more stable than SCIG. CWF is more suitable for developing countries. The second type is based on a combination of modern generators DFIG and PMSG and known as modern combined wind farm (MCWF). MCWF collects the benefits of DFIG and PMSG where DFIG features by its ability to control the active power independently of reactive power while PMSG can operate used for small and medium powers. This chapter discusses the impact of CWF and MCWF on the stability of interconnected electric distribution networks during single line to ground and double lines fault as examples of unsymmetrical and during three phase fault and three phase open circuit fault as examples for symmetrical. Also, this chapter discusses the impact of CWF and MCWF on the

A. Rashad · S. Kamel
Department of Electrical Engineering, Faculty of Engineering,
Aswan University, Aswan, Egypt
e-mail: ahmedmmrashadar@gmail.com

S. Kamel
e-mail: skamel@aswu.edu.eg

A. Rashad · F. Jurado (✉)
Department of Electrical Engineering, University of Jaén, Jaén, Spain
e-mail: fjurado@ujaen.es

S. H. E. Abdel Aleem
Mathematical, Physical and Engineering Sciences Department,
15th of May Higher Institute of Engineering, Cairo, Egypt
e-mail: engyshady@ieee.org

 
A. Arefi et al. (eds.), *Electric Distribution Network Management and Control*,
Power Systems, https://doi.org/10.1007/978-981-10-7001-3_11

stability of interconnected electric distribution networks during different types of operation conditions of electric distribution networks such as voltage sage and over voltage.

**Keywords** Squirrel cage induction generator · Double fed induction generator Permanent magnet synchronous generator · Wind farm

## 11.1 Introduction

By the end of twentieth century, the utilization of wind energy system (WES) became one of the important aims of all countries. This is because WES depends on natural source so that it can keep national income by reducing the expenditure on fuel that is used in traditional method of electricity production. The importance of WES can be tangible if that the total installed capacity of WES is increased from 236,733 MW by the end of 2011 to 456,486 MW by the end of June 2016 [1]. Figure 11.1 shows the total capacity installed from the end of 2011 till the end of 2016 [1].

The utilization of WES differs from one country to another. According to statistics and information of World Wind Energy Association (WWEA) China, USA, Germany, India and, Spain, produce 67% of total global wind capacity. Figure 11.2 shows the wind power production of several countries [1].

Figures 11.1 and 11.2 show the rapid increase in wind power production in last five years. This could be due to the great and rapid progress in wind technology. The wind turbine rating can be taken as an example of this progress; where in 1980 wind turbine rating began with 50 MW and reached 10 MW by 2012 [2].

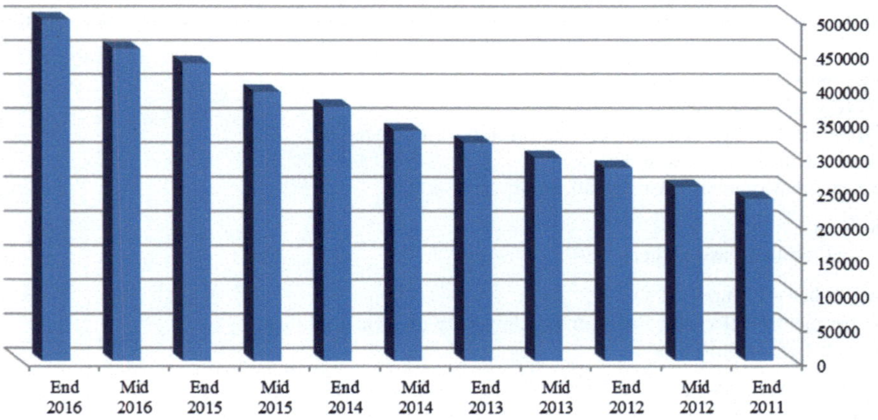


**Fig. 11.1** Total capacity installed from the end of 2011 till the end of 2016

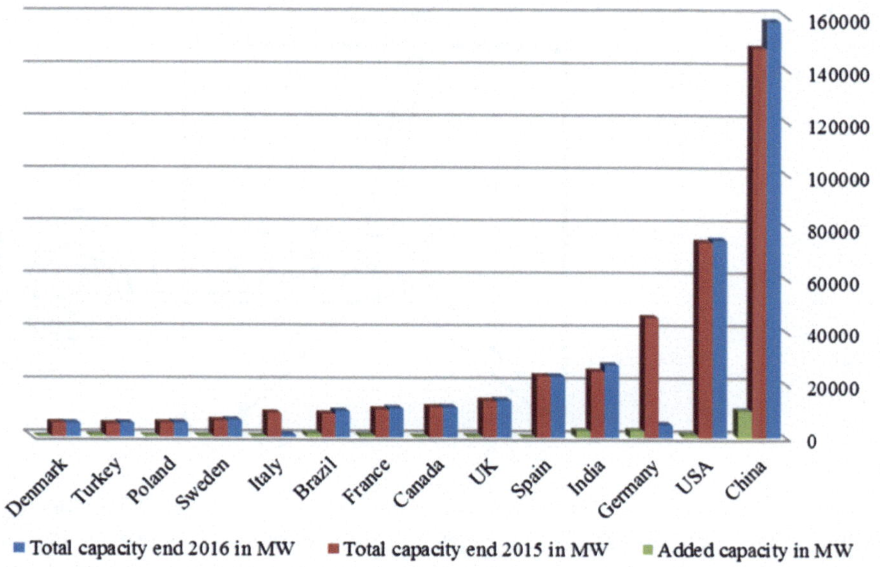


**Fig. 11.2** Wind power production of several countries

This chapter begins with modeling of SCIG, DFIG and, PMSG then modeling of two different types of combined wind farms. One of them is combine wind farm (CWF) based on combination between SCIG and DFIG while the another one is modern combine wind farm (MCWF) based on combination between DFIG and PMSG. The of chapter studies the stability of electric distribution networks inter-connected CWF and MCWF during different types of grid faults.

## 11.2 Modelling of Wind Turbines

The wind turbines compose of two main components: drive train and, generator. Drive train or wind turbine rotor consists of the hub, blade and, gear box. Figure 11.3 shows the main components of wind turbines. The blade is used to convert the wind energy to rotational motion. The hub is used to connect the blade to the gear box. The gear box is used to convert the speed of wind turbine rotor to the rated speed of generator.

The ratio between speed of turbine rotor and speed of generator rotor is called the gear box ratio ($G_r$) and is given by.

$$G_r = \frac{\omega_g}{\omega_T} = \frac{(1-s)120f}{P\omega_T} \tag{11.1}$$

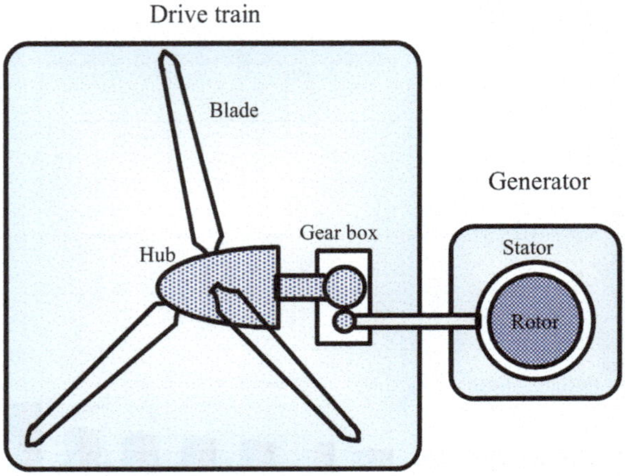


**Fig. 11.3** Main components of wind turbines

where *f* is the frequency, *P* is the number of pair, *S* is the slip of the generators, $\omega_g$ is the generator speed and, $\omega_T$ is the speed of turbine rotor. *S* it is usually 1% for induction generator and zero for synchronous generators. The drive train model of wind turbines is expressed by [2, 3].

$$T_{mec} - T_e = 2H_g \frac{d\omega_g}{dt} \tag{11.2}$$

where $T_{mec}$ is the mechanical torque of generator shaft, $T_e$ is the generator electrical torque, and, $H_g$ is inertia of generator shaft [2, 3].

From (11.1) it can be observed that any change in wind speed will change the rotor speed and hence fluctuation in the generated power. This fluctuation in generated power should be reduced as much as possible. The first step starts with controlling the speed of wind turbine rotor ($\omega_T$). $\omega_T$ is controlled by two means stall effect and pitch angle control system. The simple definition of stall effect is phenomena where aerodynamic efficiency decreased with the increase of wind speed. The good design of blade according to stall effect will save the wind turbine from damage if the wind speed exceeds the rated speed.

The pitch angle ($\beta$) can be defined as the angle between the wind speed direction in wind farm site and the blade's axis. The pitch angle is shown in Fig. 11.4. Figure 11.5 shows a simple diagram of pitch angle control system.

The output power (*P*) and rotors speed $\omega_g$ is measured the compared with reference of P and $\omega_g$. The error is injected to pitch angle actuator to generate the new pitch angle. The blade will rotate with the new pitch angle to produce the desired output power and $\omega_g$.

$\beta$ plays an important role in the value of the power that can be capture from the wind. This captured power from the wind can be given by

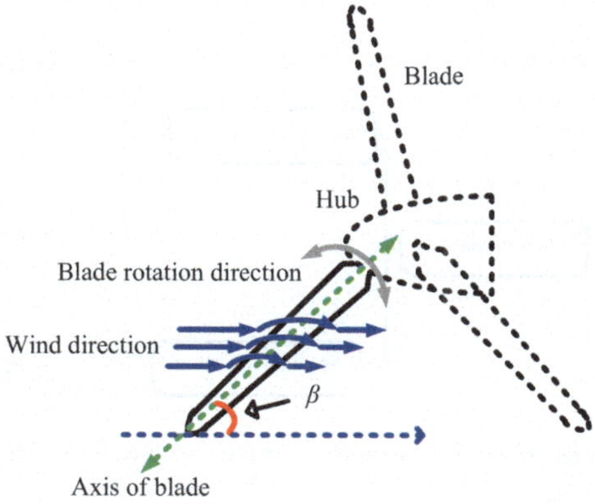


**Fig. 11.4** Angle plays an important role control system

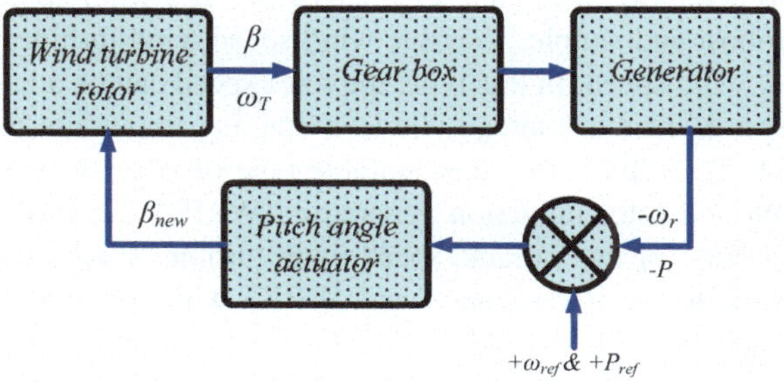


**Fig. 11.5** A simple diagram of pitch angle control system

$$P_{wt} = \frac{1}{2}\rho A v^3 C_p(\lambda, \beta) \tag{11.3}$$

where $C_p$ is the power coefficient and is function into parameters: pitch angle $\beta$ tip speed ratio $\lambda$. The tip speed ratio is the between the rotor speed and wind speed. Both of The $C_p$ and $\lambda$ can be calculated from

$$C_p(\lambda, \beta) = 0.517\left(\frac{116}{\lambda_i} - 0.4\beta - 5\right)e^{\frac{-0.0068}{\lambda_i}} + 0.0068\lambda \tag{11.4}$$

$$\frac{116}{\lambda_i} = \frac{1}{\lambda + 0.08\beta} - \frac{0.035}{\beta^3 + 1} \tag{11.5}$$

$$\lambda = \frac{\omega_r l_b}{v} \tag{11.6}$$

where $\omega_r$ is the rotor speed and $l_b$ is the blade length or rotor radius.

The wind energy market divided the wind turbines into two types according to rotational speed: fixed speed and variable speed wind turbines. The expiration fixed speed wind turbines mean that the wind turbines generate its maximum power at

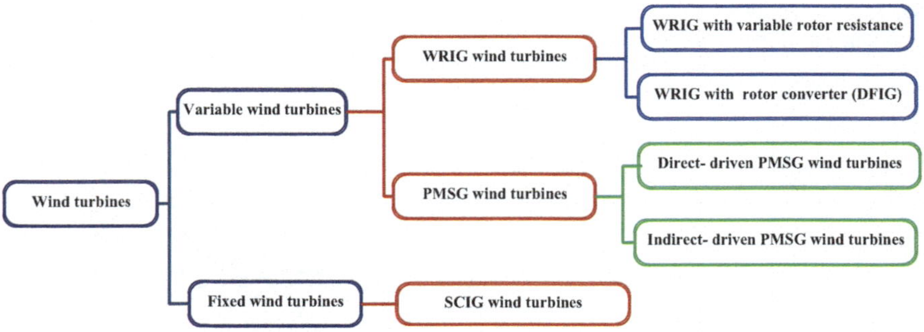


**Fig. 11.6** The division of wind turbines according to the rotational speed

certain wind speed. Over this speed, the output power of fixed wind speed will start to decrease. The main advantages of fixed wind speed are cheap easy to maintain robust and simple. Variable wind speed wind turbines can operate over wide range of wind speed. In this type, the rotor speed can be adjusted in accordance with wind speed. This advantage allows extracting maximum power at different wind speed [4, 5]. SCIG is the most suitable type of generators for fixed speed wind turbines. Wound rotor induction generators (WRIG) and PMSG are the most suitable generators for the variable speed wind turbine. WRIG include two types: WRIG with variable rotor resistance and WRIG with rotor converter (doubly fed induction generators DFIG). Also, PMSG can be divided into directly driven PMSG wind turbines and, indirectly driven PMSG wind turbines [2]. Figure 11.6 shows the division of wind turbines according to the rotational speed.

## 11.2.1 Fixed Speed Wind Turbine

As mentioned before, SCIG is the most suitable type of generators for fixed speed wind turbines. Figure 1.7 shows a Schematic diagram of SCIG wind turbines

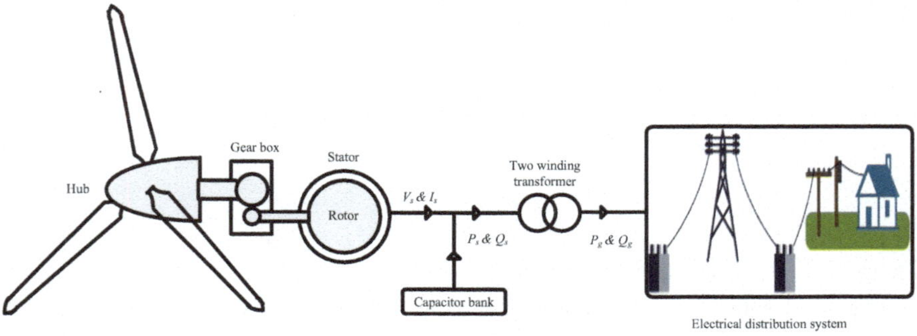


**Fig. 11.7** Schematic diagram of SCIG wind turbines (SCIG-WT) interconnected electric distribution networks

(SCIG-WT) interconnected electric distribution networks. From Fig. 12.7 it can be observed that the SCIG-WT is connected to electric distribution networks through two wind transformer. The active power is transferred from directly the stator to the system through the transformer. The capacitor bank is used for reactive power compensation. This simple construction features by reliability and cheapens. The direct connection between the SCIG-WT and the interconnected system allows exchanging the active and reactive power between them. SCIG-WT has two mean disadvantages: the active power of SCIG-WT fluctuates with the variation of wind speed, also and SCIG-WT draws a significant amount of the reactive power from the system during any system disturbances [6, 7].

## *11.2.2 Variable Speed Wind Turbine*

Variable speed wind turbine can operates at wide range of wind speed due to its construction. There are two types of generators that are suitable for this mission: WRIG and PMSG. The operation of variable speed wind turbine depends mainly on power converters. Power converters give the ability to control the active power independent of reactive power. Using power converters increase the cost of variable speed wind turbine.

### 11.2.2.1 WRIG Wind Turbines (WRIG-WT)

The main idea of WRIG-WT is controlling the rotor current in order to control the rotor speed and hence increase the range of operation speed. This aim can be achieved two means: controlling the rotor resistance or using DFIG with the power converter. Figures 11.8 and 11.9 show the Schematic diagram of WRIG with controlled rotors resistance and DFIG-WT respectively.

As shown in Fig. 11.8, the value of rotors resistance of WRIG with controlled rotors resistance is varied using variable resistance. The value of variable resistance

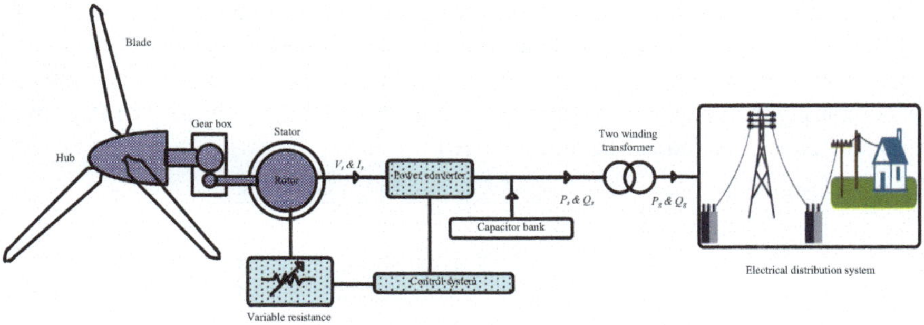


**Fig. 11.8** Schematic diagram of WRIG with controlled rotors resistance

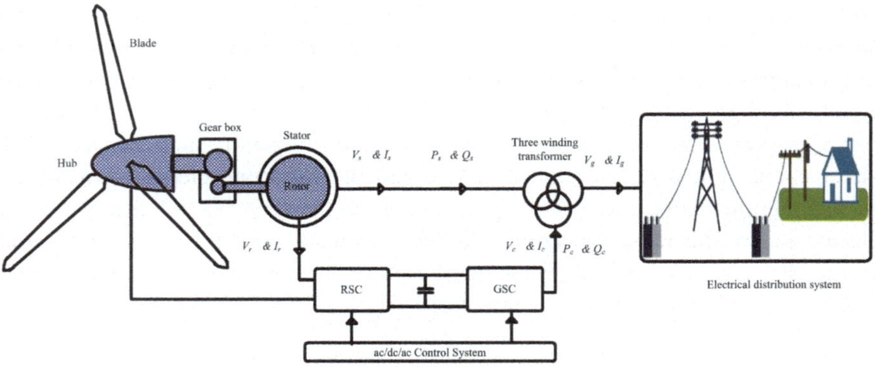


**Fig. 11.9** Schematic diagram of DFIG-WT

is controlled through a controlled converter. The main disadvantage of this type is the increased in rotor losses due to the additional resistance, the increase in cost due to the additional power converter and WRIG with controlled rotors resistance is still used capacitor bank for reactive power compensation. Figure 11.9 shows that the power of DFIG-wind turbines is transmitted to the system through three winding transformers. Two winding transfer the power from the stator wind to the system while the third winding transfer the power from the rotor to the system through ac/dc/ac converter. The ac/dc/ac converter consists of two voltage source converters (VSC) connected to each other through dc bus. One of the two VSC connects the rotor to the dc bus and called rotor side converter (RSC). The other VSC connects the dc bus to the third winding of the three winding transformers and it called grid side converter (GSC) (Fig. 11.7).

Rotor side converter (RSC) control system is shown in Fig. 11.10a. RSC is used to control the output power and voltage at the at point of common connection (PCC). The output can be controlled by adjusting the quadratic component of rotor voltage ($V_{qr}$). $V_{qr}$ can be obtained by comparing the total output power (the power at grid terminal+power losses) with the reference power ($P_{ref}$). The voltage at PCC can be controlled by adjusting the direct component of rotor voltage ($V_{dr}$). $V_{dr}$ can be obtained by comparing the grid voltage ($V_g$) with the reference voltage ($V_{ref}$). Both $V_{qr}$ and $V_{dr}$ are injected to a PI current regulator to reduce the error [8–10].

Figure 11.10b shows the grid side converter (GSC) control system. GSC is used to control or regulate the voltage of the dc bus. This function can be accomplished by comparing the voltage of the dc bus with the dc reference voltage. The PI current regulator is used to reduce the error [8–10].

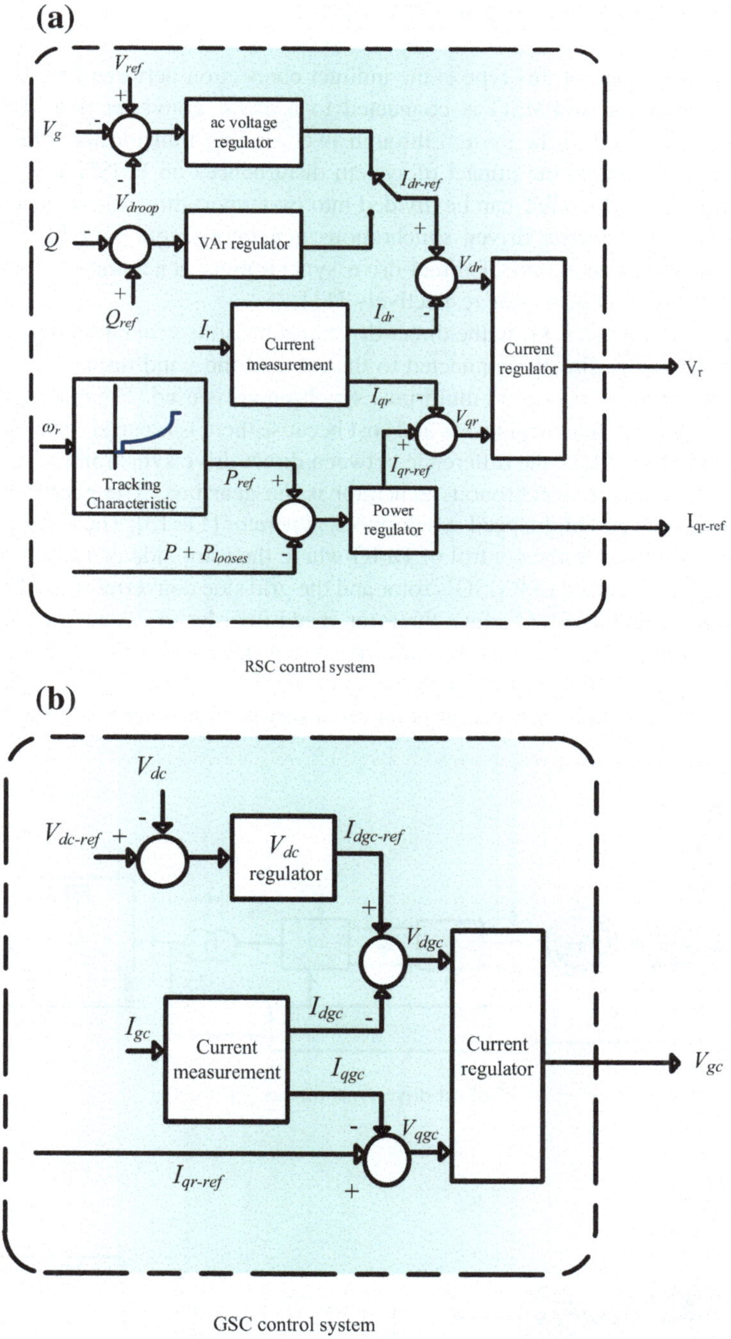


**Fig. 11.10** **a** RSC system, **b** GSC system

#### 11.2.2.2 PMSG Wind Turbines (PMSG-WT)

The main advantage of this type is the indirect connection between PMSG and the system. The stator of PMSG is connected to ac/dc/ac converter then the output power is transferred to the system through two winding transformer. This type of construction decreases the impact of system disturbances on PMSG-WT.

The rotor of PMSG-WT can be divided into two types: direct-drive synchronous generators and converter driven synchronous generator. Figures 11.11 and 11.12 show a Schematic diagram of direct-drive synchronous generators and converter driven synchronous generator respectively [11].

As shown in Fig. 11.11, in the direct-drive synchronous generators type the rotor of the generator is directly connected to the rotor of the wind turbine. Due to this direct connecting, a low-speed multi-pole synchronous is used. The main advantage is decreasing the mechanical losses and cost because there is no need to the gear box. As shown in Fig. 11.12 the difference between direct-drive synchronous generators and converter driven synchronous generator is the gear box. The gear box allows using lower pole and high-speed synchronous generator [11–13]. The control system of PMSG is similar to the control of DFIG where the rotor side converter (RSC) is used to adjust the speed of PMSG's rotor and the grid side converter is used to adjust the voltage of dc bus. RSC adjust the rotor speed in order to extract the maximum power from the wind at operating wind speed and delivered it to the dc bus. GSC adjust the dc bus voltage in order to guaranty the unity power factor and keep the voltage at the common connection point on is steady-state level [11–13].

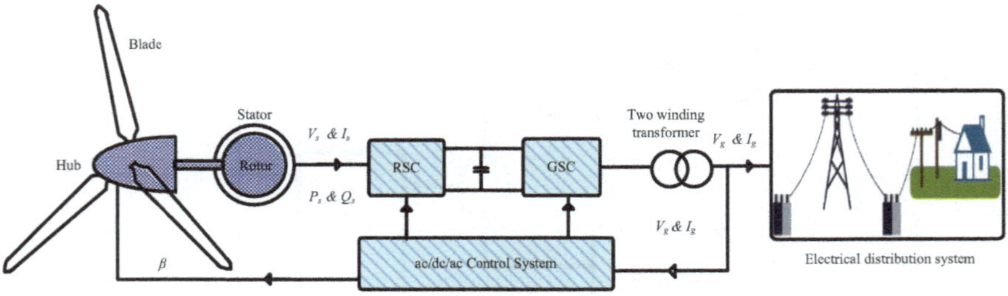


**Fig. 11.11** Schematic diagram of direct-drive synchronous generators

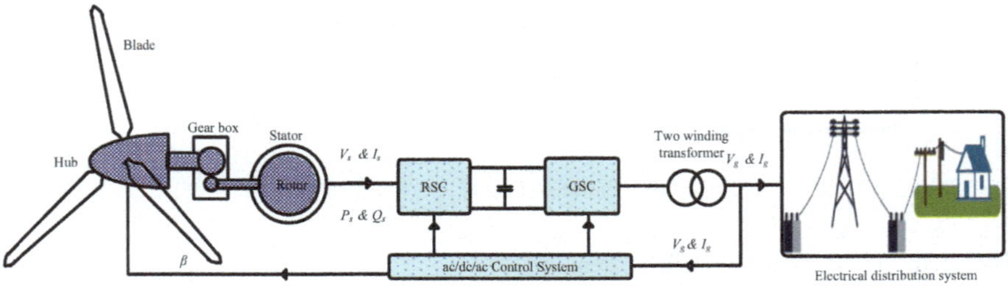


**Fig. 11.12** Schematic diagram of converter driven synchronous generator

## 11.3 Mathematical Modeling of Wind Turbine Generator

### 11.3.1 Modeling of Induction Generator

This section presents the mathematical modeling of wind turbine induction generator and wind turbine synchronous generator. The modeling of induction generator plays an important role in the controlling system of wind turbines. As it can be noted from the previous section, the control system of wind turbines depend on the direct and quadratic components of current and voltage of the system. It is already known that rotational motion of generators is resulting from the interaction between the flux of stator $\phi_s$ and flux of rotor $\phi_r$. The Schematic diagram of induction generators is shown is Fig. 11.13. The first step to conclude the mathematical modeling of induction generator, let us assume that the stator rotates in the space with speed equal to the synchronous speed $\omega_s$ and the rotor rotate in the same space, but with speed equals to $\omega_r$. The slip $S$ represents the relation between $\omega_s$ and $\omega_r$ and can be denoted by

$$s = \frac{\omega_s - \omega_r}{\omega_s} \tag{11.7}$$

$$s\omega_s = \omega_s - \omega_r \tag{11.8}$$

$\phi_s$ and $\phi_r$ rotate with the same speed of stator and rotor respectively. Both of $\phi_s$ and $\phi_r$ consist of two elements: self-inductance of their coils ($L_s$ and $L_r$) and the magnetizing inductance ($L_m$). $L_s$ and $L_r$ composed of two other parts: the leakage inductance between stator and rotor coils and the magnetizing inductance ($L_{ls}$ and $L_{lr}$ respectively). The equations describing $\phi_s$ and $\phi_r$ can be concluded by using vector representation of the parameters of stator and rotor (voltage and current) as shown [2–5]:

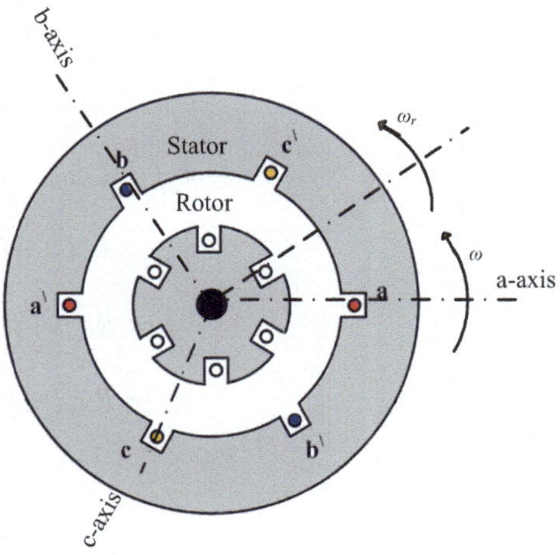


**Fig. 11.13** Schematic diagram of induction generators

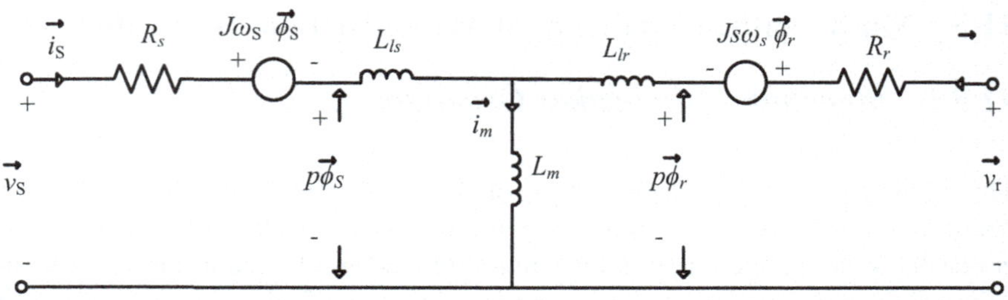


**Fig. 11.14** Equivalent circuit of induction generator according to the synchronous speed

$$\overrightarrow{\phi_s} = L_{ls}\,\vec{i_s} + L_m\,\vec{i_s} + L_m\,\vec{i_r} \tag{11.9}$$

$$\overrightarrow{\phi_r} = L_{lr}\,\vec{i_r} + L_m\,\vec{i_r} + L_m\,\vec{i_s} \tag{11.10}$$

The equivalent circuit of induction generator can be drawn (according to the synchronous speed) by using (11.7)–(11.10). Figure 11.14 shows the equivalent circuit of induction generator according to the synchronous speed [2–5].

The equivalent circuit of induction generator is shown in Fig. 11.14 can be used to determine the voltage and current of stator and rotor.

$$\vec{v_s} = R_s\,\vec{i_s} + p\,\overrightarrow{\phi_s} + j\omega_s\,\overrightarrow{\phi_s} \tag{11.11}$$

$$\vec{v_r} = R_r\,\vec{i_r} + p\,\overrightarrow{\phi_r} + js\omega_s\,\overrightarrow{\phi_r} \tag{11.12}$$

where $p = \frac{d}{dt}$ and $R$, $i$ and $v$ are resistance current and, voltage, the of the stator and rotor

The next step is converting the equation of flux and voltages from the vector for to direct and quadratic transformation from; as

$$\phi_{ds} = L_{ls}i_{ds} + L_m i_{ds} + L_m i_{dr} \tag{11.13}$$

$$\phi_{qs} = L_{ls}i_{qs} + L_m i_{qs} + L_m i_{qr} \tag{11.14}$$

$$\phi_{dr} = L_{lr}i_{dr} + L_m i_{dr} + L_m i_{ds} \tag{11.15}$$

$$\phi_{qs} = L_{lr}i_{qr} + L_m i_{qr} + L_m i_{qs} \tag{11.16}$$

$$v_{ds} = R_s i_{ds} + p\phi_{ds} - \omega_s\phi_{qs} \tag{11.17}$$

$$v_{qs} = R_s i_{qs} + p\phi_{qs} + \omega_s\phi_{ds} \tag{11.18}$$

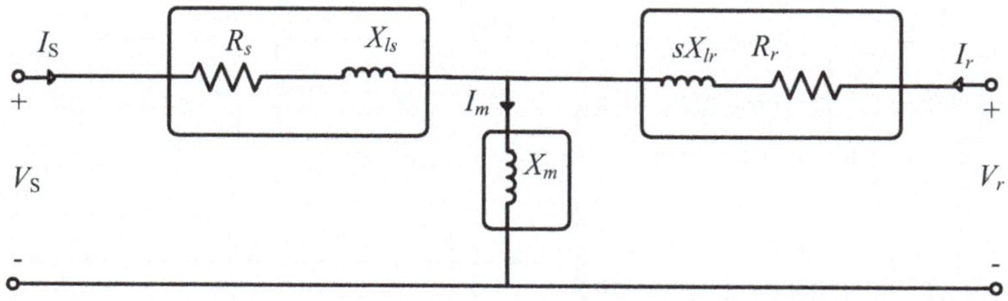


**Fig. 11.15** Equivalent circuit of induction generator according to the steady-state farm

$$v_{dr} = R_r i_{dr} + p\phi_{dr} - S\omega_s \phi_{qr} \tag{11.19}$$

$$v_{qr} = R_r i_{qr} + p\phi_{qr} + S\omega_s \phi_{dr} \tag{11.20}$$

The second step represents converting the vector model to steady-state model. This step can be obtained by applying the next steps:

1. Putting $p = \frac{d}{dt} = 0$
2. The steady stat quantities $= \frac{\text{The vector quantities}}{\sqrt{2}}$
3. Transfer the flux linkage $\phi$ to leakage reactance $X$ where $X = \omega\ \phi$

These steps can be applied to (11.9)–(11.11) and the results are as

$$V_s = R_s I_s + jI_s X_{ls} + jI_m X_m \tag{11.21}$$

$$V_r = -R_r I_r - jI_r s X_{ls} + jI_m s X_m \tag{11.22}$$

These equations can be used to draw the steady-state equivalent circuit of induction generator in general. Figure 11.15 shows the equivalent circuit of induction generator according to the steady-state farm [2–5].

The equivalent circuit that is shown in Fig. 11.15 can be used for both SCIG and DFIG. The main difference between SCIG and DFIG is the rotor of SCIG is short circuit while the rotor of DFIG is connected to power converter. The magnetizing parameters can be neglected for simplicity and to obtain the equivalent circuit of SCIG and DFIG as shown in Fig. 11.16a and b.

### *11.3.2 Modeling of Synchronous Generator*

In synchronous generators (SG), the rotating flux or magnetic field is produced by two manners: dc excitation system and permanent magnet filed as shown in Fig. 11.17.

In these two types because the flux generation depends on dc excitation system or permanent magnet field so that both of the quadratic components of rotor current

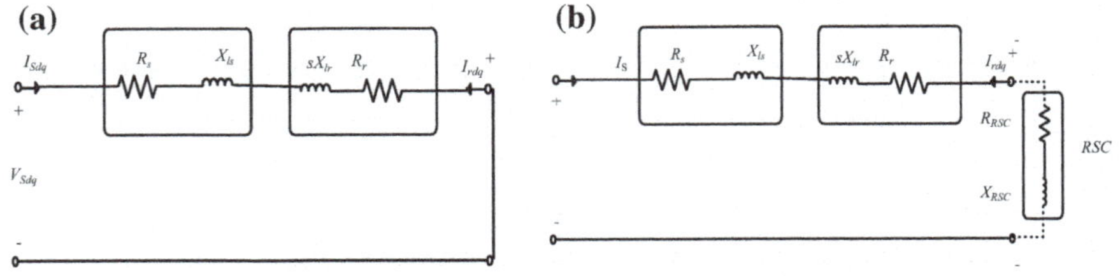


**Fig. 11.16** Equivalent circuit of induction generator in steady-state: **a** SCIG and **b** DFIG

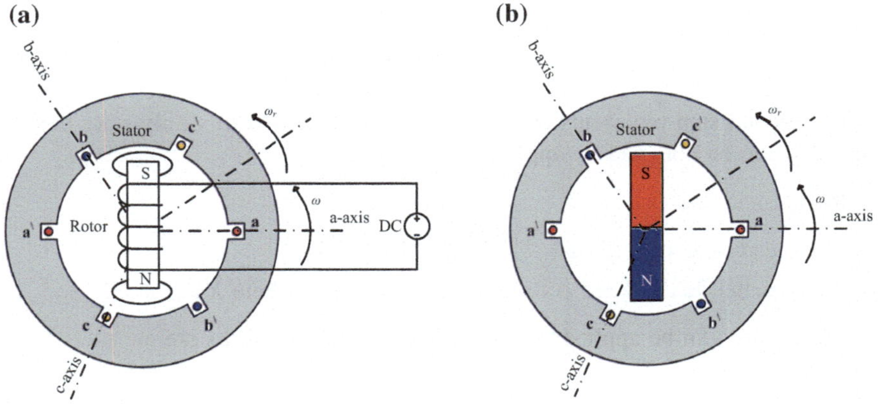


**Fig. 11.17** Schematic diagram of synchronous generators: **a** SG with dc excitation system and **b** SG with permanent magnetic field

and rotor voltage will be zero. In mathematical modeling of SG, both dc excitation system and permanent magnet field are presented by source current ($I_f$) and the rotor speed is equal to synchronous speed $\omega_r = \omega_s$. Also. the magnetizing inductance is decomposed to it direct and quadratic components $l_m = l_{dm} + l_{qm}$. The mathematical equation of SG can be obtained by rewriting the equation from (11.13) to (11.20) as [2–5, 14]

$$\phi_{ds} = -L_{ls}i_{ds} - L_{dm}i_{ds} + L_{dm}I_f = L_{dm}I_f - i_{ds}(L_{ls} + L_{dm}) = L_{dm}I_f - i_{ds}L_{ldm} \quad (11.23)$$

where $l_{ldm} = l_{ls} + l_{dm}$

$$\phi_{qs} = -L_{ls}i_{qs} - L_{qm}i_{qs} = -i_{qs}(L_{ls} + L_{qm}) = -i_{qs}L_{lqm} \quad (11.24)$$

where $l_{lqm} = l_{ls} + l_{qm}$

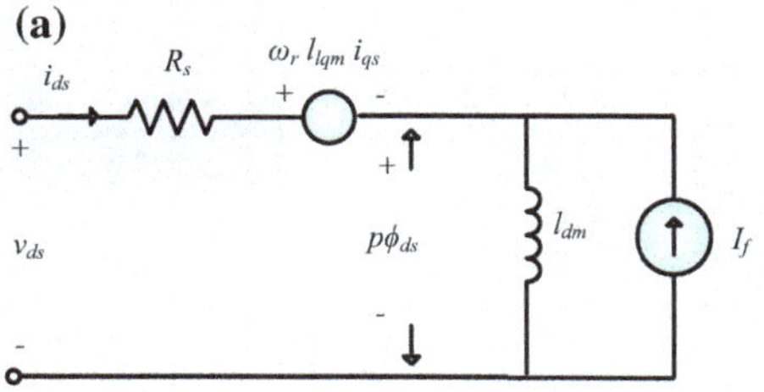


**Fig. 11.18** Equivalent circuit of SG: **a** equivalent circuit of the direct component and **b** equivalent circuit of the quadratic component

$$v_{ds} = p\phi_{ds} - R_s i_{ds} - \omega_r \phi_{qs} = p\phi_{ds} - R_s i_{ds} - \omega_r i_{qs} L_{lqm} \tag{11.25}$$

$$v_{qs} = -R_s i_{qs} + p\phi_{qs} + \omega_r \phi_{ds} = -R_s i_{qs} + p\phi_{qs} + \omega_r L_{dm} I_f - \omega_r i_{ds} L_{ldm} \tag{11.26}$$

Equations (11.25) and (11.26) can be used to investigate the equivalent circuit of SG as shown in Fig. 11.18. The equations of steady-state and the steady-state equivalent circuit of SG can be concluded by the same way that is used in induction generators. The steady-state equations of SG are given by [2–5, 14]

$$V_{ds} = -R_s I_{ds} + X_q I_{qs}$$
$$V_{qs} = X_{dm} I_f - R_s I_{qs} - X_d I_{ds}$$

where $X_q = \omega_r L_{lqm}$, $X_{dm} = \omega_r L_{dm}$ and $X_d = \omega_r L_{ldm}$

## 11.4 Performance of Electric Distribution Networks Interconnected Wind Farm

This section presents some examples of electric distribution networks interconnected to different types of wind farms. These wind farms are based on different types of generators or based on a combination of these generators. In this case, a 25 kV electric distribution network consists of 10 buses to deliver power to four loads of 1 MW. The studied wind farm is connected to bus B1. The performance of electric distribution network interconnected the studied wind farm is examined during different types such as unsymmetrical faults and, symmetrical faults. The fault occurs at Bus B6 and occurred at a time equal to 25 s and end at a time equal to 25.15 s. The studied electric distribution network is shown in Fig. 11.19.

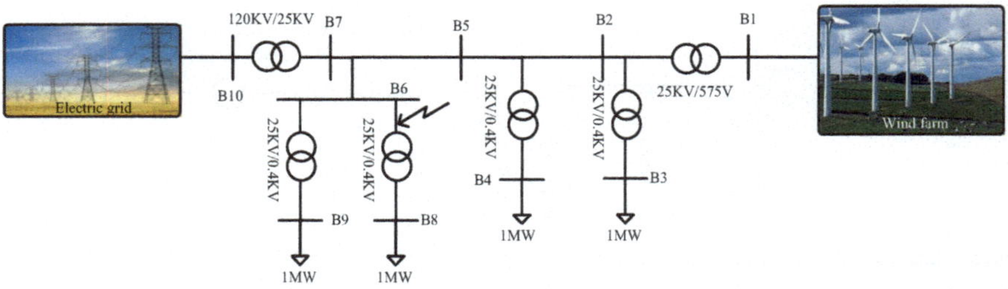


**Fig. 11.19** The studied electric distribution networks

## *11.4.1 Studied Case of Induction Generator Wind Turbines*

The studied wind farm produces 9 MW at 9 m/s wind speed. The example studied the different wind farms: SCIG wind farm (SCIG-WF), DFIG wind farm (DFIG-WF) and, combined wind farm (CWF). The combined wind farm (CWF) composed of a combination of SCIG wind turbines (SCIG-WT) and DFIG wind turbines (DFIG-WT). This combination allows collecting the advantages of SCIG and DFIG. The SCIG-WT is cheap and DFIG-WT is more stable and provides voltage sport to the system.

### 11.4.1.1 Impact of Single Line to Ground Fault

Figure 11.20 shows the impact of a single line to ground fault on the studied system. The voltage is mentored at load buses B3, B4, B8 and B9. Also, the voltage, active and, reactive power of the three wind farms is also mentored at point of common connection (PCC).

From the results, it can be observed that the three wind farms improve the voltage of load buses. The voltage of nearest load buses B3 and B4 to the wind farms has been improved more than the furthest load buses B8 and B9. The CWF represents an intermediate case between SCIG-WF and DFIG-WF where SCIG-WF has the lowest parameters' values (voltage, active power and reactive power at PCC) and DFIG-WF has the highest parameters' values. Also, it can be observed that the improvement of wind farms depend on the type of generator that is used in wind farms. The improvement of CWF is much better than the improvement of SCIG-WF and is near the improvement of DFIG-WF. This is because the CWF collects the advantage of SCIG and DFIG.

### 11.4.1.2 Impact of Line to Line Fault

In this case, the impact of line to line fault on the performance of the three wind farms (SCIF-WF, DFIG-WF and, CWF) is studied. Figure 11.21 shows the impact

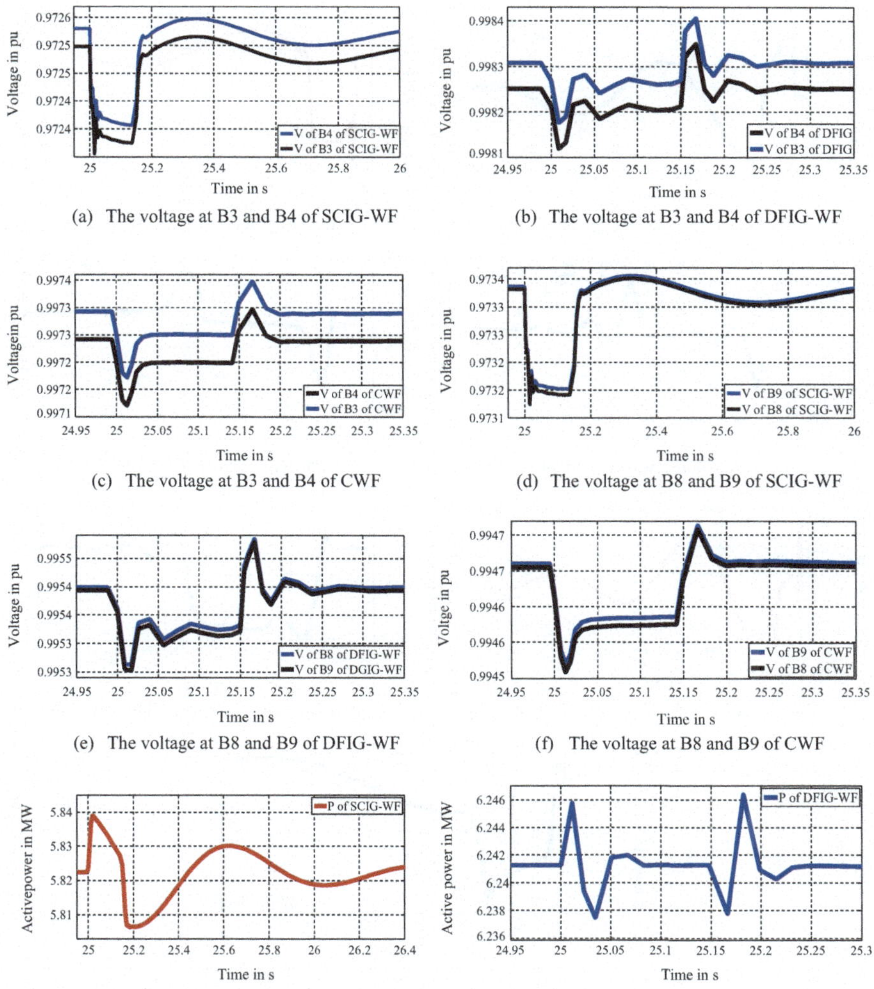


(a) The voltage at B3 and B4 of SCIG-WF
(b) The voltage at B3 and B4 of DFIG-WF
(c) The voltage at B3 and B4 of CWF
(d) The voltage at B8 and B9 of SCIG-WF
(e) The voltage at B8 and B9 of DFIG-WF
(f) The voltage at B8 and B9 of CWF
(g) The active power of SCIG-WF at PCC
(h) The active power of DFIG-WF at PCC

**Fig. 11.20** Shows the impact of single line to ground fault on the studied system

of single line to line fault on the studied system. As it can be observed from Fig. 11.21, the impact of line to line is tougher than the impact of single line to ground fault on electric distribution networks. Despite the severity of the line to line fault, three wind farms improve the voltage of load buses. The voltage of nearest load buses B3 and B4 to the wind farms has been improved more than the furthest load buses B8 and B9.

CWF shows the best performance compared with SCIF-WF and DFIG-WF. The CWF has the highest values of active power and, voltage. The CWF does not suffer from complete disconnection such as SCIG-WF or instantaneous disconnection such as DFIG-WF. The CWF remains connected to electric distribution

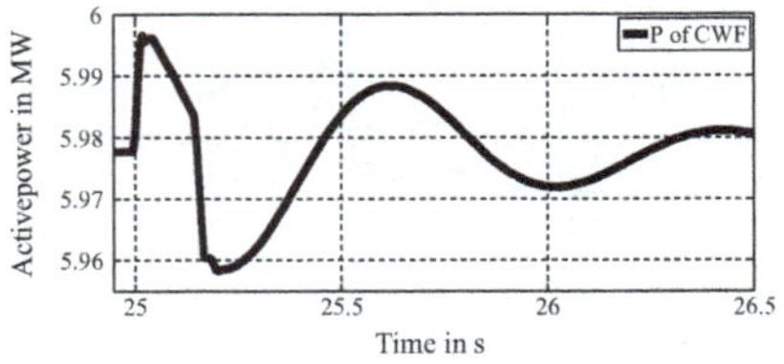


(i) The active power of CWF at PCC

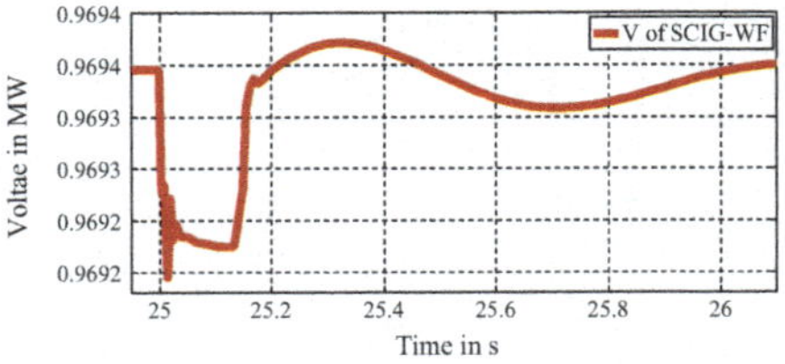


(j) The voltage of SCIG-WF at PCC

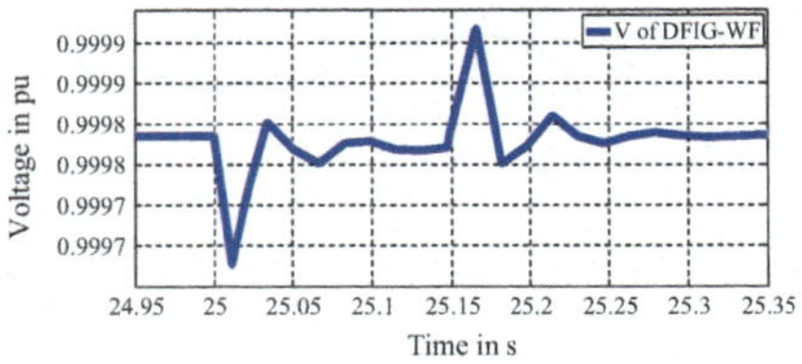


(k) The voltage of DFIG-WF at PCC

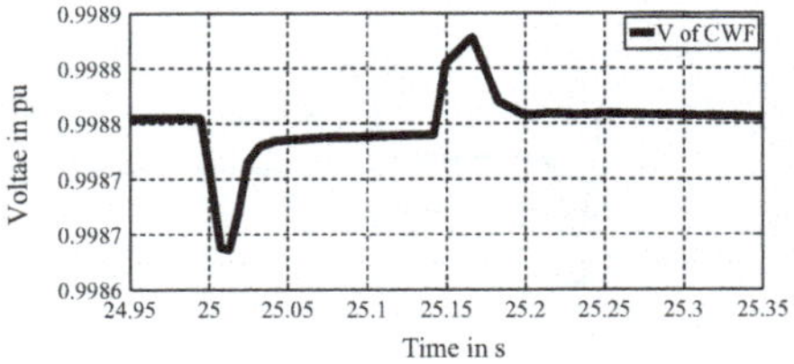


(l) The voltage of CWF at PCC

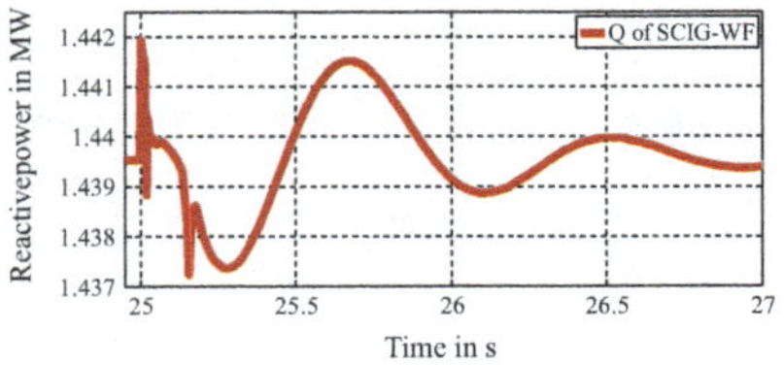


(m) The reactive power of SCIG-WF at PCC

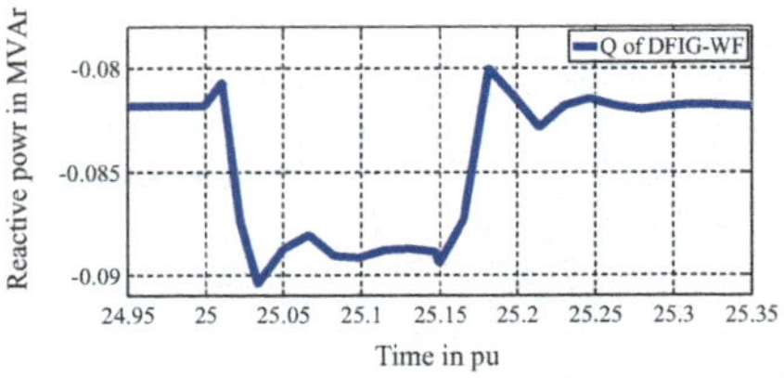


(n) The reactive power of DFIG-WF at PCC

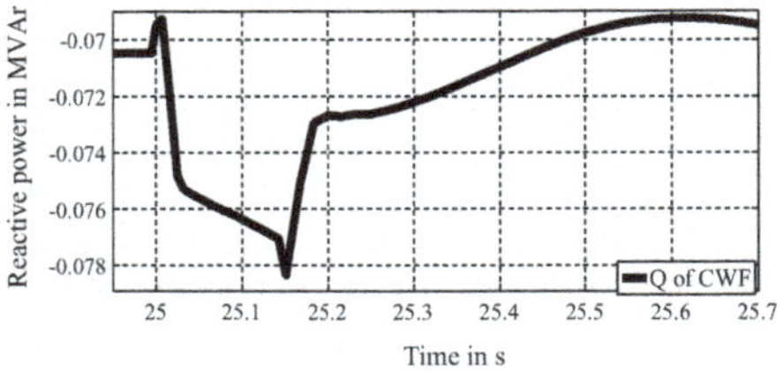


(o) The reactive power of CWF at PCC

**Fig. 11.20** (continued)

networks and injects its active power to the system during fault time. This is because the CWF collects the advantage of SCIG and DFIG.

#### 11.4.1.3 Impact of Three Phase Fault

Figure 11.22a–c show that voltage at B3 and B4 has the best improvement when the system is interconnected to CWF compared with SCIG-WF and DFIG-WF. This performance of voltage at B3 and B4 is due to the impact of generators that are

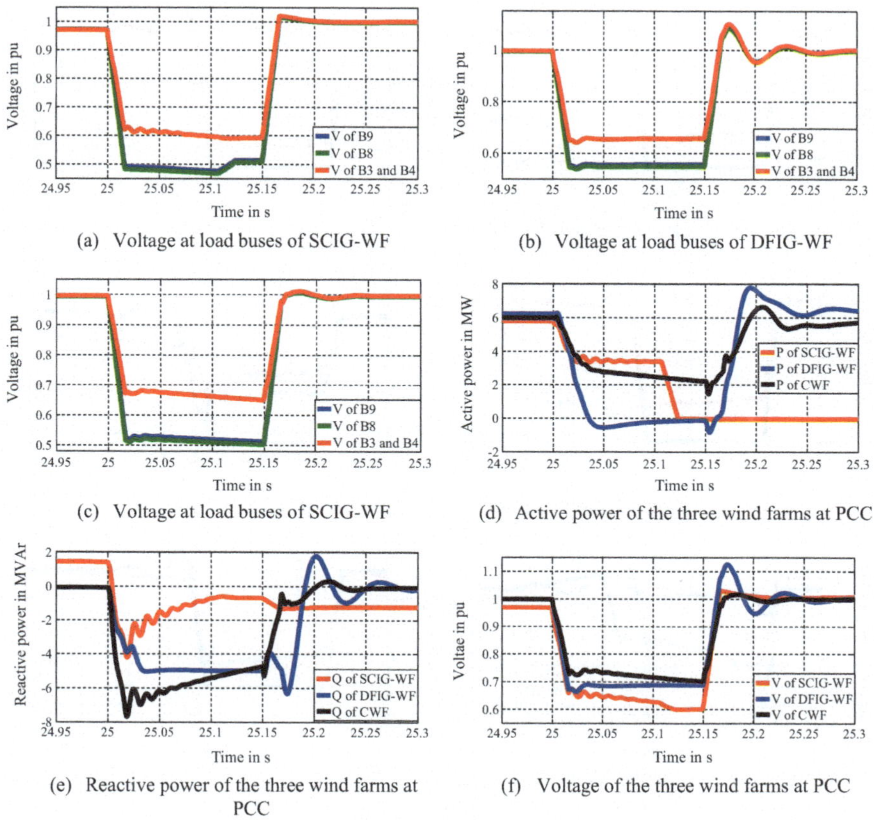


(a) Voltage at load buses of SCIG-WF

(b) Voltage at load buses of DFIG-WF

(c) Voltage at load buses of SCIG-WF

(d) Active power of the three wind farms at PCC

(e) Reactive power of the three wind farms at PCC

(f) Voltage of the three wind farms at PCC

**Fig. 11.21** The impact of single line to ground fault on the studied system

used in the wind farm on the electric distribution networks. This impact can be explained by monitoring the performance of the three wind farms at the PCC.

Figure 11.22 shows the impact of three-phase fault on the studied system. As shown in Fig. 11.22d the active power of both SCIG-WF and DFIG-WF drop to zero while the active power of CWF is still greater than zero. This means that the SCIG-WF is disconnected from the system and DFIG-WF suffers from instantaneous disconnection while CWF is never disconnected from the grid. From Fig. 11.22e and f it can be observed that CWF has the highest voltage's value at PCC because it has the highest injected reactive power at PCC. This reactive power is used to regulate the voltage and at the same time compensate reactive power demanded by SCIG wind turbines in CWF. This performance prevents CWF from complete or instantaneous disconnection from the system such as SCIG-WF and DFIG-WF.

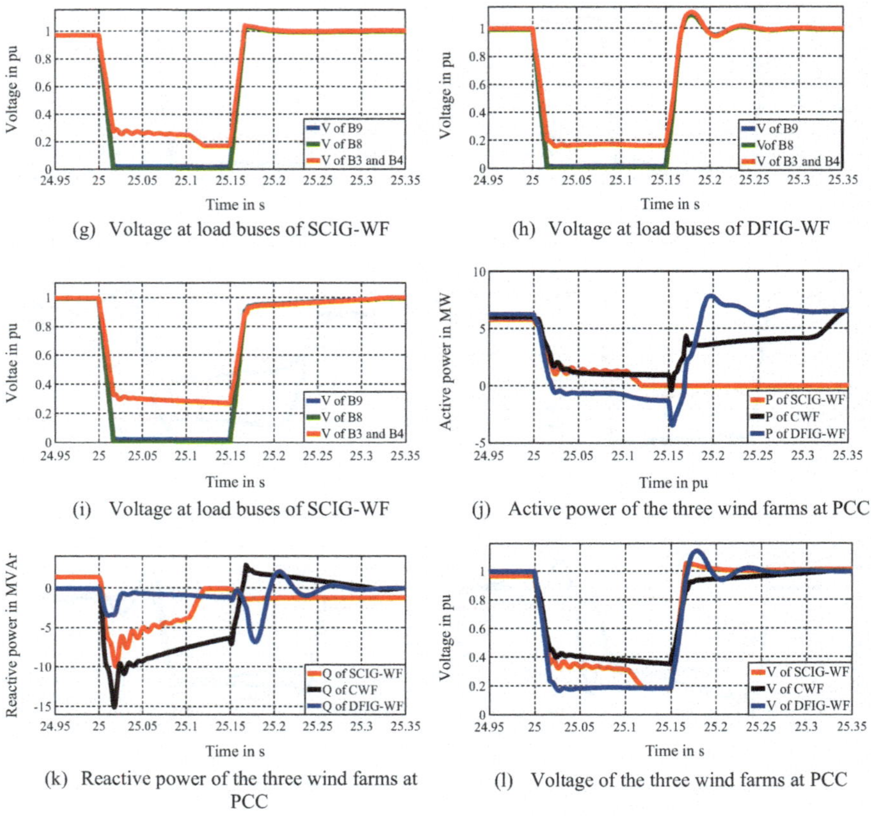


(g) Voltage at load buses of SCIG-WF

(h) Voltage at load buses of DFIG-WF

(i) Voltage at load buses of SCIG-WF

(j) Active power of the three wind farms at PCC

(k) Reactive power of the three wind farms at PCC

(l) Voltage of the three wind farms at PCC

**Fig. 11.22** The impact of three phase fault on the studied system

#### 11.4.1.4 Impact of Open Circuit Fault

In this section, the impact of the open circuit faults on the electric distribution networks interconnected different types of the wind farms (SCIG-WF, DFIG-WF and CWF) is discussed. The open circuit faults can be occurred due to the failure of one or more phase of the circuit breaker or failure of one or more phase of cables and overhead lines. In this case, a three phases open circuit fault occurs at B8 and remains for 0.15 s. Figure 11.23 shows the impact of the open circuit on the studied system. However, the wind farms improve the voltage of buses B3, B4 and B9 during the fault duration time. Figure 11.23 also shows that the CWF can almost provide the same performance of DFIG-WF during open circuit fault.

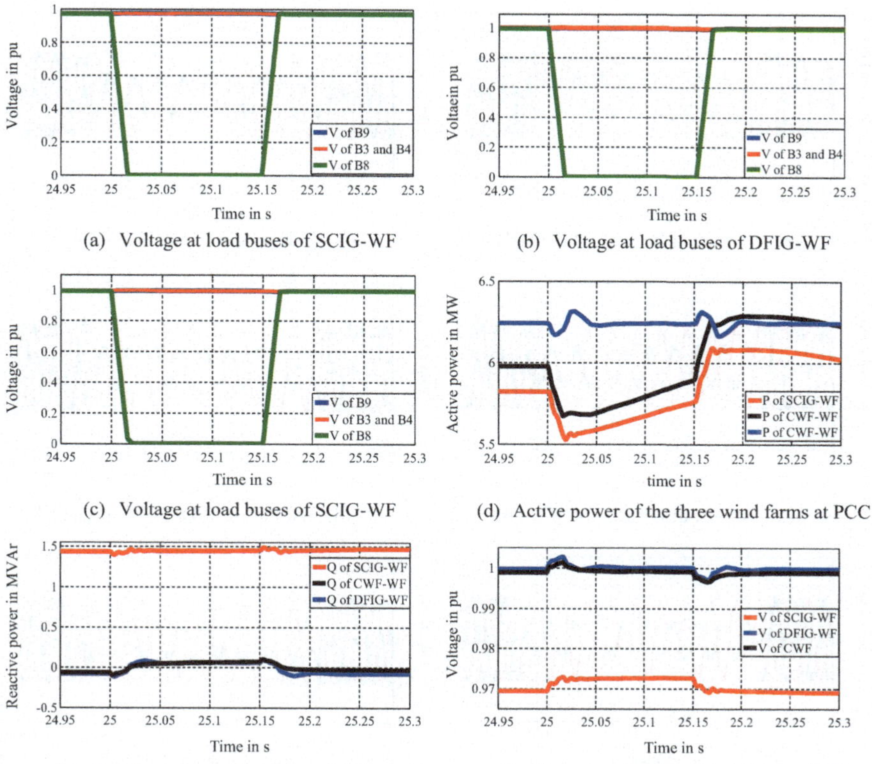


(a) Voltage at load buses of SCIG-WF

(b) Voltage at load buses of DFIG-WF

(c) Voltage at load buses of SCIG-WF

(d) Active power of the three wind farms at PCC

(e) Reactive power of the three wind farms at PCC

(f) Voltage of the three wind farms at PCC

**Fig. 11.23** The impact of open circuit on the studied system

## *11.4.2 Studied Case of Modern Wind Turbines*

Nowadays DFIG and PMSG wind turbines represent the modern technology in wind energy market. In this case, the impact of these modern types on the previous electric distribution networks is studied (Fig. 11.19). In the previous case, the combination has been accomplished between SCIG and DFIG to collect their advantage. In this case, the combination is between PMSG and DFIG to collect their advantage. The wind farm obtained from this combination is known as modern combined wind farm (MCWF). The impact of the three wind farms (PMWF, DFIG-WF and, MCWF) on studied electric distribution networks is examined during different types of faults.

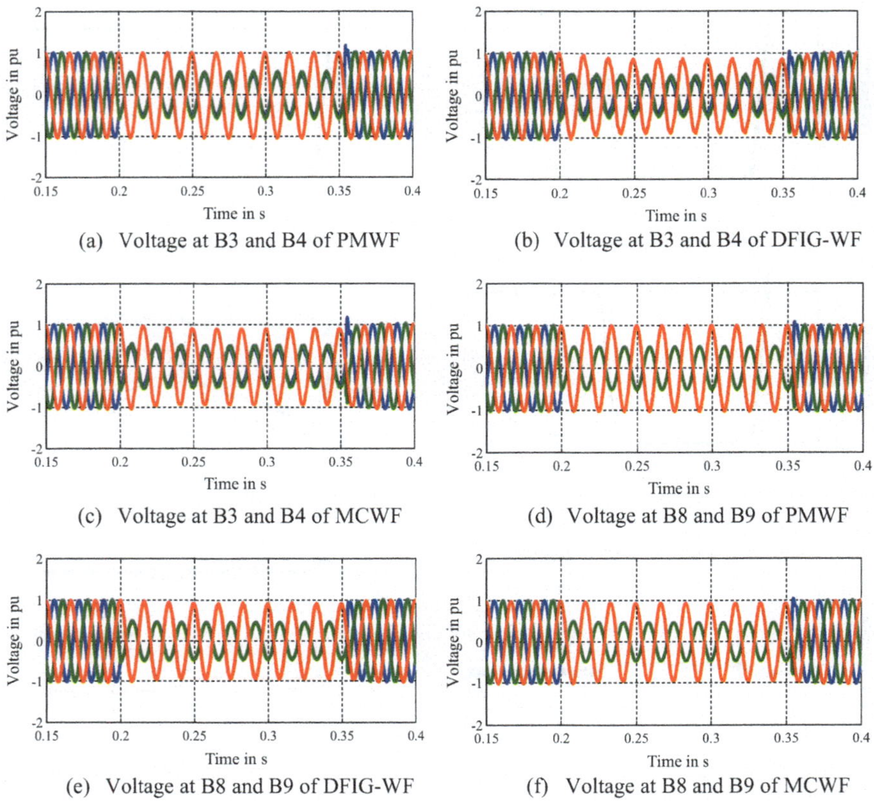


(a) Voltage at B3 and B4 of PMWF

(b) Voltage at B3 and B4 of DFIG-WF

(c) Voltage at B3 and B4 of MCWF

(d) Voltage at B8 and B9 of PMWF

(e) Voltage at B8 and B9 of DFIG-WF

(f) Voltage at B8 and B9 of MCWF

**Fig. 11.24** The voltages at the load buses during line to line fault

#### 11.4.2.1 Line to Line Fault

Figure 11.24 shows the voltages impact of line to line on the electric distribution networks during line to the line fault. As shown in Fig. 11.24, three wind farms improve the voltage of load buses especially buses B3 and B4. Also, the active power of MCWF at PCC is greater than the active power of PMWF and lower than the active power of DFIG-WF. The voltage of the voltage of MCWF at the PCC is greater than the active power of DFIG-WF and lower than the active power of PMWF. The MCWF presents an intermediate stage between PMWF and DFIG-WF and collects the advantages of both of them.

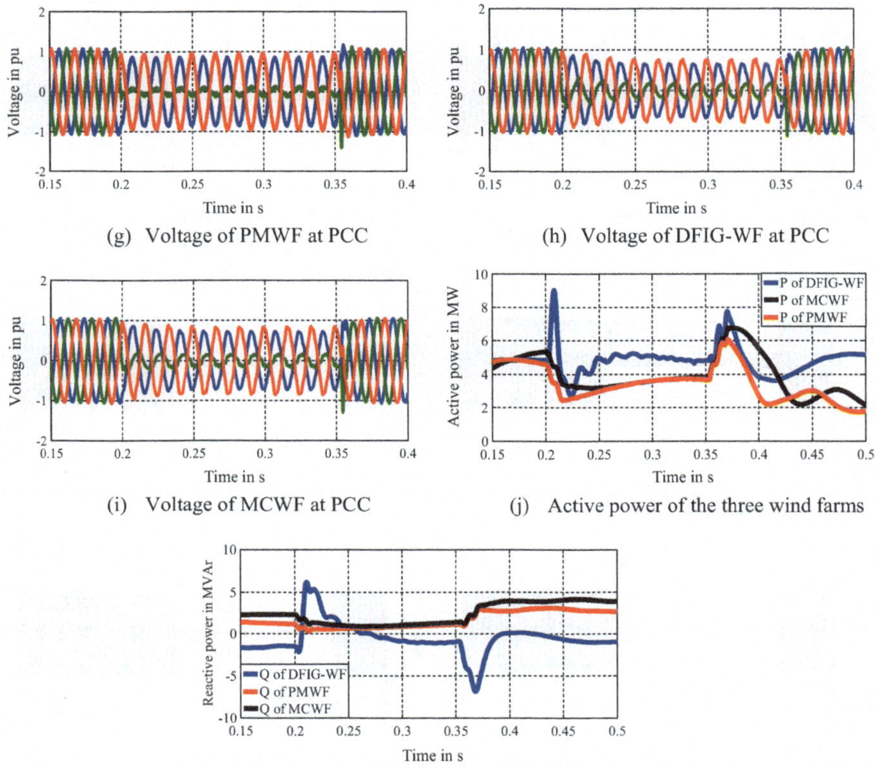


(g) Voltage of PMWF at PCC

(h) Voltage of DFIG-WF at PCC

(i) Voltage of MCWF at PCC

(j) Active power of the three wind farms

(k) Reactive power of the three wind farms

**Fig. 11.24** (continued)

#### 11.4.2.2 Three Phase Fault

The impact of the three wind farms (PMWF, DFIG-WF and, MCWF) on studied electric distribution networks is examined during three phase fault. The three phase fault occurs at a time equal to 0.2 s, has duration times equal to 0.15 s and is cleared at a time equal to 0.35 s. The results are shown in Figs. 11.25 and 11.26.

Figure 11.25 shows the voltages at the load buses during the three phase fault. The voltages of B3 and B4 are almost the same. Also, the voltages of B8 and B9 are almost the same. As shown in Fig. 11.25, the three wind farms improve the voltage

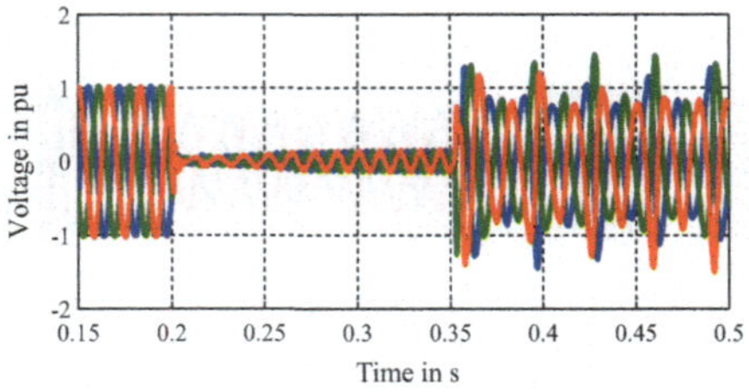


(a) Voltage at B3 and B4 of PMWF

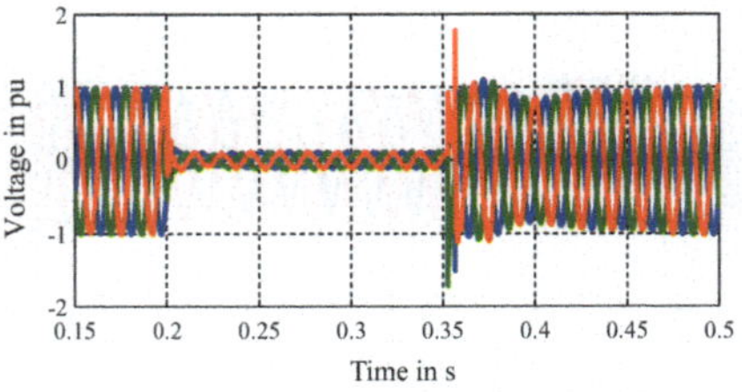


(b) Voltage at B3 and B4 of DFIG-WF

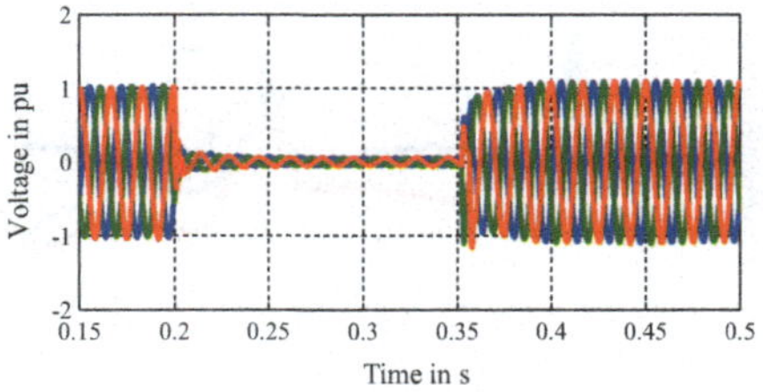


(c) Voltage at B3 and B4 of MCWF

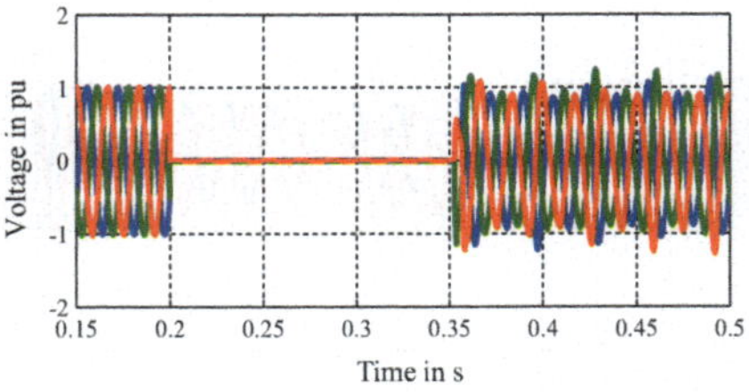


(d) Voltage at B8 and B9 of PMWF

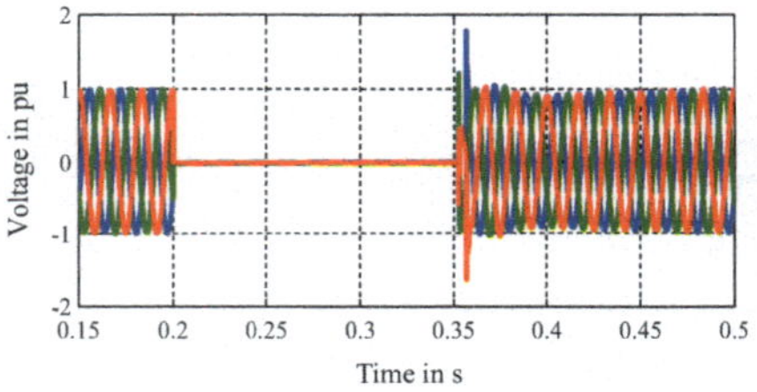


(e) Voltage at B8 and B9 of DFIG-WF

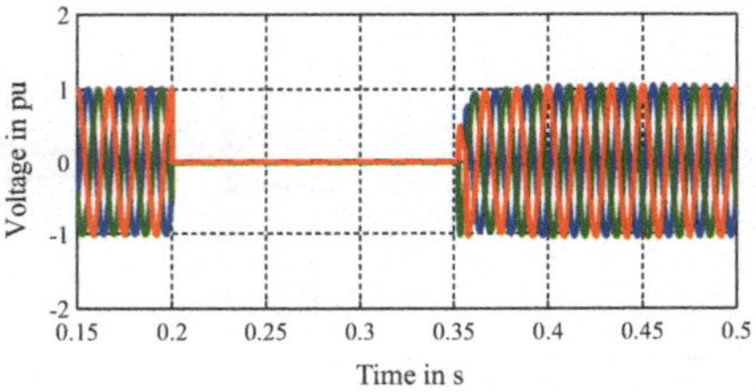


(f) Voltage at B8 and B9 of MCWF

**Fig. 11.25** The voltages at the load buses during the three phase fault system

at the nearest load buses. Figure 11.25 shows that after fault clearance the MCWF returns to stability much faster that PMWF and DFIG-WF with lowest voltage distortion. This performance of voltage at load buses can be explained by monitoring the voltage, active and, reactive power of the three wind farms at the PCC.

Figure 11.26 shows the voltage, active and reactive power of PMWF, DFIG-WF and MCWF at the PCC during the three phase fault. From Fig. 11.26 it can be observed that the distortion in voltage at load buses in cases of PMWF and

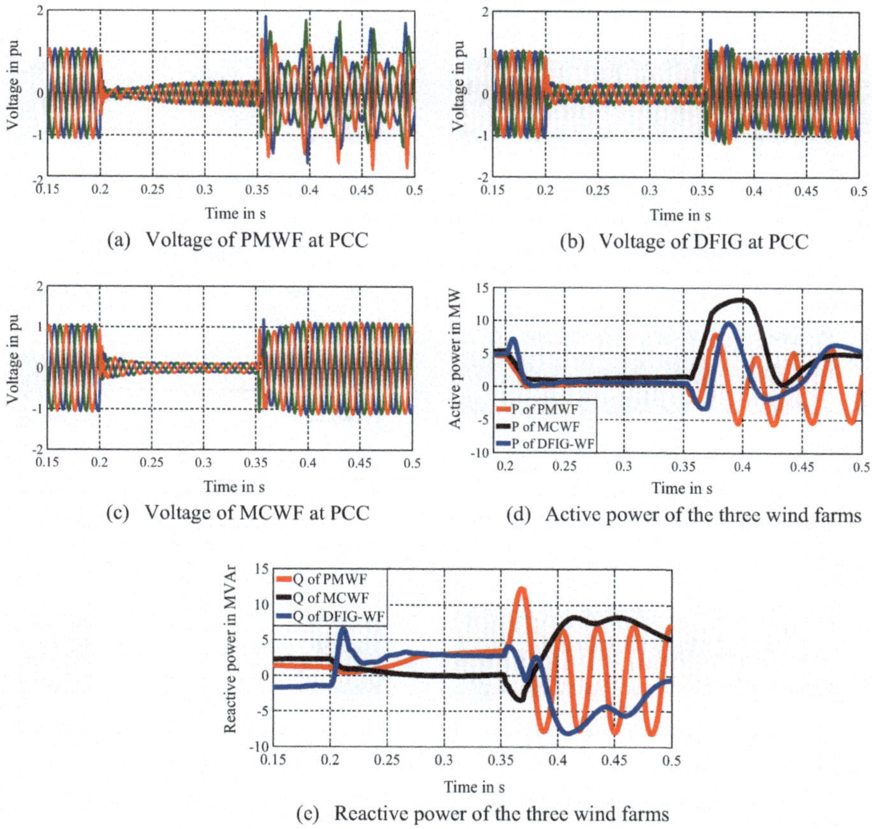


(a) Voltage of PMWF at PCC

(b) Voltage of DFIG at PCC

(c) Voltage of MCWF at PCC

(d) Active power of the three wind farms

(e) Reactive power of the three wind farms

**Fig. 11.26** The impact of three phase fault on the modern wind farms interconnect studied

DFIG-WF after fault clearance is due to the disturbance in voltage of these wind farms at PCC. As shown in Fig. 11.26 the MCWF has the highest value of active power equal to 1.3 MW and the highest value of injected reactive power during the fault. Active power of MPWF and DFIG-WF produce are equal to 0.6 MW. After fault clearance, both active and reactive power of PMWF suffers from large disturbance. This disturbance in the voltage, active and reactive power of PMWF causes the disturbance in voltage at the load buses.

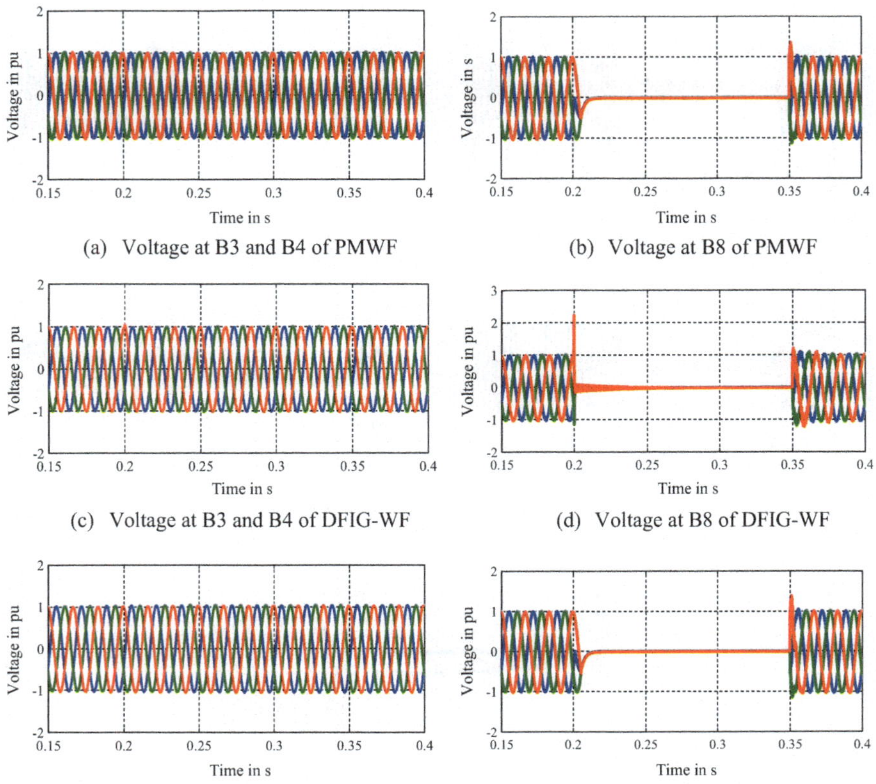


(a) Voltage at B3 and B4 of PMWF

(b) Voltage at B8 of PMWF

(c) Voltage at B3 and B4 of DFIG-WF

(d) Voltage at B8 of DFIG-WF

(e) Voltage at B3 and B4 of MCWF

(f) Voltage at B8 of MCWF

**Fig. 11.27** The impact of the open circuit on the studied system

#### 11.4.2.3 Impact of Open Circuit Fault

In this case, three phases open circuit fault occurs at B8 and remains for 0.15 s. Figure 11.27 shows the impact of the open circuit on the studied system. The wind farms improve the voltage of buses B3, B4 and B9 during the fault where the voltage values of these buses are equal to 1 pu without any distortion in voltage

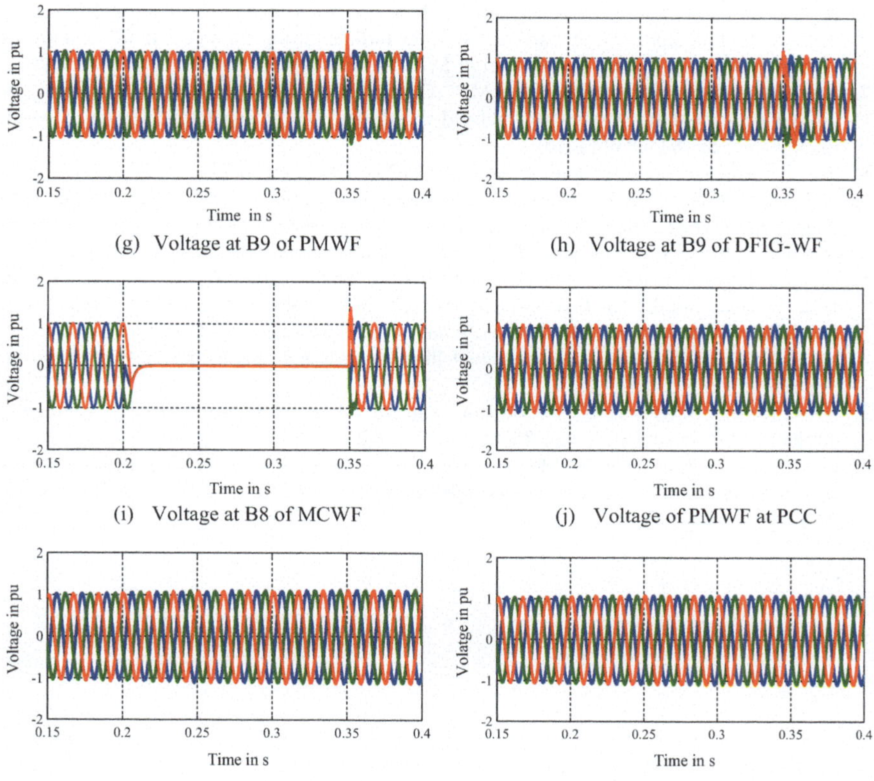


(g) Voltage at B9 of PMWF

(h) Voltage at B9 of DFIG-WF

(i) Voltage at B8 of MCWF

(j) Voltage of PMWF at PCC

(k) Voltage of DFIG-WF at PCC

(l) Voltage of MCWF at PCC

**Fig. 11.27** (continued)

profile. Figure 11.27 also shows that in the beginning of the open circuit fault and after fault clearance the voltage at B8 of MCWF is smoother than the voltage at B8 of PMWF and DFIG-WF.

## References

1. D. Ambach, Short-term wind speed forecasting in Germany. J. Appl. Stat. **43**(2), 351–369 (2016)
2. B. Wu, Y. Lang, N. Zargari, S. Kouro, *Power Conversion and Control of Wind Energy Systems* (Wiley, 2011)
3. T. Burton, N. Jenkins, D. Sharpe, E. Bossanyi, The wind resource, in *Wind Energy Handbook*, 2nd edn. (2011) pp. 9–38
4. E. Hau, H. Von Renouard, *Wind Turbines: Fundamentals, Technologies, Application, Economics* (Springer, Berlin, 2006)
5. T. Ackermann, *Wind Power in Power Systems* (Wiley, 2005)

6. H. Heydari-doostabad, M.R. Khalghani, M.H. Khooban, A novel control system design to improve LVRT capability of fixed speed wind turbines using STATCOM in presence of voltage fault. Int. J. Electr. Power Energy Syst. **77**, 280–286 (2016)
7. A. Moghadasi, A. Sarwat, J.M. Guerrero, A comprehensive review of low-voltage-ride-through methods for fixed-speed wind power generators. Renew. Sustain. Energy Rev. **55**, 823–839 (2016)
8. W.T.I. Generator, Matlab/Simulink help, in *SimPower-Systems Blocks*
9. S. Li, T.A. Haskew, K.A. Williams, R.P. Swatloski, Control of DFIG wind turbine with direct-current vector control configuration. IEEE Trans. Sustain. Energy **3**(1), 1–11 (2012)
10. N. Luo, Y. Vidal, L. Acho, *Wind Turbine Control and Monitoring* (Springer, 2014)
11. I. Al-Bahadly, A. Petersen, in *A Ducted Horizontal Wind Turbine for Efficient Generation, Wind Turbines*, ed. by Dr. I. Al-Bahadly (InTech, 2011). ISBN: 978-953-307-221-0
12. R. Melicio, V.M. Mendes, J.P. Catalão, Wind turbines with permanent magnet synchronous generator and full-power converters: modelling, control and simulation, in *Wind Turbines* (InTech, 2011)
13. H.-W. Kim, S.-S. Kim, H.-S. Ko, Modeling and control of PMSG-based variable-speed wind turbine. Electr. Power Syst. Res. **80**(1), 46–52 (2010)
14. R. Melício, J.P. Catalão, V.M. Mendes, *Wind Turbines with Permanent Magnet Synchronous Generator and Full-Power Converters: Modelling, Control and Simulation* (INTECH Open Access Publisher, 2011)

# Chapter 12
# Advanced Metering Infrastructure and Graphics Processing Unit Technologies in Electric Distribution Networks

**Zhao Li and Fang Yang**

**Abstract** The advanced metering infrastructure (AMI) has been recognized as a key communication mechanism in the modern distribution grid. As a result, integrating AMI with distribution management system (DMS) has become the focal point of distribution utilities during the past several years with the objective of enabling new applications and enhancing existing ones. In addition, with influx of massive real-time and near real-time measurements, speed up electric distribution network applications using graphic processing unit (GPU) technologies becomes attractive. Hence, the purpose of this chapter is two-fold: First it reviews a unified integration solution that enables DMS systems to flexibly adapt to various AMI systems with different communication protocols and meter data models. The feasibility and effectiveness of the integration solution are demonstrated through practical test scenarios. Second, it discusses GPU technologies and explores their applications in terms of state estimation and power flow computations. It concludes that GPU has significant potentials in improving the performance of distribution network applications. However, to unleash its power, the applications in distribution network need to be re-architected toward a GPU friendly architecture.

**Keywords** Smart grid · Advanced metering infrastructure · Distribution management system · Meter data

---

Z. Li (✉)
LY Grid Innovation, Menasha, WI, USA
e-mail: leezhao@gmail.com

F. Yang
Department of Electrical and Computer Engineering,
University of Wisconsin at Platteville, Platteville, USA
e-mail: yangfan@uwplatt.edu

 
A. Arefi et al. (eds.), *Electric Distribution Network Management and Control*,
Power Systems, https://doi.org/10.1007/978-981-10-7001-3_12

## 12.1 Advanced Metering Infrastructure

The distribution management system (DMS) [1] is an online tool running in the utility control center for the distribution system monitoring, analysis, and control. To enable a DMS system, a two-way communication network is required to connect DMS applications and monitored/controlled devices in the field. Traditionally, DMS mostly depends on the distribution supervisory control and data acquisition (SCADA) as the communication net-work. However, limited measurements (usually from the feeder head) collected by distribution SCADA inevitably constraint the DMS applications.

Developed from automatic meter reading (AMR) [2], advanced metering infrastructure (AMI) offers a two-way communication network, connecting millions of smart meters and sensors to utility control centers, transporting substantial data between them in real-time or near real-time. In the past few years, motivated by fast developing of smart grid, deploying AMI has been a critical event in utilities. The estimated installations of smart meters is about to reach 1 billion worldwide by 2020 [3].

With increased deployment in the distribution utilities, AMI has been gradually recognized as a primary two-way utility communication network. Therefore, the integration of AMI and DMS is expected to bring promising advancement in various DMS applications [4–12]. However, integration of AMI and DMS faces the following challenges: (1) adaptation of various communication protocols (Table 12.1) and information models (Table 12.2); (2) influence of substantial AMI load overwhelming the normal execution of DMS.

This chapter reviews an AMI and DMS integration solution and the evaluation of the integration solution under practical load conditions and realistic test scenarios. The overall simulation system that simulates the AMI meter load generated by millions of smart meters is discussed and test results have been presented, each focusing on individual quality attributes from the software architecture's aspect. The rest of this chapter is organized as follows: Sect. 12.2 introduces DMS

**Table 12.1** Transportation protocols and interfaces adopted by major AMI vendors

| Company | AMI data server | Transportation protocol | Transportation interface |
|---|---|---|---|
| Elster [35] | EnergyAxis Metering Automation Server (MAS) | SOAP over HTTP | Proprietarily defined XML format |
| Itron [36] | OpenWay Collection Engine | SOAP over WCF | Proprietarily defined XML format |
| Trilliant [37] | Unity Application Suite | Web Service | Interface defined by Web 2.0 AJAX |
| Sensus [38] | FlexNet | Standard or Customized | Regional Network Interface (RNI) |
| Ecologic [39] | Ecologic MDMS | JMS | Common Information Model (CIM) |

**Table 12.2** Major meter data models

| Name | Functionalities | Format | Domain | Market |
|---|---|---|---|---|
| ANSI C12.19 [40] | Model and transport meter data and control in tables | Standard Tables and procedures defined by TDL and EDL (XML) | Electricity, gas and water | North American |
| IEC62056-62 [41] | Model and transport meter data and controls in a series of interface classes | Interface classes | Electricity, gas and water | European |
| IEC61968-9 [42] | Transportation meter data between utility applications | XML based schema of meter measurement, format of messages and event | Electricity | N/A |
| MultiSpeak [43] | Data Exchange and integration in the utility enterprise | XML based model and event message interface | Electricity | North American |

applications enhanced by real-time or near real-time AMI measurements. Section 12.3 briefly presents the design of the ADI layer. Section 4 evaluates the ADI layer design.

### *12.1.1 Smart Distribution Management System*

To enhance distribution system monitoring, analysis, and control, a traditional passive operation needs to evolve into a more proactive pattern through various smart applications. The transition becomes feasible with the availability of large amount of real-time or near real-time measurements and controls across the distribution grid brought by AMI technologies. This section briefly illustrates these advanced DMS applications and their dependency on the AMI system.

#### 12.1.1.1 Distribution State Estimation (DSE)

The awareness of grid states not only facilitates the timely and accurate monitoring of the distribution system operating conditions but also provides an essential foundation to implement various other smart grid technologies that would rely on a statistically accurate snapshot of the system states. Voltage/Var control, feeder reconfiguration, and post-disturbance service restoration are some examples.

Practically, due to financial and/or technical constraints, metering devices can only be installed in very limited locations in the field. Most system states have often been estimated through the load allocation and power flow applications, which are vulnerable to measurement errors and suffered from the lack of measurement redundancy. Even though residential level AMI measurements reflect a detailed state of the distribution grid, hindered by deficiencies of measurement facilities and errors in the data transmission, the collected measurements are sometimes inaccurate, unreliable, and delayed.

Based on a three-phase system network model and a set of redundant distribution system measurements from SCADA, AMI meters, and other sensors, DSE effectively eliminates above defects and provides best estimates of the system states. In particular, DSE implements following functionalities: (1) determining whether the system state can be computed from the existing measurements; (2) detecting, identifying, and rejecting bad measurements; (3) filtering out measurement errors and computing system state with minimum error; (4) evaluating the quality of the state estimation results.

#### 12.1.1.2 Advanced Outage Management

As a fundamental function in the outage management system (OMS), outage scoping analysis (OSA) determines the efficiency and effectiveness in dispatching crew for fault reparation and service restoration tasks. Traditionally, due to the very limited real-time SCADA information available in the distribution network, the main outage information source for conducting the OSA is customer trouble calls, which usually results in prolonged OSA procedure and inaccurate OSA result due to the absence of customers and limited number of trouble calls.

For an electric distribution network equipped with AMI, smart meter data can be used as an additional outage information source for OSA. Specifically, the precise outage information carried by last gasp, sent by smart meters before they lose voltage during an outage, can be very beneficial in detecting the outage root and scope. In addition, the communication network of the AMI system also enables the on-demand polling of meter status for the purpose of outage/restoration confirmation. During an outage event, a large amount of meter data load is pushed to the OSA in a very short period of time. The DMS system is expected to be able to timely handle such a load.

#### 12.1.1.3 Demand Response

Demand response (DR) is a semi-emergency preventive action taken by the utility (or the demand response service provider) that is normally executed during peak load hour whenever the system capacity gets close to being fully utilized. The result of DR is a change in the level of energy consumption of the end customers in response to the changes in electric distribution network loading and/or energy prices. The energy consumption can be either reduced or shifted (postponed) to a different time to reduce the total demand at peak load hour. As a result, the utility can avoid purchasing power from the spot market or having to impose forced outages on customers [13].

Successful implementation of DR requires an efficient two-way communication system between the utility and the individual customers. The utility would need to poll meter data from the customers, process it in the DR engine, and generate signals accordingly to be transmitted to the appropriate customer meters. The timeframe required for demand response varies from one event to another. While there are cases where the event is scheduled well in advance, there are cases where demand needs to be reduced immediately (i.e., emergency demand response). Nonetheless, the meter poll is usually performed over a very large geographical area, retrieving data from thousands of customers at once. All these underline the importance of an efficient and reliable communication network between the customers and the control center.

In summary, all the advanced applications discussed above and many other typical applications that facilitate a smart distribution system have one common feature: their implementation is largely dependent on real-time or near real-time meter information provided by the AMI system. Regardless of the type of the application and its timeframe, it is easy to see that the dependency of these applications on the AMI system imposes a substantial challenge on the AMI communication system and its interface with the DMS. Finding an efficient way to handle this large amount of meter data in DMS becomes a major task. This challenge and its solution will be further discussed in the next section.

### *12.1.2 The ADI Layer—The Solution for AMI and DMS Integration*

The purpose of AMI and DMS integration is twofold: (1) to exchange meter data and control information between the two systems and (2) to architecturally minimize the impact of the information exchange on both systems. As the mid-layer between AMI and DMS, the ADI layer, as shown in Fig. 12.1, implements the above two objectives.

DMS and AMI are two separate systems, which have different business goals and architectures. The purpose of AMI and DMS integration is to exchange meter data and control information between the two systems, at the same time to minimize the influence of the integration on both the AMI and DMS system in terms of performance and engineering cost.

To reach the above goals, an AMI and DMS integration solution, called the AMI and DMS integration layer (ADI), is described in this section. The ADI layer can be viewed as a middleware between the AMI and DMS systems. Some major business considerations of the ADI layer are summarized below:

- Different DMS applications use different approaches to import external data. For example, some DMS applications utilize Enterprise Service Bus (ESB), and others rely on the SCADA system. In this sense, the ADI layer should be easily adapt to both ESB and SCADA interfaces.

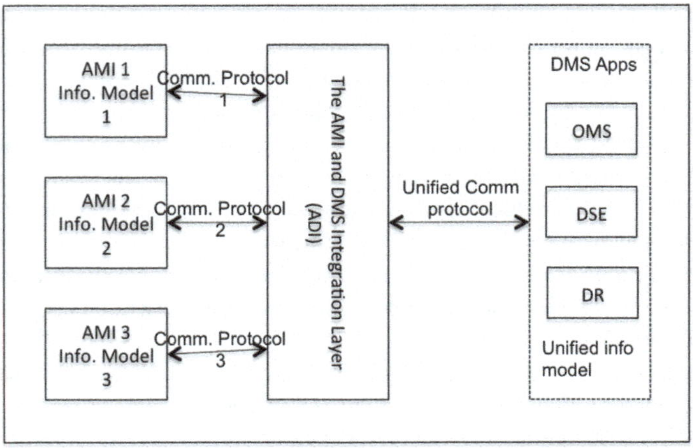


**Fig. 12.1** The context of AMI DMS integration

- Generally, utilities have deployed (or are currently developing) AMI systems with different types of meter data servers, such as MDMS, meter data collection systems. These AMI systems are usually built by different AMI vendors. The ADI layer should be adaptable to the diverse AMI systems.
- Meter data models in AMI systems are designed for applications in different domains (i.e., electricity, water, and gas); however, DMS applications primarily require a dedicated meter data model (i.e., IEC61968-9). The ADI layer should consider information gaps between a general domain meter data model and a distribution network specific meter data model, and match these gaps while exchanging meter data between an AMI system and a DMS system.
- As of today, the commercial requirement for regular meter reading is at an interval of 15 min. With the development of AMI technologies, the interval is getting shorter and shorter. However, even handling meter data generated by millions of smart meters in every 15 min can pose substantial challenges to DMS systems. The original architecture of most legacy DMS systems was not designed for heavy AMI meter data load conditions. Therefore, the ADI layer has to minimize the influence of the meter data load on a DMS system by caching the meter data and adjusting the meter data stream throughput to a level that can be accepted by the DMS system when necessary.
- AMI and DMS systems are usually developed by different vendors. Thus, it is likely that each uses a different dialect in both meter models and meter transportation protocols to describe even the same grid network. The ADI layer should harmonize these dialects.

Major requirements for the ADI layer are listed as follows:

- As a unified solution, the ADI layer should adapt to the communication protocols used by different DMS systems.
- The ADI layer should adapt to various AMI communication protocols.

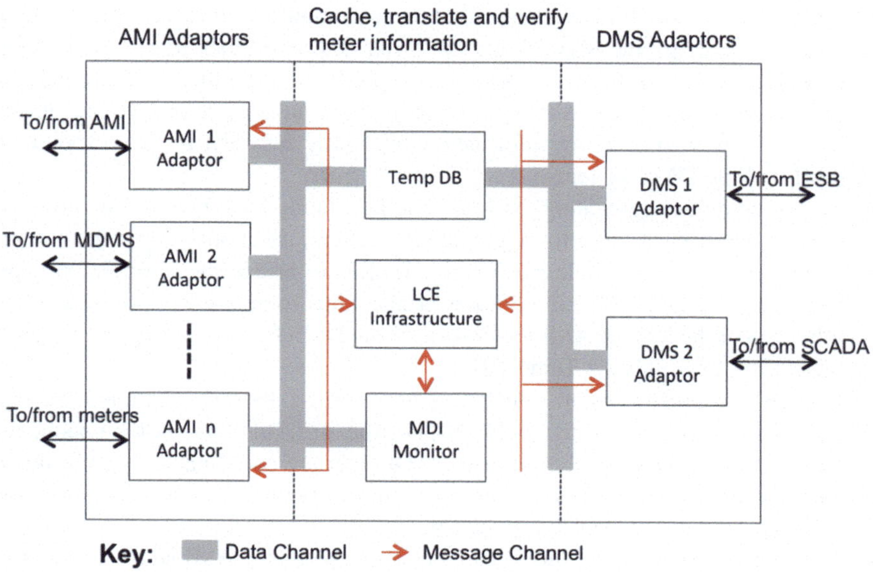


**Fig. 12.2** The architecture design of ADI

- The ADI layer should bridge gaps between the data model used in AMI and DMS when exchanging information between them.
- The ADI layer should architecturally minimize the impact of the heavy AMI meter data load on the DMS system, which was not originally designed for handling such type of heavy load.

To design the ADI layer, the following challenges should be addressed from the software architecture's aspect: Performance: The ADI layer's capabilities in timely processing both scheduled (expected) meter reads and burst occurred (unexpected) outage reports, especially the worst case scenario, where a regular meter reading session coincides with an outage session consisting of large-scale outage-reporting events.

Scalability: The ADI layer's capabilities in adapting to a wide range of load conditions without losing its performance by only adjusting its hardware configurations rather than modifying its source code.

Adaptability: The ADI layer's capabilities in adapting to different AMI systems deployed by utilities in the current and the future, which potentially use different meter data models and communication protocols.

The components of the ADI layer, as shown in Fig. 12.2, can be classified into four categories: the AMI adaptors, the AMI information translation and verification infrastructure, the loosely coupled event (LCE) infrastructure, and the DMS adaptors.

- AMI adaptors. AMI adaptors are used to accommodate different types of AMI data sources (e.g., the MDMS and the meter data collector), transferring meter data streams either from the AMI to the DMS or the other way around. In practice, each integrated AMI system should have an associated AMI adaptor in the ADI layer, processing information bilaterally between the corresponding AMI and the DMS systems.
- The structure of aggregating, caching, and translating information. The primary goal of the infrastructure for aggregating, caching, and translating information, as illustrated in the middle part of Fig. 12.2, is to bridge the information gaps between AMI and DMS. In addition, it also aims to architecturally isolate the influence of the load of AMI meter data on the DMS by temporarily caching the meter data load into the Temp DB.
- Loosely coupled event infrastructure. The loosely coupled event (LCE) infrastructure is the publish/subscribe infrastructure. All components (e.g., AMI and DMS adaptors) in the ADI layer are coordinated by publishing and subscribing messages to the LCE infrastructure. The coordination by messages makes the ADI layer scalable.
- DMS adaptor. Generally, most design considerations for AMI adaptors are also applicable to those of the DMS adaptors. The unique feature of the DMS adaptors is their capability of adapting to the throughput limitation of DMS data channels when delivering meter data to the DMS system.

Dynamically, the ADI layer supports the following three types of activities or events: (1) the AMI pushing meter data (e.g., outage report and meter reads) to the DMS, (2) the DMS pushing meter control commands to the AMI, and (3) the DMS polling meter data from the AMI (e.g., verifying outage and requiring meter measurements from certain meters).

- AMI pushing meter data to the DMS. The workflow of the ADI layer processing meter data pushed by the AMI is as follows: Upon receiving a meter data package from the AMI, an AMI adaptor parses and temporarily caches the parsed meter data to the Temp DB for further translation and verification; Then it publishes a message to the LCE infrastructure and notifies subscribers (e.g., a DMS adaptor) that the AMI meter data has arrived. After receiving this notice, the DMS adaptor picks up the verified and translated meter data from the TempDB, packs it based on the message format required by the DMS system and delivers to it.
- DMS pushing meter control commands to the AMI. The work-flow of processing meter control commands pushed by the DMS to the AMI is the same as the workflow of processing meter measurements pushed by the AMI but the direction of information flow is from the DMS to the AMI.
- DMS polling meter data from the AMI. The workflow of DMS polling meter data from the AMI consists of the following processes: (1) The DMS pushes meter control commands to the AMI and (2) the AMI pushes the meter data back to the DMS. The details of two processes were discussed earlier in this section.

### 12.1.3 The Evaluation of the ADI Layer

This section develops a simulation system and five test scenarios to comprehensively evaluate the quality of the ADI layer architecture.

The simulation system consists of two AMI simulators and two ADI layers, deployed to four physical servers as shown in Fig. 12.3. To make the simulation more practical, the authors have used the 100 Mbps network, instead of the 1 Gbps network, to link these servers. The AMI simulator has the following functionalities: (1) Simulate various meter loads (e.g., regular meter reads and large scale outage report), (2) Respond to the control commands sent by the DMS (e.g., meter poll commands, outage verification request, and demand response request).

The AMI simulator supports two communication interfaces to exchange data with external systems: a standard web service (i.e., SOAP) and a Java Message Service (i.e., JMS). Accordingly, to interface with the AMI simulators, the web service adaptor and the JMS adaptor are developed in the ADI layer. In addition, to clarify the analysis, all the messages are packaged into the format of IEC61968-9.

The AMI simulator, as shown in Fig. 12.4, consists of the meter farm, the meter management layer and various load generators. The meter farm manages the 63,445 self-spinning objects of smart meters, simulating various smart meter events (e.g., regular meter reads, outage and demand response). The meter management layer

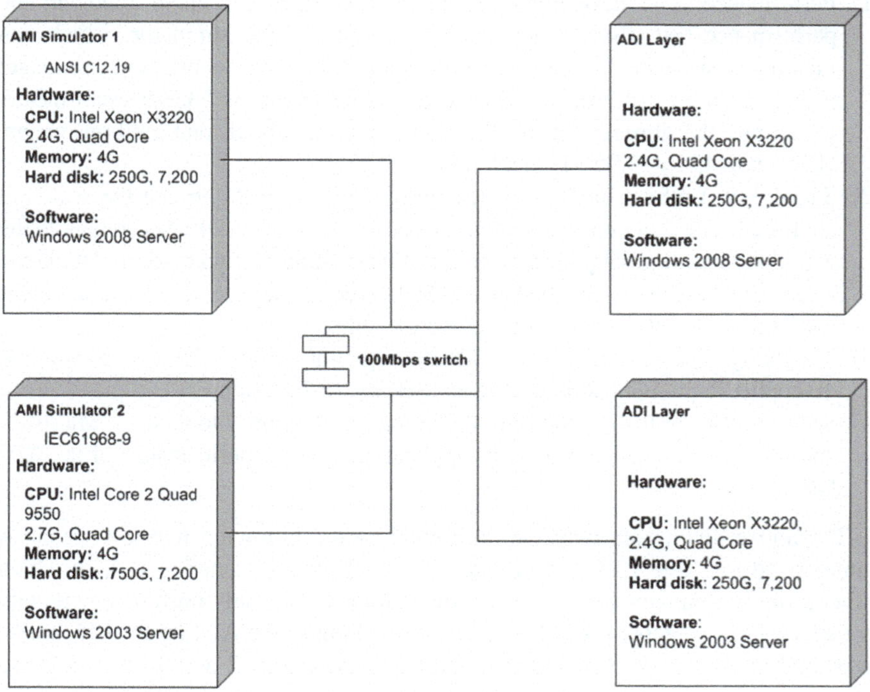


**Fig. 12.3** The evaluation system with two AMI simulators and two ADI layers

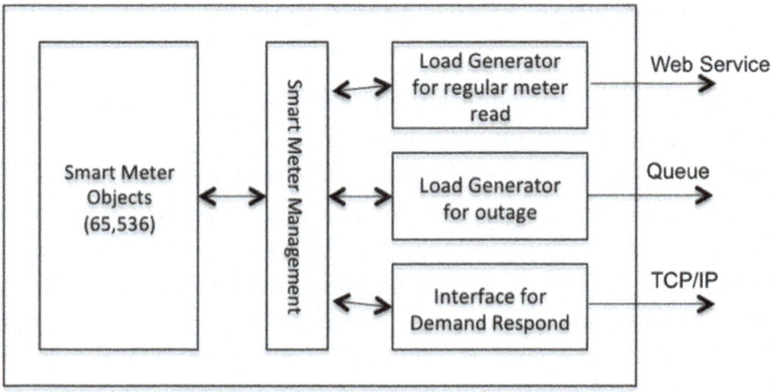


**Fig. 12.4** The infrastructure of the AMI simulator

provides interfaces for external applications to access meter objects. The meter read load generator, the outage generator, and the interface for demand response simulate the data required by related DMS applications.

#### 12.1.3.1 Test Strategies

(1) Performance. The throughput of the ADI layer is the major focus in the performance test. Constrained by the capacity of the simulator, which can maximally simulate 63,445 smart meters, the following assumptions are made: if the ADI layer can process meter data load generated by 63,445 smart meters in 1 min, the throughput of the ADI layer is about one million meters ($15 \times 63{,}445 = 0.95$ M) in 15 min.
(2) Scalability. Without the loss of generality, both the scale up and the scale out are tested. The scale up test is to verify capability of the ADI layer to handle the extra load by improving the utilization of its existing hardware setup. While the scale out test is to verify whether the ADI layer can be extended to multi-server configuration without revising the source code.
(3) Flexibility. The flexibility is to verify the capability of the ADI layer to adapt to diversified communication protocols used by AMI systems.
(4) Security. The security is to verify the influence of security tools (e.g., Microsoft and McAfee Enterprise Anti-virus applications) on the performance of the ADI layer.

The above quality attributes are examined by the following five test scenarios summarized in Table 12.3. Performance is tested by the scenarios 1–4, in which either a single simulator serially or two simulators concurrently push outage reports or meter measurements to ADI layer. The throughput of the ADI layer in 15 min is measured in each case. Scale-up is tested by the scenarios 2 and 4, in which two simulators concurrently sending outage reports and/or meter measurements to the ADI layer. The CPU utilization are measured. Scale-out is tested by the test

Table 12.3 The quality attributes and the designed test cases

| Test scenario | Performance | Flexibility | Scale-up | Scale-out | Security |
|---|---|---|---|---|---|
| 1 | ✓ | | | | ✓ |
| 2 | ✓ | | ✓ | | ✓ |
| 3 | ✓ | | | | ✓ |
| 4 | ✓ | ✓ | ✓ | | ✓ |
| 5 | | | | ✓ | ✓ |

scenario 5, which extends a system with one ADI layer to a system with two ADI layer without changing the source codes. The flexibility is tested in the scenario 4, in which two types of communication protocols are selected to trans-port meter data. Finally, as industrial standard security tools (e.g., McAfee Enterprise Anti-virus tools) are installed in the ADI layer, the performance overhead of these security tools is automatically examined through the test cases 1–4.

#### 12.1.3.2 Test Results

The permutations and combinations of the AMI simulators and the ADI layers, as shown in Fig. 12.3, produce the following five test scenarios:

Test Scenario I: one AMI simulator pushes meter outage reports formatted as JMS to one ADI layer.

The objective of this test case is to verify if the ADI layer can handle outage load pushed by one million smart meters in 15 min. The outage reports, containing meter id, the service transformer id, and the timestamp, are simulated by AMI simulators 1 and 2 separately. On the ADI layer, to handle the incoming outage reports, the ADI layer conducts the following two tasks: (1) unpack and parse the outage reports sent from the AMI simulator, and (2) validate the integrity of the carried meter information against the current DMS meter asset.

The AMI simulator 1, which has more advanced hardware configuration than the AMI simulator 2, on average takes about 27 s to generate and send 63,445 outage reports to the ADI layer, the throughput is 2.1 million per 15 min; while the AMI simulator 2 takes about 67 s to transport the same amount of data, the through-put is 0.8 million per 15 min.

The test scenario 1 demonstrates that the ADI layer server can process 0.8–2.1 million outage reports in 15 min.

Test Scenario 2: Two AMI simulators concurrently push meter out-age reports in the JMS format to one ADI server.

The objective of the test scenario 2 is to test the scale-up capability of the ADI layer design. To fulfill this test, AMI simulators 1 and 2 concurrently push 2 × 63,445 outage reports to the ADI Layer server. The test results demonstrate that AMI simulator 1 spends 27.8 s to process the meter data; while AMI simulator 2 spends 61.2 s to do the same amount of work. Compared to the test scenario 1, two AMI simulators concurrently pushing the meter data to the ADI layer barely influence its performance. This demonstrates that AMI adaptors are well isolated in terms of performance and can concurrently receive data from various data sources.

Comparing to the test scenario 1, in which the CPU utilization is 40–50%, the test scenario 2 has higher CPU utilization rates, which is about 60–70%. This demonstrates the well-designed scale up capability by the fact that the ADI layer handles the outage reports pushed by 3.1 million smart meters in 15 min with scaling up the utilization of the hardware resource in the ADI layer.

Test Scenario 3: the AMI simulator 1 pushes meter measurements to the ADI layer through web service remote call.

Test scenario 3 examines the capability of the ADI layer in receiving the regular meter measurements through the web service. The regular meter measurement package contains: AMI meter id, AMI transformer id, measurement time, frequency, voltage (Va, Vb, Vc), current (Ia, Ib, Ic), phase angle (ANGa, ANGb, ANGc), real power (kWa, kWb, kWc), reactive power (kVAa, kVAb, kVAc), power factor, and energy (kWha, kWhb, kWhc).

On average, it takes 31.4 s to transport and process the 63,445 m measurement packages. The throughput is about 1.8 million packages per 15 min. Even though the time spent on the test case 3 is similar to that of the test scenarios 1 and 2, as the size of the meter read package is much larger than the size of the out-age package, the transportation through web service over HTTP is much faster than that through the WebSphere MQ regardless other advanced features of WebSphere MQ (e.g., security and transaction based transportation). The test scenario 3 shows that the ADI layer can process the regular meter reads, carrying major engineering measurements generated by 1.8 million smart meters in 15 min.

Test Scenario 4: AMI Simulator 1 pushes meter measurements and AMI Simulator 2 pushes the outage reports to the same ADI layer simultaneously.

Test scenario 4 simulates the worst-case scenario, in which the regular meter reads and outage reports happen simultaneously: AMI simulator 1 pushes the regular meter reads and AMI simulator 2 pushes the outage packages to the same ADI layer. The ADI layer takes about 41 s to process the 63,445 regular meter measurement packages sent from simulator 1 and takes about 68 s to handle the 63,445 outage reports sent from simulator 2. Hence, in 15 min, the ADI layer can concurrently handle 1.4 million smart meter reads and 0.8 million outage reports.

Test Scenario 5: Scale out test.

The test scenario 5 is intended to examine the scale out attribute, focusing on extending a system with one ADI layer to a system with two ADI layers with no changes of source codes.

**Table 12.4** The measurements of test scenarios 1–4

| Scenario | Simulator | Load type | Throughput (M) | CPU utilization (%) |
|---|---|---|---|---|
| 1 | 1 | Outage reports | 2.1 | 40–50 |
| 2 | 1 + 2 | Outage reports | 3.1 | 60–70 |
| 3 | 1 | Meter reads | 1.8 | 40–50 |
| 4 | 1 + 2 | Meter reads + Outages | 2.2 | 60–70 |

In order to build the above system, the following tasks are con-ducted: (1) in the ADI layer 1, configure the AMI adaptor 1 hosted in Internet Information Service (IIS) and connect it to the AMI simulator 1 through web service over HTTP; (2) set up a new queue in IBM WebSphere MQ 7 and in the ADI layer 2, configure and connect the AMI adaptor 2 to the AMI simulator 2 through the new queue configured in the WebSphere MQ 7.

The above procedure verifies that constructing a system with two AMI simulators and two ADI layers can be achieved by only updating the hardware/software configuration and demonstrates that the ADI layer can be well scaled out by adding more ADI servers without modifying the source codes (Table 12.4).

## 12.2 Graphics Processing Unit (GPU) Technologies

As the states of distribution network change continuously, its operations require constant monitoring and control to balance the power production and consumption. The deployment of smart grid technologies exacerbates this situation because of the increasing uncertain behaviors in both power generation (e.g., the intermittency of renewable resources) and the power consumption (e.g., the randomly charging behavior of electric vehicles). To effectively manage these uncertainties, smart grid needs continuously executing distribution network monitoring, simulation, control, and analysis applications such as the short-term planning, state estimation, and voltage var control. These applications are generally computation intensive and inevitably incur substantial amount of computational load.

In literature, large computation load is commonly handled by mainframe or specially designed high performance computation (HPC) facilities, which are equipped with hundreds or even thousands of CPUs and a large amount of memory. These HPC facilities are mostly deployed in universities and national research labs [14]. However, the high cost of constructing and maintaining these facilities prevent them from being adopted by a distribution network control center. Hence, identifying affordable HPC solutions suitable for deploying practical distribution network applications becomes attractive.

The HPC solutions in the affordable personal computer (PC) market fall into two categories: the multi-core CPU and the many-core GPU (Graphic Processing Units) [15, 16]. The latter, which was originally designed to exclusively boost graphic functionalities in the PC platform, has been accessible for general computation through C++ APIs (e.g., CUDA or OpenCL) [17] since 2007. From the technical standpoint, the number of computational cores is the primary criterion to distinguish the above two HPC platforms: a GPU has thousands of computational cores; while a multi-core CPU has less than 20 cores. As of today, a CPU with four cores and a GPU with 2700+ cores (e.g., NVIDIA GTX Titan [18]) dominate the current market.

Equipped with excessive computational cores, GPU offers more computation power than multi-core CPU does, particularly suitable for parallelizable massive computations (e.g., matrix calculations). For instance, the main stream GPU can concurrently run thousands of computation tasks, fulfilling more than one trillion floating-point operations per second. Its computation capability is equivalent to that of a mainframe computer could offer a few years ago. The recent literature is seeing that GPU tremendously boosts the performance of applications in various domains [19, 20].

Motivated by the GPU's performance potentials, considerable research efforts have been devoted to the application of GPU technologies to power system. For example, in [21], the GPU was applied to accelerate transient stability simulation of the large-scale power system and achieved a speed improvement of 345 times for a power system with 320 generators and 1248 buses; in [22], authors comprehensively studied the shared memory computational infrastructure and implemented a GPU based iterative solver, which is around 50 times faster than its CPU counterpart. The above results theoretically demonstrate the potentials of using GPU technologies to improve the performance of power system applications.

This chapter investigates the potential enhancement of the GPU technologies on power grid analytic tools along both the theoretical direction and the practical direction. The rest of this chapter is structured as follows: Sect. 12.2 takes a closer look at the computation architecture of multi-core CPU and many-core GPU. Section 12.3 prototypes the GPU version of CGNR algorithm, an iterative solver that has been widely applied to distribution network applications, and evaluate it against the data sets abstracted from large scale estimation problems and power flow problems. Section 4 studies the potentials of applying the GPU technologies on accelerating a real-world application.

### 12.2.1 Potentials of GPU on Accelerating Sparse Linear Solvers

This section reports the experience of using the main stream computation architecture to improve performance of solving a system of linear equations, the key part of most power system applications, using iterative methods. Since Conjugate Gradient (CG) algorithms [23] have been applied to power system applications in the literature with a suggested benefit from parallelization, they are selected and evaluated against the mainstream computation architectures (i.e., multi-core CPU and many-core GPU) in the context of both power system state estimation and power flow applications. The evaluation results show that solving a system of linear equations using iterative methods is highly memory bonded and multi-core CPU and GPU computation architecture have different impacts on the performance of such an iterative solver: unlike multicore CPU, GPU can greatly improve the performance of CG-based iterative solver when matrices are well conditioned as typically encountered in the DC power flow formulation.

#### 12.2.1.1 A Closer Look at GPU Architecture

From an architecture's aspect, both multi-core CPU and many-core GPU package several cores into one die, sharing the same memory channel. For this type of computation architecture, having a high-speed memory bandwidth, through which data can be quickly fed into computational cores, is necessary to keep the computation structure in high performance.

To measure the performance of computation architecture, two concepts, the computational power and speed of data channel, are introduced. The former are measured by GFLOPS, billions of floating point operations per second that a processor can perform; while the latter are evaluated by memory bandwidth, the data transportation rates between computation cores and main memory.

Figure 12.5 illustrates the major differences between multi-core CPU and GPU: a multi-core CPU has a larger cache system and stronger control capabilities; while GPU has significant number of computational cores (ALUs). Table 12.5 further explains the above differences. Overall, a multi-core CPU has total 10 M cache (L1 cache + L2 cache + L3 cache), 6–7 times larger than the capacity of the GPU's cache system. On the other hand, GPU has numerous computational cores (i.e., GTX480 that has 480 cores), which is far more than the number of cores of multi-core CPU has.

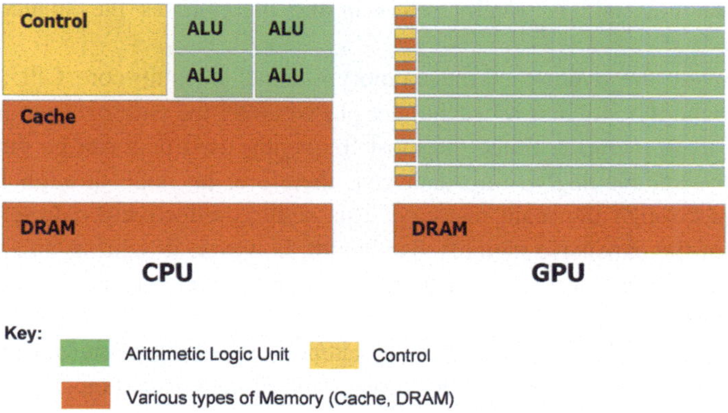


**Fig. 12.5** The models of multi-core CPU and GPU

**Table 12.5** Multi-core CPU versus many-core GPU

| | Number of cores | L1 cache | L2 cache | L3 cache | Memory bus (GB/s) |
|---|---|---|---|---|---|
| Intel Bloomfield Core i7 950 (June, 2009) | 4 | 32 KB(Instruction) 32 KB Data | 256 KB/Core | 8 M for all cores | 25 |
| Intel Yorkfield Core 2 Quad Q9400 (Aug. 2008) | 4 | 4 × 32 (Instruction) 4 × 32 (Data) | 2 × 3 M shared by two cores | 0 M | 10 |
| AMD Phenom II 1090T (Apr. 2010) | 6 | N/A | 3 M | 6 M shared | 37 |
| AMD Phenom X4 9950 (Mar. 2008) | 4 | 64 KB + 64 KB Data + Instructions | 2 M | 2 M | 33 |
| Femi NVDIA GTX480 (Apr. 2010) | 480 | 64 KB (16 + 48) | 768 KB | 768 KB | 180 |

*Key* I/O bus includes front side bus (FSB), quick path structure and hyper transport bus

As more transistors are devoted to data processing rather than caching and flow control, the computation power offered by GPU is significantly larger than the one offered by a multi-core CPU. For instance, the GTX480 from NVIDIA offers 576 GFLOPS computing power, 20 times more than the cotemporary six-core Intel CPU Westmere does.

Increasing the computing power by assembling many computational cores into one die, however, escalates memory I/O requests and overuses the existing memory channel. The increased memory I/O requests blocks the memory channel, makes computational cores starving for data and eventually eliminates the benefit of additional computation power. Hence, memory bandwidth is the key factor that influence the performance of both multi-core CPU and many core GPU computation architectures and is usually refers in the literature as the "memory wall" problem.

To mitigate the overuse of the memory channel in multi-core CPU, different types of cache (L1, L2 and L3 cache) are placed along the path between individual cores and main memory. Since the most frequently used data can be put into the caching system, reading/writing data can, therefore, be done through the cache rather than through the main memory. This strategy effectively reduces the consumption of the expensive memory bandwidth, however, it requests an application having a regular memory usage pattern in order to efficiently utilize the caching infrastructure.

Unlike CPU, GPU hides the memory latency by driving many threads to concurrently access multiple memory locations. This strategy is particularly suitable for the data parallel computation, in which the same operation is executed on many data in parallel. Most matrix calculations (i.e., Matrix multiplication) that dominate power system calculations fortunately fall into this category.

Technically, it is easier to implement the GPU's strategy of hiding memory latency through concurrent accesses. This leads to GPU having higher memory bandwidth growth rates than multi-core CPU does in the past decade. For instance, from 2003 to 2009, the GPU memory bandwidth has been improved from 20 to 180 GB/s; while during the same time frame, the multi-core CPU's bandwidth has just improved from 10 to 25 GB/s, which is six times slower than the growth rates of the memory bandwidth in GPU.

In summary, efficient utilizing the limited memory bandwidth is the key to unleash the computation power of both multi-core CPU and many-core GPU computation architectures. An application with a regular memory access pattern is more suitable for the multi-core computation architecture, in which most memory footprints can be cached; on the other hand, an application with simple logic controls, but bulk concurrent data calculations, fits in the many-core computation architecture, in which concurrently transportation of data effectively hides the memory latency.

#### 12.2.1.2 Features of the Numerical Methods Applied to Distribution Network Calculations

In this section, the serial implementation of Conjugate Gradient Normal Residual (CGNR) against the power system state estimation application is evaluated as an instance to identify the features of distribution network calculations in terms of GFLOPS and memory bandwidth. The CGNR method is a general technique to solve a Weighted Least Squares (WLS) based system of linear equations and it is built upon on the concepts of the Conjugate Gradient (CG) method. Both CG and CGNR involve similar computational operations such as matrix-vector multiplication, vector dot products and vector updates.

A. **CGNR and CGNR + Jacobi preconditioner**

Consider the problem where the solution to the system of linear equations $Ax = b$ is sought, A is an $n \times m$ real matrix, $b$ is a known real vector of dimension $n \times 1$ and x is an unknown real vector of dimension $m \times 1$. Also, $n > m$ since the state estimation problem is over-determined (more equations than unknowns). For the state estimation problem, matrix $A$ is the measurement Jacobian.

In the Weighted Least Squares (WLS) Normal Equations formulation of the state estimation problem, $A^T A$ describes the Gain matrix that is Symmetrical Positive Definite (SPD) by structure. This makes the Conjugate Gradient (CG) method applicable. Note that the Gain matrix requires a matrix-matrix multiplication of the measurement Jacobian A with its transpose. This can be avoided by using CGNR (Fig. 12.6).

Given an initial guess $x^{(0)}$

Compute $r^{(0)} = b - Ax^{(0)}$, $z_0 = A^T r_0$, $p_0 = z_0$

For i = 1, 2, ... until convergence Do:

$w_i = Ap_i$

$temp = z^T z$

$\propto_i = \frac{temp}{w^T w}$

$x_{i+1} = x_i + \propto_i p_i$

$r_{i+1} = r_i - a_i w_i$

$z_{i+1} = A^T r_{i+1}$

$\beta_i = {z_{i+1}}^T z_{i+1} / temp$

$p_{i+1} = z_{i+1} + \beta_i p_i$

EndDo

Given an initial guess $x^{(0)}$

Compute $r^{(0)} = b - Ax^{(0)}$, $\dot{r_0} = A^T r_0$, $z_0 = M^{-1}\bar{r_0}$, $p_0 = z_0$

For $j$ = 1, 2, ... until convergence Do:

$w_j = Ap_j$

$temp = (z_j, \bar{r_j})$

$\propto_j = \frac{temp}{w^T w}$

$x_{j+1} = x_j + \propto_j p_j$

$r_{j+1} = r_j - a_j w_j$

$\dot{r}_{j+1} = A^T r_{i+1}$

$z_{j+1} = M^{-1} \dot{r}_{j+1}$

$\beta_i = {z_{j+1}}^T \dot{r}_{j+1} / temp$

$p_{j+1} = z_{j+1} + \beta_j p_j$

EndDo

**Fig. 12.6** CGNR (left) and CGNR with Jacobi preconditioner (right)

CGNR with Jacobi preconditioning method results when the diagonal of the Gain matrix is used as the preconditioner matrix *M* (in the right part of Fig. 12.6). The inverse of *M* consists of only the reciprocal of the diagonal of *M*. This makes the Jacobi preconditioning quite attractive for iterative methods as it involves very few additional computations compared with other preconditioning methods [24].

B. **Serial Implementations of CGNR and CGNR + Jacobi preconditioner**

The CGNR and CGNR with Jacobi preconditioner as presented in Fig. 12.6. have been implemented using C on the Windows platform. In order to study the behavior of iterative methods on the state estimation application, they are executed using a group of state estimation data matrices listed in Table 12.6 and the following measurements are recorded: the number of iterations to convergence and the time spent during iterations. Iterations are declared as converged when the 2-norm of a residual vector is less than a predefined threshold value—1e−3.

C. **The test environment**

*Hardware*

- Intel Core 2 Quad Q9400 (2.66 GHz)
  - L1 Cache: 4 × 32 KB instruction caches, 4 × 32 KB data caches
  - 2 × 3 M 12-way set associative caches (each L2)
  - caches is shared between 2 cores
  - Seagate 1T hard drive, 7200 rpm
  - 4G DDR2 memory
  - PCI-E × 16 (GPU interface)

**Table 12.6** The test data sets abstracted from the estimation problem

| Data set (SE measurement Jacobian H) | Size | Non zero elements | Condition number | Ratio of non-zero elements (%) |
|---|---|---|---|---|
| 3 bus system | 22 × 6 | 88 | 26.2 | 66.7 |
| 14 bus system | 73 × 27 | 329 | 118.78 | 16.7 |
| 14 bus system diagonal extension | 36500 × 13500 | 164,500 | 137.87 | 0.033 |
| 14 bus system non-diagonal extension | 36500 × 13500 | 205,296 | 1571.9 | 0.041 |
| PSSE2 | 28,634 × 11,028 | 115,262 | $1.24 \times 10^6$ | 0.036 |

- NVIDIA GTX480 GPU
  - GPU GF100
  - Memory 1.5G (DDR5)
  - Bus width (384 bit)
  - Bandwidth 182.4 GB/s

*Software*

- Windows 7 enterprise edition
- Visual Studio 2008 professional

D. **The test data set**

Table 12.6 describes the test matrices for the state estimation problem using their dimension, number of nonzero elements, condition number and percentage of the nonzero elements. The condition number is specified for the index matrix $A$ that represents the measurement Jacobian. It should be emphasized that for CGNR, we are dealing with the Gain matrix that has a squared condition number, even though it accepts matrix $A$ as an input, its convergence behavior depends upon the condition number of the Gain matrix ($A^TA$).

The 3 bus system and the 14 bus system test matrices are generated from a simple 3 bus system and the standard IEEE 14 bus system. To test the scalability of the algorithm, the 14 bus system measurement Jacobian is extended to form a large sparse matrix, in which the original 14 bus matrix is duplicated 500 times along the diagonal. From a practical viewpoint, such a diagonally extended matrix stands for 500 independent 14 bus systems and its condition number is the same as of the 14 bus system measurement Jacobian. In order to obtain a test case having a worse condition number, a few off-diagonal elements are introduced. The matrix so obtained is referred to as the 14 bus system diagonal extension with inter-area dependencies. The condition number of this matrix is 1571, which is 9–10 times larger than the original one. In addition, it considers another large state estimation matrix PSSE2 [25].

### E. Sparse matrix storage scheme

The chosen matrix storage scheme is a combination of Coordinate (COO) Format [26] and compressed sparse row Format [26]. The combination of COO and CSR efficiently uses both the search based on individual locations (COO) and the search based on rows, resulting in the following data structure as illustrated in Fig. 12.7.

### F. Test results and analysis

Table 12.7 lists the test results, which include the number of iterations and time taken until convergence by CGNR and CGNR with Jacobi preconditioner when executed using the test matrices in Table 12.6. For the termination criterion, the residual thresholds is assumed as 1e−3. The data type used is the double precision float type. The observations are:

- For a linear system $Ax = b$, the degree of ill-conditioning of the matrix $A$ influences the rate of convergence of an iterative method. An ill-conditioned matrix A results in more iteration to converge.
- The computational time spent by the iterative algorithms is not as affected by the size of the problem as by the matrix conditioning. In this evaluation, the 14 bus system with diagonal extension only takes a reasonable time (<0.1 s) to

$$A = \begin{bmatrix} 1 & 7 & 0 & 0 \\ 0 & 2 & 8 & 0 \\ 5 & 0 & 3 & 9 \\ 0 & 6 & 0 & 4 \end{bmatrix}$$
$$row = [0\ 0\ 1\ 1\ 2\ 2\ 2\ 3\ 3]$$
$$col = [0\ 1\ 1\ 2\ 0\ 2\ 3\ 1\ 3]$$
$$data = [1\ 7\ 2\ 8\ 5\ 3\ 9\ 6\ 4]$$
$$ptr = [0\ 2\ 4\ 7\ 9]$$

**Fig. 12.7** The combination of COO and compressed sparse row

**Table 12.7** Number of iterations and time spent by CGNR, CGNR with Jacobi preconditioner under different test systems

| Residual = $10^{-3}$ | | 3 bus system | 14 bus system | 14 bus system diagonal extension | 14 bus system diagonal extension with inter-area dependencies | PSSE2 data |
|---|---|---|---|---|---|---|
| CGNR | Iteration | 5 | 28 | 33 | 1369 | 188,271 |
| | Time (s) | 0.000159 | 0.000318 | 0.077579 | 3.801 | 271.45 |
| CGNR + Jacobi | Iteration | 3 | 24 | 25 | 845 | 27,605 |
| | Time (s) | 0.000007 | 0.000117 | 0.064539 | 2.428825 | 40.844019 |

converge, even though its size is 36500 × 13500 (larger than PSSE2). Note that this is a well-conditioned matrix.

- The Jacobi preconditioner greatly reduces the number of iterations to converge and therefore effectively reduces the computation load. In this test, CGNR with Jacobi preconditioner using PSSE2 data and residual threshold of 1e−3 as stopping criterion takes 27,605 iterations to converge, which is 14.4% of the number of iterations taken by CGNR without any preconditioning.

G. **Performance evaluation**

To roughly estimate GFLOPS and memory bandwidth, the number of double precision operations, the number of accessing memory and time consumed by each iterative algorithm are calculated manually (Table 12.7). The estimation assumes the data is directly picked up from memory with no caching capability. As a simple case, the statement $c = a + b \times c$ can be counted as two double precision operations (double precision add and multiplication), and four memory operations (three reading operations and one writing operation). In the estimation process, only the double precision operations and memory operations that occur during the iterative loop are counted. Table 12.8 illustrates the overall performance of the serial implementations of CGNR and CGNR + Jacobi preconditioning in terms of GFLOPS and memory bandwidth, The CGNR with Jacobi preconditioner takes 0.36 GFLOPS, which is far below the peak GFLOPS of the Intel Quad Core 9400 (42.56 GFLOPS (Double Precision)) [27]. The average memory bandwidth is around 6.2 GB/s, which is more than half of the peak memory bandwidth of the Intel Core 2 Quad Core system (8.5 GB/s [28]).

The above test results show that the serial implementations of iterative methods consume a small part of CPU computation power (less than 1% of the peak CPU computation power), but a large portion of memory bandwidth (around 60% of peak memory bandwidth) to transport double precision data back and forth between CPU and memory. Consumption of a large portion of the memory bandwidth demonstrates that an iterative method in the context of state estimation application is highly memory bounded.

**Table 12.8** Performance valued by GFLOPS and memory bandwidth

| | Total GFLOP | Total mem Op | Spent time | GFLOPS | Memory bandwidth (GB/s) |
|---|---|---|---|---|---|
| CGNR | 1.5G | 3.11G × 8 | 3.83 | 0.39 | 6.4 |
| CGNR + Jacobi | 0.932G | 1.98G × 8 | 2.53 | 0.36 | 6.2 |

*Key* DP: Number of Double Precision Float Point Calculations
Test Data: 14 bus system diagonal extension with inter-area dependencies
Residue: $10^{-3}$

#### 12.2.1.3 Parallelization of Iterative Methods

This section discusses the parallelization of iterative methods using multi-core and many-core computation architectures. CGNR + Jacobi preconditioning and CG + Jacobi preconditioning are selected for parallelization based on the test results in the former section.

A. **The parallelization strategy**

As is discussed, both of CG + Jacobi and CGNR + Jacobi preconditioning have a common framework, which includes an initialization part and an iterative part, and is constructed by several common operations, including the matrix vector multiplication, vector inner products and vector updates. Here, the vector update operation is defined as $u_i = v_i + \alpha w_i$, where *u, v, w* are $n \times 1$ vectors, and $\alpha$ is a scalar.

Table 12.9 illustrates the time spent by above common operations, (i.e., the matrix-vector multiplication, the vector inner product, and the vector update) in two test scenarios: CG + Jacobi against a power flow test data set and CGNR + Jacobi against a state estimation data set. The fact that the matrix-vector multiplication takes 66.1, 82.3% of total time separately in above two test cases demonstrates the importance of matrix vector multiplication. Hence, improving the performance of a matrix-vector multiplication is the first priority of the parallelism.

B. **Multi-core implementation**

This section aims to boost the performance of matrix-vector multiplication operations using multi-core technologies. To simplify the analysis, only the matrix-vector multiplication of $w_i = Ap_i$ (in the left part of Fig. 12.8) is parallelized using a series of operations that can be concurrently executed: $A_0p_i, A_1p_i, \ldots, A_np_i$ ($A_0, A_i, \ldots, A_n$ are the individual rows of matrix $A$). Other parts of the algorithm use their serial implementations.

The multi-core CG + Jacobi has been implemented using C++ on the windows platform. To reduce the overhead of creating and deleting a thread, the thread pool provided by the Microsoft platform is used. Managed by the thread pool, the overhead of creating and destroying a thread is greatly reduced since the thread pool creates a certain number of threads and reuses them. The thread synchronization is done by a shared variable and an event handler.

**Table 12.9** Time spent by three basic operations

| | Total time (s) | Matrix × Vector (s) | Vector × Vector (s) | Vector update (s) | Iterations |
|---|---|---|---|---|---|
| CG + Jacobi | 2.557 | 1.69 (66.1%) | 0.46(17.9%) | 0.40 (15.6%) | 715 |
| CGNR + Jacobi | 2.544 | 2.0945 (82.3%) | 0.2272(8.9%) | 0.2(7.8%) | 845 |

Test data set

CG + Jacobi: 1138 bus system with 56 times extension for power flow

CGNR + Jacobi: 14 bus system diagonal extension with inter-area dependencies for state estimation

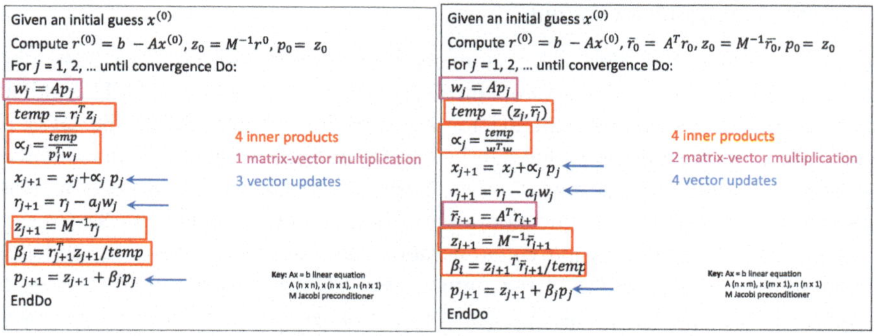


**Fig. 12.8** CG with Jacobi preconditioner and CGNR with Jacobi preconditioner

**Table 12.10** The performance and CPU utilization of multi-core CG and single core CG

| | CPU utilization (%) | Consumed time (s) |
|---|---|---|
| CG multi-core version | 94 | 32.168 |
| CG single-core version | 25 | 3.39 |

*Key* The test is against 1135_56 data set

The workflow of multi-core CG + Jacobi is as follows: the main function enqueues individual operations $A_0P_i$, $A_1P_i$, …, $A_nP_i$ into the buffer of the thread pool, and then idles for an event that indicates all the operations $A_0P_i$, $A_1P_i$, …, $A_nP_i$ are finished. Once the main function receives such an event, it finishes its idling status and continually conducts rest of other part of the algorithm. For each thread, a shared variable is used to synchronize its execution: once a thread finishes the calculation of $A_jP_i$, it adds one to the shared variable through the atomic add operation, and then check if the shared variable reaches $n$ (the size of the problem). The shared variable reaching $n$ will trigger the event, telling the main function that calculations for all threads are done. The above multi-core CG + Jacobi preconditioner is tested against the data set 1138_56, a DC power flow Jacobian test data extended from the 1138 system provided by NIST [29].

Demonstrated by Table 12.10, it is apparent that the multi-core version CG + Jacobi has a higher CPU utilization rate (94%) than the single-core version CG + Jacobi does, however, the actual performance of multi-core CG is 10 times slower than CG + Jacobi.

Due to the irregular sparsity pattern of the input matrix, it is hard to be cached by the caching infrastructure of the multi-core CPU in the test system. Most of the data have to be picked up from the main memory. The observations in above section demonstrate that a serial version iterative method already consumes >60% memory bandwidth. The experiment results here confirm the above observation: four cores concurrently consuming the same memory bandwidth blocks the memory data channel and makes individual cores idling and starving for data.

C. **Many-core implementation**

As is discussed in Sect. 12.2, a many-core system consists of a host—the original CPU system, and a device—the GPU system. A function in GPU is called kernel. The CPU system is suitable for the workflow control, while the GPU is suited for bulk computations that can be run on numerous data elements simultaneously in parallel. To let each type of hardware do the work it does best, it is important to partition the application and offshore a suitable load to GPU.

Concurrent execution of individual iterations are not possible because the latter iterations rely on the results of the former ones. Hence the parallel strategy is to parallelize individual iterations by utilizing parallelism of GPU threads and synchronize the execution of GPU threads using a global barrier to maintain the execution sequence between individual iterations.

As common operations of an iterative method (Fig. 12.8) are GPU friendly, in an actual implementation, these operations are offloaded to GPU for high performance computation. Meanwhile, the CPU is utilized to coordinate the execution of the operations that are offloaded to GPU.

Figure 12.9 illustrates the implementation of CGNR + Jacobi preconditioning under the CPU-GPU computation architecture, in which the CPU and GPU collaborate with one another to maximize hardware utilization of both CPU and GPU. To minimize the overhead of data transportation, the data is transported at the beginning and the end of the calculation. During the calculation, only the synchronization control signal is transported, the data stays in GPU.

D. **GPU kernels implementation**

As the vector-update operation is straightforward, this section focuses on the matrix-vector multiplication and the vector-vector dot product.

The GPU implementation of the matrix vector multiplication is based on [30], in which the matrix vector multiplication $Ax = b$ is partitioned into a series of independent operations $Aix = bi$ (Ai refers to the row vector of A, bi is the $i$th element of the vector b), concurrently executed by individual GPU threads. In this partition, the number of currently executed threads is equal to the row number of matrix A. For example, in the test scenario of PSSE2, the Ax = b is done by 28634 threads, calculating individual Aix = bi concurrently.

The GPU implementation of the vector inner product is based on [31], where the vector inner product is defined as the sum of the elements of $c_i = a_i \times b_i$ ($a_i$ and $b_i$ are the $i$th element of the vectors $a$ and $b$). Generally, the GPU version of the vector-vector dot product has two steps:

- Concurrently conduct $c_i = a_i \times b_i$ ($a_i$ and $b_i$ are the $i$th element of the vector)
- Sum $c_i$ ($i = 1\ldots n$)

Theoretically, the parallel vector-vector dot product algorithm requires all the calculations ($c_i = a_i \times b_i$) need to be done before doing Sum $c_i$. This requires a global synchronization of the executions of $c_i = a_i \times b_i$ by CPU.

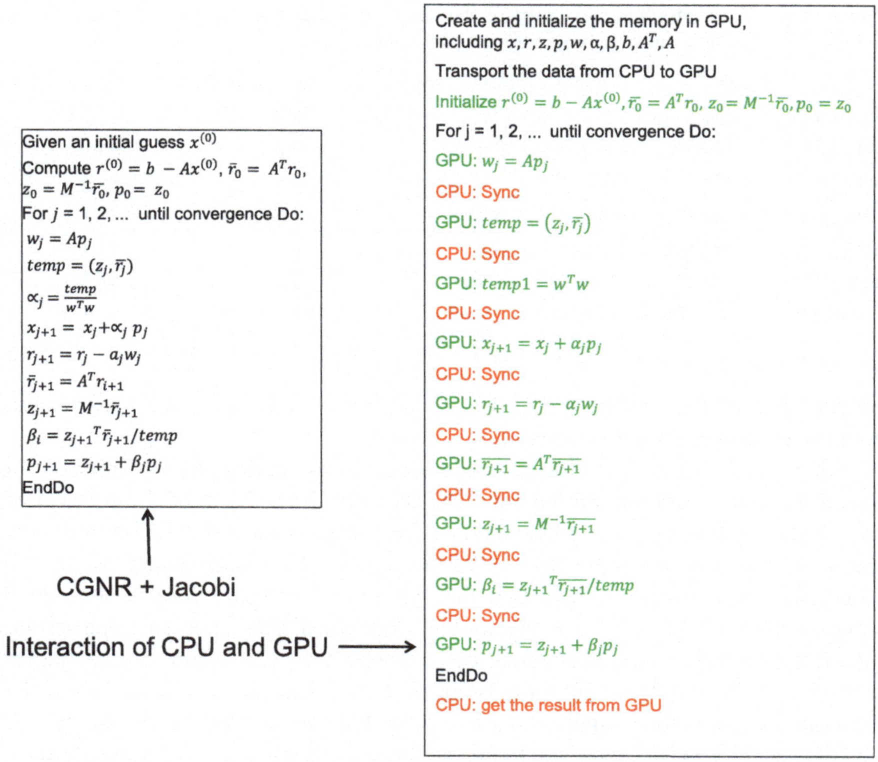


**Fig. 12.9** The CGNR with Jacobi and its CPU + GPU execution

### 12.2.1.4 Evaluations

This section evaluates the GPU implementation of CG series algorithms with Jacobi preconditioning in the context of state estimation and power flow applications. To keep the results comparable, the same test system is used in the previous sections. Table 12.11 shows the test results of CPU and GPU version CGNR with Jacobi preconditioner using the test data sets listed in Table 12.6. Irrespective of the

**Table 12.11** The performance comparison between the serial and GPU implementations of CGNR + Jacobin in the context of state estimation application

| | 14 bus system diagonal extension | | 14 bus system diagonal extension with interarea dependencies | | PSSE2 | |
|---|---|---|---|---|---|---|
| | Number of iterations | Spend time (s) | Number of iterations | Spend time (s) | Number of iterations | Spend time (s) |
| CPU | 25 | 0.063972 | 845 | 2.43 | 27,605 | 42.873 |
| GPU | 25 | 0.006327 | 845 | 0.3619 | 27,605 | 7.13 |

**Table 12.12** The performance comparison between the serial and GPU implementations of CG + Jacobin in the context of power flow application

| | Size | CPU time (ms) | GPU time (ms) | Speed Up |
|---|---|---|---|---|
| 494 bus | 494 × 494 | 7.53 | 38.8 | −5.15 |
| 494_28 | 13,832 × 13,832 | 226.6 | 40.0 | 5.6 |
| 494_56 | 27,664 × 27,664 | 451.4 | 40.8 | 11.0 |
| 494_336 | 165,984 × 165,984 | 5400 | 141.7 | 38.1 |
| 685_336 | 230,160 × 230,160 | 4030 | 124 | 32.5 |
| 662_336 | 222,432 × 222,432 | 3700 | 74.9 | 49.3 |
| 1138_336 | 382,638 × 382,638 | 25000 | 687.6 | 36.4 |

conditioning of the matrix, the GPU version has a factor of 7–10 performance improvement over the CPU version.

Table 12.12 lists various test matrices used for comparing the performance of CG with Jacobi preconditioning on CPU and on GPU. Each of these matrices represents a power flow admittance matrix [32], equivalent to the DC power flow Jacobian. The first row represents a 494 bus system that is duplicated 28, 56 and 336 times respectively in rows two through four. Similarly, rows 5 through 7 represent a 685, 662 and 1138 bus system duplicated 336 times to obtain corresponding power flow admittance matrices. The results in 12.11 demonstrate that for a large system, the GPU solution is much faster than the corresponding CPU solution; however, for a small system such as 494 bus network, GPU solution is slower than the CPU solution due to the overhead of the GPU computation platform (i.e., such as the expense of launching thousands of threads).

Matrices that are typically used in DC power flow calculations, namely the active power flow Jacobian or equivalently the system admittance matrices, are better conditioned than the ones used in other applications such as state estimation. Well-conditioned matrices are more favorable for preconditioned iterative methods such as CG using Jacobi preconditioner. Furthermore, a GPU architecture being suitable for parallelizing large-scale matrix-vector multiplications in iterative methods, it in effect appears promising for efficiently solving DC power flow problems on a large-scale system (Table 12.12). The GPU architecture may also be effective for DC power flow based contingency analysis as it involves multiple power flows for various contingency situations.

### *12.2.2 How Far Is the GPU Technology from Practical Application?*

Even though substantial efforts have been incurred, few of existing research works reported the successful or even attempted applications of GPU technologies to real-world software products that power utilities currently execute in their daily operation. To address this issue, a commercial software product widely used in the present US power industry is selected as the platform, based on which the potentials and the challenges to apply GPU technologies in power system practical applications are analyzed. For the sake of convenience, in this session, the selected power system software product is referred to as the GridSimulation rather than its real name. Although the evaluated GridSimulation is a transmission system tool, it has the same computational structure as the distribution application does, which consists of 3rd party optimization solver and pre- and post-data processing [33]. Hence, the conclusion drawn in this section can be applied to electric distribution network applications. The rest of the session is organized as follows: Session 12.2.2.1 profiles the Grid Simulation and identifies its performance bottlenecks. Section 12.2.2.2 implements the GPU counterparts of these performance bottlenecks. Section 12.2.2.3 evaluates the performance of the GPU solutions against practical power system test data set.

#### 12.2.2.1 Profile the Grid Simulation Tool

GridSimulation is a power grid planning tool, which has the following functions: (1) assessment of the exposed risks during the critical situations of power systems (e.g., fluctuation in generation output, transmission component outage, or load change); (2) prediction of important events for the power grid in the future (e.g., load forecast, the long-term, day-ahead, and near real-time market predictions) by simulating the hourly economic operations of the power market for periods from one day to multiple years.

While providing a forecast of the utilization levels of power system components and power flow patterns in the transmission grid, GridSimulation involves a massive amount of computational load. For example, an annual simulation of a power grid with 5000 buses and 20,000 constraints takes tens of hours. To reduce simulation time, customers of the GridSimulation system often have to simplify the simulation problems and cut off relatively unimportant constraints. However, the problem simplification and the constraints exclusion constantly result in approximate results. Therefore, improving the performance of GridSimulation becomes an effective solution that can obtain more accurate results while reduces the simulation time.

The core of GridSimulation is to solve an optimization problem. Therefore, it, architecturally, consists of three parts: pre-processing to formulate the optimization problem, solving the formulated problem, and post-processing to calculate results. The source code in both the pre- and the post-processing parts are accessible. The source code of the optimization solver, developed by a 3rd party vendor, is unreachable. Hence, the 3rd party solver is treated as a black box in the profiling procedure.

The used test data are from several practical power systems. The scales of these test systems are represented by the dimensions of the generator shifter factor (GSF) matrices. To thoroughly investigate the features of the computational load,

**Table 12.13** Selected test power grids

| System name | Dimension (Row × Column) |
|---|---|
| PG1 | 1943 × 3817 |
| PG2 | 2336 × 4252 |
| PG3 | 8429 × 13622 |
| PG4 | 4291 × 8155 |

**Table 12.14** The Test Environment

| | Profiling hardware configuration S1 | Profiling hardware configuration S2 |
|---|---|---|
| CPU | Intel Core 2 Q9400 | Intel Core i7-3930 |
| Memory | 4 GB DDR2 | 32 GB DDR3 |
| Hard disk | Mechanical HD | SSD |
| OS | Win 7 (64 bits) | Win 7 (64 bits) |

**Table 12.15** Performance of GridSimulation on power grids

| | | PG1 (w/Loss) | PG2 (w/o loss) | PG3 (w/o loss) | PG4 (w/Loss) |
|---|---|---|---|---|---|
| S1 | Total runtime (s) | 700 | 299 | 3930 | 7310 |
| | Ratio of solver | 6.28% | 35.05% | 64.09% | 25.17% |
| S2 | Total runtime (s) | 190 | 138 | 1860 | 2990 |
| | Ratio of solver | 8.83% | 34.32% | 69.12% | 24.37% |

*Key* Ratio of Solver − Ratio of time spent by solver over the total runtime

power grid models with and without system loss are both considered. In the case of considering loss, a loss model is included into the optimization problem. In particularly, the test cases PG1 and PG4 Table 12.13 are profiled without the loss model; PG2 and PG3 are profiled with the loss model (Tables 12.14 and 12.15).

Two types of profiling environment, referred to as S1 and S2, are designed for the purpose of studying the performance features under different hardware configurations. The S1 hardware setup has relatively outdated computer technologies, including the Intel Core 2 Quad CPU (2008), the GTX480, the Fermi based GPU (2010). In contrast, the S2 hardware setup has the most upgraded technologies, including a four-channel DDR3 1600 MHz memory system, an Intel extreme series CPU i7 3930 and the SSD hard disk.

Four identified performance bottlenecks are listed as follows: updating GSF, checkflow that retrieves power flow results, calculating Locational Marginal Price (LMP) and calculating delivery factors. To clarify the analysis, the profiling results are categorized into two groups: overall measurements and measurements on each individual sub-components. The former includes the overall performance of GridSimulation and the performance of the 3rd party solver. The latter illustrates the performance on each of the identified performance bottlenecks.

The observations on the overall performance are listed as follows:

- The overall simulation time is significantly increased when the size of the power grid is getting large. The simulation time for PG1 is 190 s versus 1860 s to that of the PG4.
- The time spent on the 3rd party solver varies case by case, ranging from 15.46% (PG1 with loss) to 82.59% (PG3 without loss) of the total simulation time. In the PG3 case, in which the loss model is excluded from the simulation, the solution time of the solver takes a large portion of runtime. On the contrary, when the power grid loss is considered, the solution time of the solver only takes less than 20% of total runtime.
- Inclusion of power grid loss in the simulation triggers substantial computational load on the pre- and post-processing parts and significantly increases the simulation runtime. In the practice of the power grid analysis software, most simulation scenarios need to consider the power grid loss.
- Because of the hardware evolution, particularly the upgrade of CPU, the S2 test environment exhibits significant performance advantages over the S1 configuration. The runtime in S2 environment reduces more than half comparing to that of S1 in some cases (e.g., PG3).

The observations on the identified performance bottlenecks are listed as follows:

- Including the power grid loss introduces two expensive operations: updating GSF and calculating the delivery factors. In regard to PG1, for instance, in the S1 setup, the above two operations take about 65.36 and 7.77% of the total runtime (Table 12.16). In the S2 setup, they take 39.36 and 13.62% (Table 12.17). In other words, 74% of the total runtime in the S1 environment and 53% of the total runtime in the S2 environment are consumed by the power grid loss related calculation.
- In cases that the power grid loss is excluded from the simulation, checkflow and calculating LMP are the major performance bottlenecks. In the simulation of the PG2 (no loss), these two operations take about 44.82% (Table 12.16) and

**Table 12.16** Test results from the S1 setup

| | CAISO (Loss) | ISO_B2021 (No loss) | EI_NERC (No Loss) | EI_EPA_Reg (Loss) |
|---|---|---|---|---|
| GSF update | 65.36% | N/A | N/A | 41.64% |
| Cal delivery factor | 7.77% | N/A | N/A | 7.68% |
| Check flow (%) | 8.15 | 29.26 | 10.40 | 3.00 |
| Calculate LMP (%) | 3.26 | 15.56 | 7.01 | 1.95 |
| Solving optimization problem (%) | 6.28 | 35.05 | 64.09 | 37.43 |
| Other functions (%) | 9.18 | 20.13 | 18.5 | 8.3 |
| Total execution time (s) | 700 | 299 | 3930 | 69110 |

*Key* The percentage refers to the ratio of runtime of the function over total runtime

**Table 12.17** The test results from the S2 setup

| | CAISO (Loss) | ISO_B2021 (No loss) | EI_NERC (No Loss) | EI_EPA_Reg (Loss) |
|---|---|---|---|---|
| GSF update | 39.36% | N/A | N/A | 33.48% |
| Cal delivery factor | 13.62% | N/A | N/A | 11.88% |
| Check flow (%) | 11.63 | 27.16 | 9.66 | 2.58 |
| Calculate LMP (%) | 7.66 | 21.27 | 6.45 | 2.59 |
| Solving optimization problem (%) | 8.75 | 34.32 | 69.12 | 33.57 |
| Other functions (%) | 18.98 | 17.25 | 14.77 | 15.9 |
| Total execution time (s) | 190 | 138 | 1860 | 27,550 |

*Key* The percentage refers to the ratio of runtime of the function over total runtime

48.43% (Table 12.17) of the total runtime in S1 and S2 configurations, separately.

In a nutshell, GridSimulation is a computationally intensive application; especially when including power grid loss in the simulation procedure, the annual simulation could take tens of hours. The loss calculation generally adds on heavy computational load to the pre-processing and post-processing parts when loss is considered.

#### 12.2.2.2 GPU Solutions for Identified Performance Bottlenecks

A. **The partial and full GPU optimization**

Technically, GPU optimization falls into two categories: the partial and the full optimization. In the partial optimization, both CPU and GPU participate in the calculation: CPU controls the workflow and conducts some lightweight computational tasks; the heavy computational load is offshored to GPU. In the full GPU optimization, CPU only controls the workflow and GPU performs all the calculation. By keeping all the data in GPU, the full GPU optimization maximally reduces the data transportation overhead. Since implementing the full GPU optimization requires the re-architecture of the existing software and unavoidably triggers a large amount of efforts, only the partial GPU optimization is studied in this work.

B. **GPU implementation for the performance improvement**

In regards to the GPU implementation, both the existing GPU APIs and new GPU functions developed from scratch are utilized.

(1) GPU kernels implemented by cuBLAS [17]

Essentially, the calculation of delivery factors and checkflow are both matrix-vector multiplication; the calculation of LMP is matrix-matrix

```
//Matrix vector multiplication
void GPUFAccelerator::gpu_mv(float* h_x, float* h_b)
{
    1. Transport vector x to GPU
    2. Transport vector b to GPU
    3. cublasSgemv(x, b)  //the matrix vector multiplication function in cuBLAS lib
    4. Transport results back to GPU
}

Key: cublasSgemv is the matrix vector multiplication function in the cuBLAS library
```

**Fig. 12.10** The GPU implementation of matrix vector multiplication

multiplication, the corresponding APIs in the GPU cuBLAS library are deployed directly to implement the required multiplications between matrices and vectors.

In particular, the matrix-vector multiplication can be expressed by $b = Ax$, where $A$ is the GSF matrix. The GSF matrix is transported to GPU first before the calculation begins and reused by other bottleneck calculations in GPU. During calculation, only the vector x and b are transported between CPU and GPU. Sharing the GSF matrix among several performance bottlenecks greatly reduces the data transportation overhead. Figure 12.10 shows the example of GPU implementation for matrix-vector multiplication.

(2) Developed GPU solutions

Since no API is offered for the matrix update, the GPU function for GSF update is prototyped from scratch. Figure 12.11 illustrates the theory of the matrix update. Its GPU counterpart is developed (Fig. 12.12), which can fully take advantage of the GPU's capability in concurrently launching thousands of threads. As a result, the calculations of the GSF update can be implemented in one step in GPU instead of multiple steps required in CPU implementation such as moving data to CPU and conducting calculations serially.

$$\begin{bmatrix} b_0 \\ \vdots \\ b_n \end{bmatrix} \begin{bmatrix} a_{00} & \cdots & a_{0m} \\ \vdots & \ddots & \vdots \\ a_{n0} & \cdots & a_{nm} \end{bmatrix}$$

$$\begin{bmatrix} c_0 & \cdots & c_m \end{bmatrix}$$

$$\Downarrow$$

$$\begin{bmatrix} a_{00} * b_0 * c_0 & \cdots & a_{0m} * b_0 * c_m \\ \vdots & \ddots & \vdots \\ a_{n0} * b_n * c_0 & \cdots & a_{nm} * b_n * c_m \end{bmatrix}$$

**Fig. 12.11** The element update in the GSF update

```
__global__ void VecMulti(double* A, const double* B, double* C, double* D, int N, int Column)
{
    int i = blockDim.x * blockIdx.x + threadIdx.x;

    if (i < N)
    {
        D[i] = A[i] * (B[i % Column]) * (C[i / Column]);
    }
}
```

**Fig. 12.12** GPU Implementation of the GSF update

#### 12.2.2.3 Evaluate GPU Solutions

A. **Overall performance measurements on bottlenecks**

The measurements of the identified performance bottlenecks collected from both the S1 and S2 hardware configurations are listed in (Tables 12.18 and 12.19), separately. To be precise, the data transportation is further classified into synchronized transportation and synchronized transportation.

The test results demonstrate that synchronized transportation effectively reduces the data transportation time and significantly improves the performance. For example, in the synchronized transportation (Tables 12.18 and 12.19), the checkflow GPU solver is about 9 times and 1.7 times faster than its CPU counterpart in the S1 and S2 test environment. In contrast, in the synchronized transportation, the above two performance gains are 148 and 49.7, separately.

**Table 12.18** The runtime of GPU and CPU solvers (S1 test system, PG1)

| | | GPU (s) | CPU (s) | Performance gain |
|---|---|---|---|---|
| Syn | Checkflow | 0.001156 | 0.010384 | 8.98 |
| | Delivery factor | 0.00284 | 0.0527 | 18.54 |
| | Update GSF | 0.000332 | 0.40165 | 1315 |
| Asyn | Checkflow | 7.12e−5 | 0.105 | 148.3 |
| | Delivery factor | 0.00294 | 0.0579 | 19.709 |
| | Update GSF | 7.27e−5 | 0.377 | 5184 |

**Table 12.19** The runtime of GPU and CPU solvers (S2 test system, PG1)

| | | GPU (s) | CPU (s) | Performance gain |
|---|---|---|---|---|
| Syn | Checkflow | 0.001319 | 0.002291 | 1.7367 |
| | Delivery factor | 0.001925 | 0.0246 | 12.81 |
| | Update GSF | 0.000185 | 0.0444 | 241.03 |
| Asyn | Checkflow | 5.76e−5 | 0.002419 | 41.97 |
| | Delivery factor | 0.0021 | 0.0221 | 10.48 |
| | Update GSF | 5.20e−5 | 0.044 | 847.68 |

**Table 12.20** The scalability of GPU solver versus CPU solver (checkflow, the S2 test system)

| | PG1 | PG2 | PG4 |
|---|---|---|---|
| CPU | 0.002228 | 0.00254 | 0.004653 |
| GPU | 0.000112 | 0.000155 | 0.000163 |

In GSF update, as GSF matrix stays in GPU, the measurements only consider transportation of the vector x and b from CPU to GPU and the calculations in GPU. Since GPU can concurrently launch thousands of threads, updating all elements in one step, the GPU version of GSF update has significant performance advantage over its CPU counterpart. The test results demonstrate that the GPU solver of GSF update is about 5184 and 847 times faster than its CPU counterpart in the S1 and the S2 test systems, separately if the data transportation is not considered.

B. **Scalability of GPU solver versus CPU solver**

Table 12.20 illustrates the performance variances of the GPU solver versus the CPU counterpart with the size of calculated data increased. In the measurements, only calculation time (excluding the transportation time) is considered. The test results demonstrate that time spent on the GPU solver increases in a flat ratio (0.000163/0.000112 = 1.03) with the increased size of the simulated system. On the contrary, the ratio of the CPU solver is not flat (0.004653/0.002228 = 2.08). In other words, performance of GPU solver is insensitive to the size of the system, which makes GPU solver suitable for calculating large systems.

C. **Measurements on calculation and transportation**

The detailed measurements on data transportation and calculations are listed in Tables 12.18, 12.19, 12.20, and 12.21. Some observations are listed as follows:

- The transportation from CPU to GPU is much faster than that of the opposite direction because GPU memory (DDR5) has faster read/write speed than the CPU memory (DDR3).
- The calculations on two GPUs (GTX480 versus GTX680) take almost the same amount of time. The reasonable explanation is that the problem is not large enough to distinguish capability of the two GPUs.
- For a GPU solver, the ratios of transporting data between CPU and GPU over the total runtime are significant. The round-trip of data transportation takes about 95% + of the total runtime; while calculation only takes 5% of the total runtime. Hence, reducing the transportation time becomes important in designing high performance GPU solver.

**Table 12.21** The runtime of GPU solver for transportation and matrix-vector calculation (PG1)

| | S1 (s) | S2 (s) |
|---|---|---|
| CPU to GPU | 2.03e−5 | 1.59e−5 |
| GPU to CPU | 0.001 | 0.001 |
| Calculation | 3.1e−5 | 3.1e−5 |

**Fig. 12.13** Three bus system network

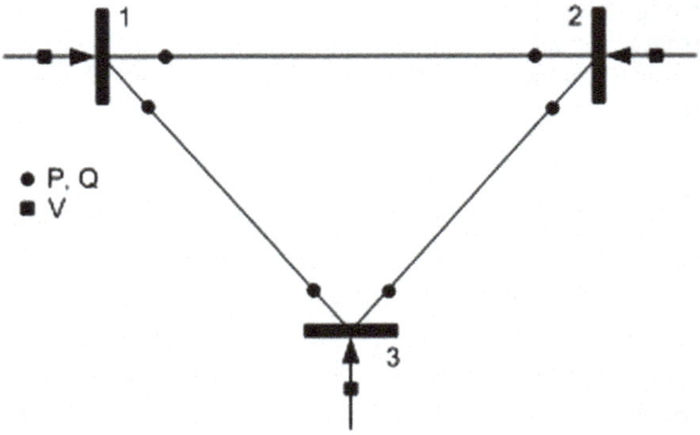


## Appendix

Referring to Table 12.6, system networks used for experimentation include a 3-bus system, IEEE 14 bus system [34] and PSSE2 system from [25]. These are described in this appendix (Fig. 12.13).

The 3-bus system is shown in where the solid circles denote the flow and injection measurements used for state estimation. Similarly, Fig. 12.14 shows the measurement placement for IEEE 14-bus system. Measurements are obtained from the load flow solution with noise added.

The 14 bus system is duplicated several times to obtain larger systems. The sparsity pattern of the measurement Jacobian matrices in Table 12.6 (rows 3, 4 and 5) is shown in Fig. 12.15.

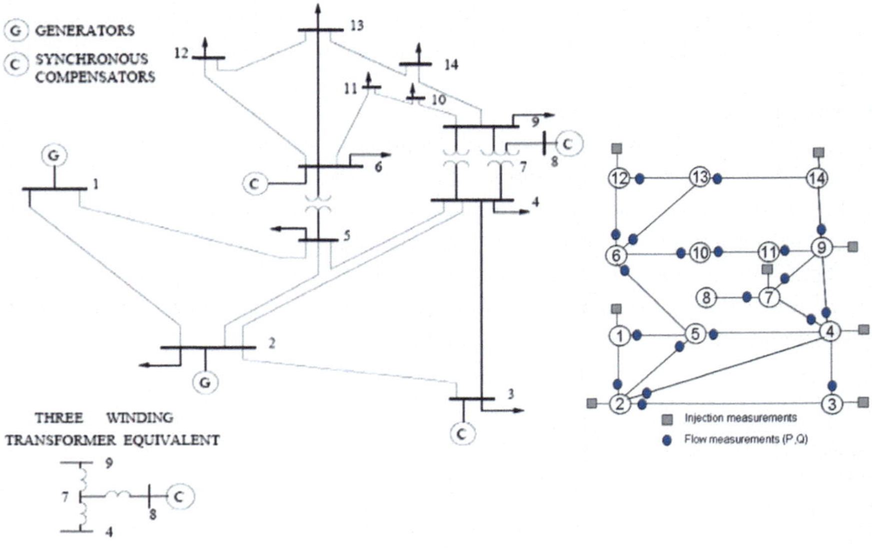


**Fig. 12.14** IEEE 14-bus system and measurement placement

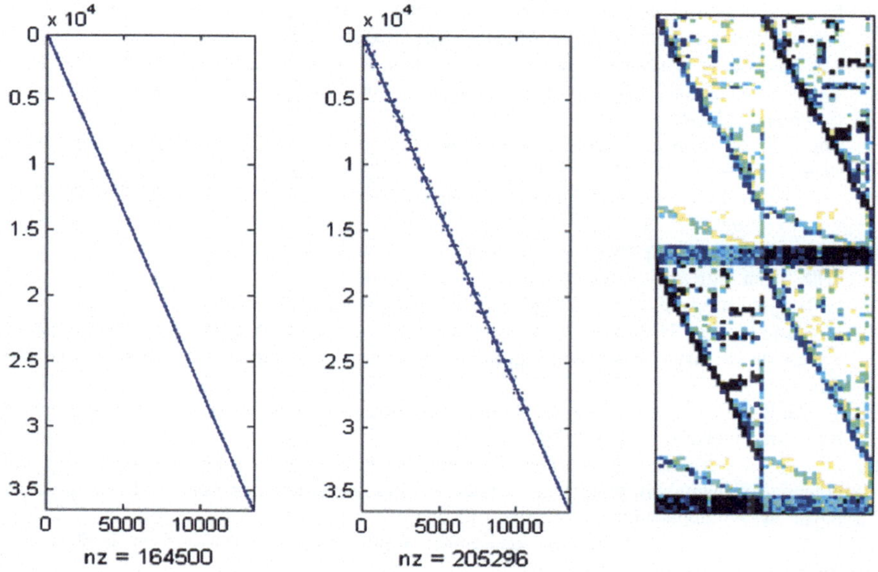


**Fig. 12.15** Sparsity pattern of the test matrices: 14-bus with diagonal extension (left) 14-bus with diagonal extension and inter-area tie lines (middle), PSSE2 system (right)

## References

1. W.R. Cassel, Distribution management systems: functions and payback. IEEE Trans. Power Syst. **8**(3), 796–801 (1993)
2. H. Dorey, Advanced metering in old and new worlds. Power Eng. J. **10**(4), 146–148 (1996)
3. "Smart Meter Installation Progress", The Pike Research Report (2012)
4. S. Mak, D. Radford, Design considerations for implementation of large scale automatic meter reading systems. IEEE Trans. Power Deliv. **10**(1), 97–103 (1995)
5. V.C. Gungor, D. Sahin et al., SmartGrid technologies: communication technologies and standards. IEEE Trans. Ind. Inf. **7**(4), 529–539 (2011)
6. S. Depuru, L. Wang, V. Devabhaktuni, Smart meters for power grid: challenges, issues, advantages and status. Renew. Sustain. Energy Rev. **15**(6), 2736–2742 (2011)
7. R.W. Uluski, Interactions between AMI and DMS for efficiency/reliability improvement at a typical utility, in *IEEE PES GM* (2008)
8. Z. Li, F. Yang, S. Mohagheghi, Z. Wang, J.C. Tournier, Y. Wang, Toward smart distribution management by integrating advanced metering infrastructure. Electr. Power Syst. Res. **105**(3), 51–56 (2013)
9. Z. Li, F. Yang, Governing Millions Meters Data, in *IEEE PES GM* (July 2015)
10. J. Zheng, Z. Li, Speeding up processing data from millions of smart meters, in *Proceeding of the 5th ACM/SPEC International Conference on Performance Engineering*, pp. 27–37, New York, NY, USA (2014)
11. Z. Li, J. Zheng, F. Yang, Unity AMI information models to support diversified smart grid systems and applications, in *IEEE PES SmartGridComm* (Oct. 2013)
12. Z. Li, F. Yang, D. Ishchenko, The standardization of distribution grid communication networks, in *IEEE PES GM* (July 2012)

13. Demand response: design principles for creating customer and market value, Technical Report, Peak Load Management Alliance (Nov. 2002)
14. J. Nieplocha, D. Chavarría-Miranda, V. Tipparaju, H. Huang, A. Marquez, A parallel WLS state estimator on shared memory computers, in *Proceedings of Power Engineering Conference* (2007)
15. The TITAN Super Computer, http://www.theregister.co.uk/2012/10/29/oak_ridge_titan_supercomputer/
16. J.D. Owens, M. Houston, D. Luebke, S. Green, GPU computing, in *Proceedings of the IEEE*, vol. 96, issue 5 (2008)
17. CUDA, NVIDIA, http://www.nvidia.com/object/cuda_home_new.html
18. GTX Titan, http://www.nvidia.com/titan-graphics-card
19. GPU applications, http://www.nvidia.com/object/gpu-applications.html
20. D. Castano-Diez, D. Moser, A. Schoenegger, S. Pruggnaller, A.S. Frangakis, Performance evaluation of image processing algorithms on the GPU. J. Struct. Biol. **164**(1), 153–160 (2008)
21. V. Jalili-Marandi, V. Dinavahi, Large-scale transient stability simulation on graphics processing units, in *PES GM* (2009)
22. Z. Li, V.D. Donde, J.C. Tournier, F. Yang, On limitations of traditional multi-core and potential of many-core processing architectures for sparse linear solvers used in large-scale power system applications, in *PES GM* (2011)
23. H. Dag, A. Semlyen, A new preconditioned conjugate gradient power flow. IEEE Trans. Power Syst. **18**(4) (Nov. 2003)
24. J. Nieplocha, D. Chavarría-Miranda, V. Tipparaju, H. Huang, A. Marquez, A parallel WLS state estimator on shared memory computers, in *Proceedings of Power Engineering Conference* (2007)
25. Y. Cheng, PSSE2, http://www.cise.ufl.edu/research/sparse/matrices/YCheng/psse2
26. N. Bell, M. Garland, Implementing sparse matrix-vector multiplication on throughput-oriented processors, in *Proceedings of the Supercomputing 09* (Nov. 2009)
27. The Computation Power of Intel Core 2 Quad Core Q9400, http://www.intel.com/support/processors/sb/CS-023143.htm#3
28. The memory bandwidth of a general computation architecture, http://en.wikipedia.org/wiki/Front-Side_bus
29. NIST, http://math.nist.gov/MatrixMarket/
30. N. Bell, M. Garland, Implementing sparse matrix-vector multiplication on throughput-oriented processors, in *Proceedings of the Supercomputing 09* (Nov. 2009)
31. M. Harris, Optimizing parallel reduction in CUDA, NIVIDIA Technical Report (2008)
32. NIST, http://math.nist.gov/MatrixMarket/
33. X. Feng, F. Yang, W. Peterson, A practical multi-phase distribution state estimation solution incorporating smart meter and sensor data, in *IEEE PES GM* (July 2012)
34. Power Systems Test Case Archive, http://www.ee.washington.edu/research/pstca/pf14/pg_tca14bus.htm
35. Elster, "EnergyAxis MAS 6.2 Web Service Interface", Revision L (May 29, 2009)
36. Itron, "OpenWay Collection Engine Web Service Reference Guide", CEv3.0 SRv2.0 (Jan. 22, 2009)
37. "Trilliant – A trusted Solution Partner", Solution Brief, Trilliant Incorporated (2009)
38. FlexNet, Sensus metering systems—Advanced Metering Infrastructure Technology for Utilities
39. Ecologic MDMS, http://www.ecologicanalytics.com/
40. ANSI C12.19-2008, "American National Standard - Utility Industry End Device Data Tables", Approved February 24, 2009, American National Standards Institute, Inc.

41. IEC 62056-62 The interface class for Electricity metering Data exchange for meter reading, tariff and load control, Second edition, 2006-11
42. IEC 61968-9 Ed. 1 Part 9: Interface for meter reading and control, 2009-08-14, IEC/TC 57
43. G.A. McNaughton et al., Comparison of the multispeak distribution connectivity model and the IEC common information model network data set. National Rural Electric Cooperative Association (2009)

The manufacturer's authorised representative in the EU is Springer Nature Customer Service Centre GmbH, Europaplatz 3, 69115 Heidelberg, Germany. If you have any concerns regarding our products, please contact ProductSafety@springernature.com

Printed and bound by CPI Group (UK) Ltd, Croydon, CR0 4YY

30/09/2026

02225448-0014